P9-CDZ-098

Advance Praise for *Churchill's Bomb*

"This is a fascinating book. Graham Farmelo offers a fresh and thoroughly researched history of the development of atomic weapons in his insightful and engaging account of Winston Churchill's failure to forge a partnership of equal exchange between Great Britain and the United States in the development of the bomb. Farmelo offers vivid vignettes of political and scientific personalities, with special attention to the widely disliked Oxford physicist Frederick Lindemann, who became Churchill's science and technology guru in the 1920s."

—Mary Jo Nye, Professor of History Emerita, Oregon State University, and author of *Michael Polanyi and His Generation*

"An excellent book. Graham Farmelo draws on many sources to show how Churchill, his scientific adviser Frederick Lindemann, and a host of other scientists and politicians developed the atomic bomb. *Churchill's Bomb* brings these characters back to life with anecdotes, quotations, and personal sketches. But Farmelo's book does more than unfold the hopes, doubts, and fears engendered by the bomb: it illuminates the relationship between big science and modern democracy."

—James W. Muller, University of Alaska, Anchorage

"What a brilliant and compelling book! Graham Farmelo sensitively and eloquently deconstructs the twists and turns of Winston Churchill's involvement with nuclear weapons over nearly half a century, setting this unfamiliar tale in the context of the turbulent times. At its heart are the ambiguities of the World War II relationship between a scientifically innovative but economically weakened Britain and the inexhaustibly energetic USA with unlimited resources."

—Sir Michael Berry, University of Bristol

"A nicely detailed and balanced record of the British ambivalence toward building an atom bomb in favor of the American effort. . . . A tremendously useful soup-to-nuts study of how Britain and the U.S. embraced a frightening atomic age."

—*Kirkus Reviews*

CHURCHILL'S BOMB

CHURCHILL'S BOMB

How the United States Overtook Britain in the First Nuclear Arms Race

---

GRAHAM FARMELO

BASIC BOOKS

A Member of the Perseus Books Group
New York

First published in the United States in 2013 by Basic Books,
A Member of the Perseus Books Group

Published in Great Britain in 2013 by Faber and Faber Ltd.

Books published by Basic Books are available at special discounts for bulk purchases in the United States by corporations, institutions, and other organizations. For more information, please contact the Special Markets Department at the Perseus Books Group, 2300 Chestnut Street, Suite 200, Philadelphia, PA 19103, or call (800) 810-4145, ext. 5000, or e-mail special.markets@perseusbooks.com.

A CIP catalog record for this book is available from the Library of Congress.

ISBN: 978-0-465-02195-6 (hardcover)
ISBN: 978-0-465-06989-7 (e-book)

LCCN: 2013940827

10 9 8 7 6 5 4 3 2 1

For Lindsey Jones, F.S.

In the next fifty years mankind will make greater progress in mastering and applying natural forces than in the last million years or more. And the first question we must ask ourselves is: 'Are we fit for it? Are we worthy of all these exalted responsibilities? Can we bear this tremendous strain?'

—WINSTON CHURCHILL, 14 November 1937

Scientists on the whole are a very docile lot. Apart from their own particular job they do just what they are told and are content to sit down and be very minor entities.

—MARK OLIPHANT, 20 April 1940

Devil: 'In the arts of life Man invents nothing; but in the arts of death he outdoes Nature herself . . . his heart is in his weapons. This marvellous force of Life of which you boast is a force of Death: Man measures his strength by his destructiveness.'

—GEORGE BERNARD SHAW, *Man and Superman*, 1903

Contents

PROLOGUE

FEBRUARY 1955

Churchill, his nuclear scientists and the Bomb

> 'I do not pretend to be an expert or to have technical knowledge of this prodigious sphere of [nuclear] science. But in my long friendship with [Frederick Lindemann] I have tried to follow and even predict the evolution of events.'
>
> WINSTON CHURCHILL to the Commons, 1 March 1955[1]

His swansong was sure to have a nuclear theme. In February 1955, when Churchill was eighty years old and inching reluctantly towards his resignation as Prime Minister, he set his heart on making one last great speech in the Commons. The hydrogen bomb, his obsession, supplied the perfect theme – it made all the other business of the day look trifling. As he had told his doctor a few months before: 'I am more worried by [the H-bomb] than by all the rest of my problems put together.'[2]

The H-bomb was, Churchill believed, the greatest threat to civilisation since the Mongols began their conquests three-quarters of a millennium before.[3] This threat had become a monomania for him, driving his final great diplomatic initiative: to bring the Soviet Union and the United States together to ease the tensions of the Cold War and so minimise the risk that H-bombs would be used.[4] He was certainly going to mention that campaign in his speech, but his main task was to argue that the UK must acquire the weapon he feared so much, as a deterrent to the Soviet Union. This argument was almost certain to win the day in the Commons – his main challenge was to give his country a sense of hope at a time when the world seemed to be careering towards a nuclear holocaust.

He threw himself into the speech, researching the nuclear story and his role in it, all the way back to the articles he had written in the 1920s and 1930s about the potential of nuclear energy to change the world. Among the best of the pieces was 'Fifty Years Hence', a four-thousand-word speculation on the effects science might have on life in the future, first published in late 1931. In it, he drew attention to the likely advent of nuclear weapons and the challenges their invention would pose. He even glimpsed the destructive power of the H-bomb, which would be detonated for the first time twenty-one years later:[5]

> High authorities tell us that new sources of power, vastly more important than any we yet know, will surely be discovered. Nuclear energy is incomparably greater than the molecular energy which we use today . . . If the hydrogen atoms in a pound of water could be prevailed upon to combine together and form helium, they would suffice to drive a thousand horse-power engine for a whole year . . . There is no question among scientists that this gigantic source of energy exists . . .

Churchill had based the article on a draft by his scientific Grand Vizier, Frederick Lindemann, an acid-tongued professor of physics at the University of Oxford. Lindemann was 'one of the best scientists and best brains in the country', in Churchill's opinion, a view not shared by many leading academics.[6] To most of them, 'the Prof', as Churchill called him, was a distinguished scientist with a gift for summarising complex arguments simply and accurately, but not a deep or imaginative thinker and certainly not an expert on nuclear science.

One of the services the Prof rendered to his admiring friend was to nourish his inquisitive mind with briefings on the latest advances in basic science. In the spring of 1926, when the new and revolutionary quantum theory of matter was the talk of physicists, Lindemann sent Churchill – then Chancellor of the Exchequer – a book on how the structure of atoms can be understood using basic quantum ideas. The text grabbed Churchill's attention so completely that, for a few hours, he was incapable of concentrating on his Budget.

A few years later, Lindemann kept Churchill abreast of the headline-making advances in nuclear physics made by Ernest Rutherford and his colleagues at Cambridge University, including the first artificial splitting of the atom. Soon afterwards, Churchill marvelled at the scientists' achievements and said so after chairing one of Lindemann's non-specialist talks on nuclear physics: 'Here is this great study of science proceeding.'[7] The Prof ensured that Churchill had been aware of the opportunities and threats of nuclear technology for longer than any other leading politician, living or dead. In return, Churchill made his friend one of the most politically influential scientists ever to serve in government.

The speech Churchill was preparing in late February 1955 was part of his final bid for a glorious place in Britain's post-war history, having positioned himself as a link between the reigns of the British Empire's two most recent queens.[8] During the closing years of Victoria's reign, he had read about the widely publicised discovery of radioactivity, which involved the release of nuclear energy, as scientists later understood. Now, in the new Elizabethan era, he was commissioning a weapon that would release this energy with a destructiveness he had first fully appreciated only a year before, when he read a front-page article in the *Manchester Guardian*, 'Devastation and the Hydrogen Bomb'. It almost made the eyes stand out of his head, as he told President Eisenhower a few months later.[9]

His colleagues in Parliament were now expecting a great speech from him, to round off his second premiership. Although they all knew of his obsession with the H-bomb, few of them appreciated the full extent of his involvement in the development of nuclear weapons. The handful in Whitehall who were familiar with the details of his record knew that it had been not been especially distinguished by his standards. He had almost always responded to events rather than shaping them, had shown poor judgement in his choice of advisers, and demonstrated none of his fabled vision and imagination until it was too late.

It was without doubt a misfortune for him that he had to think about the possibility of nuclear weapons when he was also deeply involved in the tumult of a global war. The news from Birmingham that two 'enemy aliens' – as the government classified them – had discovered a viable way of making a nuclear bomb arrived in Whitehall less than two months before he first became Prime Minister in May 1940. During most of the next two years, Churchill's pool of nuclear advice was too narrow and too shallow. Most damagingly, he froze out Henry Tizard, Britain's leading expert on the application of science to military problems – a decision that dismayed many leading scientists. The computer pioneer and former radar engineer Sir Maurice Wilkes later remembered: 'Scientists offered the Prime Minister the man best able to give their consensus, but he chose a maverick.'[10] Churchill discussed the new 'explosives', as he usually called nuclear weapons before they became a reality, only with Lindemann and with their colleague Sir John Anderson, keeping it secret from almost the entire Cabinet for most of the war. He demonstrated neither his usual sure-footedness nor any of his habitual enthusiasm for innovative new weapons, such as – during World War I – the tank.[11]

In August 1941, when Churchill endorsed plans to build the Bomb, he had not grasped the transformative qualities of a weapon that could be delivered by a single aeroplane and wipe out a city in seconds. British nuclear scientists, then far ahead of their American colleagues in this field, had given him a high-value bargaining chip to play in his dealings with Roosevelt, who wrote to suggest that they embark on an equal-harness collaboration to develop the Bomb. Churchill as good as threw the chip away. He did not reply to the President's generous note for several weeks and even then appeared unenthusiastic about a nuclear collaboration. By that time, the United States had entered the war and was gearing up to begin its gargantuan Manhattan Project, which it pursued with a self-interest so ruthless that it left Churchill floundering. It seems that he first appreciated the strategic significance of the nuclear project only in the early spring of 1943, some eighteen months after Roosevelt. One consequence of the myopia Churchill shared with his closest advisers was that British physicists played only a minor role in the leadership of the project, and the influence they had on the application of their pioneering ideas was limited.

Churchill's lack of vision about the Bomb was embarrassingly clear in May 1944, when he met the Danish theoretical physicist Niels Bohr in 10 Downing Street. By common agreement, Bohr was the world's most accomplished nuclear scientist and a man of exceptional wisdom, though not an articulate speaker. When Bohr mumbled his suggestion that the US and Britain should share the secret of the Bomb with their Soviet allies to help build trust and avert a post-war arms race, Churchill was dismissive, having shown him none of the respect and attentiveness he gave to Lindemann. Roosevelt also had no time for the Dane's ideas. Had the leaders thought more deeply about his views, it is at least possible that the worst excesses of the post-war arms race might have been averted.

Of all the wartime agreements Churchill made with the American administration, he was especially proud of the one he struck on the Bomb when he met Roosevelt in Quebec during the summer of 1943. This agreement brought British scientists into the Manhattan Project after almost a year of exclusion and enshrined an undertaking that Britain and the US would not use the Bomb against another country without each other's consent. The problem was that this was not a treaty but a private agreement that both Churchill and Roosevelt withheld from all but a tiny number of their colleagues. The leaders regarded the Bomb as an essentially private matter, but after the war the plan predictably backfired, with serious consequences for Britain. Churchill's successor as Prime Minister, Clement Attlee, discovered that Truman and his administration had no wish to continue with the Quebec Agreement: in 1946, the American government passed a brutally self-interested Act forbidding collaboration on nuclear matters with any foreign country.[12] Attlee eventually decided to cut his losses and set up a team of nuclear scientists to build the British Bomb, using the skills and scraps of information retrieved from the Manhattan Project, with virtually no assistance from the United States for several years. Rarely had the relationship between the US and Britain, so special to Churchill, been so devoid of practical value.

It was inevitable that the Soviets would have the Bomb soon after the war, as Churchill knew. Appalled by their military adventurism and their repressive regimes in Eastern Europe, he made the astonishing argument that if there was no rapprochement – his preferred option – then America should stage a pre-emptive nuclear attack on Russia.[13] President Truman wanted nothing to do with this, and Churchill quickly changed his line when the Soviets tested their first nuclear weapon in August 1949. The arms race foreseen by Niels Bohr was now well under way and the world appeared to be sliding into an age of mutually assured destruction.

Beginning in October 1951, the prospect of an imminent nuclear war gradually became the great theme of Churchill's second premiership. He spent most of his final two years in office trying to avoid such a catastrophic conflict, believing that he could bring the Soviets and Americans to the conference table and talk them into a more rational approach to living with the 'frightful' H-bomb.[14] Churchill pursued his perhaps quixotic cause with all the tenacity and courage he had shown in 1940, against widespread derision and after a stroke that, he boasted, 'would have killed most men'.[15] Only when it was clear that there was no chance that either the Americans or the Soviets would cooperate, and that his hopes of becoming a latter-day global saviour were over, did he finally throw in the towel. His failure to make headway in what was – at that time – a hopeless cause was one of the tragedies of his political career, though its prosecution did him credit and helped to erase his reputation as a warmonger. This was the defeat of a statesman years ahead of his time – he was trying too early to hurry along the détente agenda that brought such credit to later leaders, notably Ronald Reagan and Mikhail Gorbachev.

One curiosity of Churchill's second term was that at first he showed no interest in developing nuclear power, which he had foreseen and discussed in widely read articles decades before. As usual, he trusted in the goodwill of the Americans – he wanted his country to piggy-back on their technology, but was persuaded to change his mind by Lindemann, who became one of the godfathers of the nuclear industry in Britain. By that time, Churchill felt comfortable in the company of several senior scientists other than the Prof, even with a few leading nuclear physicists. In the four months before he began to prepare his valedictory speech on the H-bomb, he had talked at length three times with Sir John Cockcroft, one of the duo that first artificially split the atom. Two of these discussions on nuclear policy were held over long, bibulous lunches, with the Prime Minister in fine form.

By late February 1955, Churchill was spending most mornings polishing the text of his speech, sitting up in his silk dressing gown. He was still an imposing figure, though he was small in stature and looked like an outsize doll, with skin as smooth and shiny as pink celluloid.[16] Usually holding a cigar, he dictated for hours on end to his secretary Jane Portal, later Lady Williams of Elvel, who sat a respectful distance away with her pen and notepad.[17] She now remembers that 'he was absolutely determined to go out on a high, to prove that he was still on top of his job, dealing with the biggest threat to the world'. He was in no doubt that he was better equipped than any other international leader to deal with the crisis.

Near the beginning of his speech, he intended to quote a long passage from 'Fifty Years Hence' to underline how far ahead of his time he had been – almost a quarter of a century before – in appreciating how close scientists were to tapping huge reservoirs of nuclear energy. This was sure to impress his audience. One of the other far-sighted sections of the essay that he did not quote, about the demands new science would place on future democracies, would probably not be welcomed so favourably. So great were the challenges, he had written in 1931, that the current generation's leaders would probably not be up to the task:

> Great nations are no longer led by their ablest men, or by those who know most about their immediate affairs, or even by those who have a coherent doctrine. Democratic governments drift along the line of least resistance, taking short views, paying their way with sops and doles, and smoothing their path with pleasant-sounding platitudes.

This censorious passage may well have given him pause and led him to ask himself two obvious questions. How well had he risen to the nuclear challenge, having foreseen it so long in advance? And how effectively had he worked with the scientists who had created it?

I

TOWARDS THE NUCLEAR AGE

1894–1925

Wells and his liberating 'atomic bombs'

> 'Wells is a seer. His *Time Machine* is a wonderful book . . . one of the books I would like to take with me to Purgatory.'
>
> WINSTON CHURCHILL, 7 December 1947[1]

Winston Churchill almost certainly first heard about 'atomic bombs' from his friend and irritant H. G. Wells, who gave the weapons their enduringly inaccurate name. The term first appeared in Wells's novel *The World Set Free*, published in January 1914, a few months before the outbreak of the First World War. Churchill probably purchased the book shortly after it was published, as he was exceptionally interested in Wells's work. Almost two decades later, he wrote that he had 'shouted for joy' after wolfing down *The Time Machine* and afterwards read every book Wells wrote, twice.[2]

Later, neither Churchill nor Wells could remember their first meeting. It probably took place at one of the garden parties or gentlemen's clubs they frequented in the summer of 1900, cultivating their most talented and influential peers. Both were celebrity socialites, relatively new to the limelight and enjoying every minute of it in their different ways. Churchill, a twenty-five-year-old scion of one of the country's wealthiest political families, had trained as a soldier and fought in active service in Cuba, India, the Sudan and South Africa. In the Boer War, still raging in South Africa, he had been – in modern parlance – an 'embedded reporter'. After his recent return, he had been feted internationally as a hero, having escaped from prison and come home with a twenty-five-pound bounty on his head. Already well known as a lively writer, his work showed the influence of Wells – in Churchill's modest debut novel, *Savrola*, his description of the universe ending as 'cold and lifeless as a burnt-out firework' echoed a passage in Wells's *Time Machine*.[3] Intent on a political career, Churchill had earlier told his mother that he was 'a Liberal in all but name', but was so fiercely opposed to their policy of granting self-rule to Ireland that he chose to stand as a Conservative.[4]

Wells, eight years older, was at the forefront of the new wave of novelists. A son of struggling shopkeepers, he was a proselytising Socialist and in his student days had sported a plain red tie to underline his political allegiance, though now his wardrobe was more discreet, even dapper. By 1900, he had several commercial successes under his belt, including his scientific romances *The Island of Dr Moreau* and *The Invisible Man*, works that showed him to be the kind of energetic, forward-looking thinker that Britain needed in the new century. Dismissing Thomas Carlyle's lament that the modern age had sacrificed its spirituality to machines and materialism, Wells looked forward to an age when scientists and engineers would sweep away the moth-eaten brocade of sentimentalism, replacing it with a sturdy infrastructure of new inventions and innovative methods of production.

He had been a talented student of science, taking a good combined honours degree in zoology and geology, albeit at the second attempt.[5] His principal scientific talent, however, was the one that shone through in his writings and impressed even the best and most conservatively minded scientists – his ability to see where their new theories might take society. Some of the period's finest writers admired him, too, including Henry James, who told him, 'You are for me . . . the most interesting "literary man" of your generation.'[6] Oscar Wilde had described him as 'a scientific Jules Verne'.[7] Although these literary luminaries knew that Wells was no great stylist, they acknowledged him to be the new era's secular priest, praising science and materialism in prose that, though often pedestrian, had an appealing undertow of optimism. This quality is likely to have appealed strongly to Churchill.

In November 1901, ten months after Queen Victoria's death, Wells published his first work of non-fiction, *Anticipations*, a rambling rumination on the future of technology, the Western economy, education and warfare. The book was studded with exciting predictions and had irresistible verve, but was not without flaws – some of its more opinionated passages read as if they had been dictated from the top of a soap box. Less than a week after *Anticipations* appeared in bookstores, Churchill received a copy from the publishers. Only six months before, he had spoken thoughtfully on the future of warfare, commenting that in modern conflicts 'the resources of science and civilisation sweep away everything that might mitigate their fury'.[8] He found plenty in the book to nourish his military thinking, especially Wells's point that warfare was then being waged using strategies long out of date but 'is being drawn into the field of the exact sciences'.[9] It was time, Wells believed, for governments to stop thinking that wars could be won by drunken armies led by ignorant generals who were proud of their old-fashioned ways. Rather, conflicts should be run by technical experts, supported by aerial intelligence. He predicted the invention of aeroplanes ('very probably before 1950') and foresaw the crucial importance in war of dominating the sky, imagining civilians far below the coming aerial battles: 'Everybody, everywhere, will be perpetually looking up, with a sense of loss and insecurity.'

He was rather less convinced of the strategic importance of submarines. 'I must confess that my imagination, in spite even of spurring, refuses to see any sort of submarine doing anything but suffocate its crew and founder at sea.' Predictions like that did not much trouble him in the coming years – he preferred to be specific and wrong rather than vague and correct.

Wells received an eight-page letter on his new book from Churchill, who had read it within days of its arrival.[10] 'I read everything you write,' Churchill began, before launching a thoughtful critique of Wells's technocratic view of government, demonstrating that he was not just another of the scientifically illiterate dullards Wells despised. One of Churchill's fundamental objections was that Wells seemed to assume that the advent of new technology would be accompanied by a concomitant improvement in human nature. 'It is the nature of the beast that counts,' Churchill insisted: 'You may teach a dog all kinds of tricks . . . but you can't improve the breed of a dog in a hurry.'

Churchill was stung by the suggestion that politicians should not be bumbling generalists, learning as they went along, but should come to their posts armed with a technical training. His response summarises a point of view that he held for the rest of his life, as politicians and scientists who worked with him would find out in the decades to come:

> Expert knowledge is limited knowledge: and the unlimited ignorance of the plain man who knows only what hurts is a safer guide than any vigorous direction of a specialised character. Why should you assume that all except doctors, engineers etc. are drones or worse? . . . Is not government itself both an art and a science? To manage men, to explain difficult things to simple people, to reconcile opposite interests, to weigh the evidence of disputing experts, to deal with the clamorous emergency of the hour; are not these things themselves worth the consideration and labour of a lifetime? . . . Wherefore I say, from the dominion of all specialists (particularly military specialists) good Lord deliver us.

That last line hit home with Wells.[11] He replied immediately, saying that he agreed on that point, adding that he should have said that 'the predominating people to come' should be properly educated, not necessarily technically trained. Wells did not agree, however, that he had overestimated the speed at which humans could progress, telling Churchill why he had got this wrong: 'You belong to a class that has scarcely altered internally in a hundred years. I really do not think that you people who gather in great country houses realise the pace of things.'

Soon afterwards, Wells accepted an invitation from Churchill to meet for dinner, replying with more than a touch of condescension, 'To me you are a particularly interesting & rather amiable figure,' adding that he expected humanity to have to face great challenges in the coming years, although he fancied that Churchill was 'a little too inclined towards the Old Game' to be able to deal with them.[12] It was not until the following year that they were able to get together, eventually agreeing to meet at 8 p.m. in the lobby of the House of Commons on 6 March 1902.[13] They then headed out into the fog of London, the horse-drawn carriages only rarely encountering one of the new-fangled automobiles on the city's reeking streets.

They will have looked an odd pair. Little more than five feet tall, Wells was a weedy man with fiery eyes and a hirsute moustache. It never seemed quite fitting that the author of such astringent prose spoke in a hoarse squeak, which contrasted comically with Churchill's arresting baritone, marred slightly by a lisp and mild stutter.[14] Churchill was taller by six inches, already slightly stooped and with red hair that was starting to recede. Both men oozed ambition, especially Churchill, who had already made it plain that he wanted to be Prime Minister.[15]

No record of the conversation remains, but it is a fair bet that the two men explored each other's geopolitics, Churchill as proud of the British Empire as Wells was ashamed of it. They will also have discovered that they had in common a restless confidence, an impatience for fame and a broad-mindedness that made it natural for them to befriend even some of their political opponents. The first seeds of their unlikely companionship had been sown. Wells next wrote to Churchill in the autumn of 1906, sending a copy of his latest book, *A Modern Utopia*. This explored how humanity might best function as a one-party state after it had solved its material problems, mainly by making intelligent use of new science and technology.

The new volume was not to Churchill's taste. He replied appreciatively, tactfully pointing out that the book's main weakness was its lack of a good story: 'I am always ready to eat your suet . . . but I must have the jam, too.' [16] For all its shortcomings, *A Modern Utopia* does appear to have encouraged Churchill to think about where technical developments were taking society – a subject that became one of his favourite themes. By then, he was recognised in Westminster as a conviction politician, unafraid of challenging party bosses. Two years earlier, the rise in support for Protectionism had led him to dramatically cross the floor of the Commons to join the Liberal Party, which supported free trade. In 1908, he married the radical Liberal Clementine ('Clemmie') Hozier – charming, attractive, loyal and firmly supportive of her husband, even though her political instincts went against his.[17] With her support, he became a leader of popular radicalism, introducing the first proposals for unemployment insurance, and minimum-wage rates in industries whose workers were especially vulnerable to exploitation.

Wells was so impressed with his friend's talent that he supported him in a by-election in April 1908, giving his reasons in a controversial newspaper article, 'Why Socialists Should Vote for Mr Churchill'.[18] Churchill was soon back in the House of Commons and making swift progress, becoming a Cabinet minister as President of the Board of Trade when he was only thirty-three and, two years later, the youngest Home Secretary for almost a century.

Around this time, Wells appeared to turn his back on science fiction in favour of novels with social themes. Perhaps as a result of hints from his friend Joseph Conrad that he was squandering his talent, Wells returned to science and introduced 'atomic bombs' to his huge readership in the novel *The World Set Free*.[19] By early 1913, when he started writing the book, Wells was much talked about in literary circles as a self-styled feminist Lothario. The traffic of his bed comprised two wives and dozens of lady friends, most of whom served as muses, a role that seemed essential to maintaining his creative flow. He worked on his story during a stay in the Swiss Alps with a new mistress, the diminutive widow Elizabeth von Arnim, in her gorgeously situated chalet, built using the proceeds of her popular novels and plays.[20] Even by his standards, their relationship was intensely physical, he later recalled.

The World Set Free imagined the consequences of harnessing the energy released in radioactivity. The process had been discovered by the French physicist Henri Becquerel seventeen years earlier and had been the last global scientific sensation of the nineteenth century. Readers of newspapers, magazines and novels had long been gripped by stories of scientists uncovering secrets that would eventually bring the human race to a grisly end.[21] Radioactivity supplied rich material for authors attracted to the long-established Armageddon genre, and it was only a matter of time before it caught the eye of Wells. His interest was piqued by the book *The Interpretation of Radium*, written in 1909 by the English chemist and radioactivity pioneer Frederick Soddy, who based his account on popular public lectures he had given in Glasgow. The book supplied Wells with just the type of raw material he loved to mould into fiction – exciting new science with the potential to revolutionise the way humans live.

Soddy pointed out that radium, a new chemical element, is unusual in 'giving out heat and light like Aladdin's lamp'.[22] If this energy could somehow be harnessed, then 'We stand today where primitive man first stood with regard to the energy liberated by fire.'[23] He foresaw some of the prizes awaiting societies that could capture this energy – they 'could transform a desert continent, thaw the frozen poles, and make the whole world one smiling Garden of Eden'. The problem was that radium and other radioactive elements are stubborn in the extreme – they give out their energy at the same rate, regardless of any attempt to change them. However they are treated – warmed, crushed or stretched – they decay at exactly the same rate and so slowly that it is not feasible to use the energy to drive turbines or do anything useful. Soddy surmised that if it were possible to utilise this energy, there would be huge benefits. He mentioned one consequence of this on the book's fourth page, in a phrase that captured Wells's interest: releases of radioactive energy 'could with effect be employed as an explosive incomparably more powerful in its activities than dynamite'.[24]

Wells read Soddy's account in the early spring of 1913, near the beginning of his stay with Elizabeth von Arnim. His imagination on fire, he asked friends for more information about radioactivity[25] and around May began a novel that he provisionally entitled *The Atom Frees the World*. It seems he was unaware that he was not quite the first to write about the idea of using radioactive energy to make weapons: five years earlier, the French writer Anatole France had published the satirical novel *Penguin Island*, featuring terrorists who make explosives using a gas from which 'radium evolves'.[26] Wells's vision was, however, more graphic, more powerful and ultimately more influential.

Wells and von Arnim wrote in the mornings and then went on long mountain walks in the afternoons, often pausing to make love alfresco on beds of 'sun-flecked heaps of pine needles'.[27] He also read parts of the book to her on the slopes above her chalet, on one occasion offending her delicate sensibilities: punching him with her fur-gloved hands, she complained that he actually '*liked* smashing up the world'.[28]

In his story, Wells as usual let loose hostages to posterity by making absurdly precise predictions, this time about the future of nuclear physics. He imagined scientists in 1933 discovering how to make some chemical elements radioactive and, as a result, releasing large amounts of energy.[29] Twenty years later, a special engine brings 'induced radioactivity into the sphere of industrial production', making energy available at negligible cost and rendering fossil fuels such as oil and gas too expensive to bother with. Out of the economic chaos, incompetent governments wage war with the new 'atomic bombs'. Although quite small – three of them would fit into a coffin[30] – they are powerful enough to reduce a city the size of Chicago to a pile of radioactive rubble. This wipes the Earth's slate clean, enabling Wells to spell out his latest vision of Utopia: people finally realise that war is pointless, nations and races become obsolete, conventional politics ends while a new age of leisure begins, and the entire world becomes a single state that speaks only English. One measure the government takes is to keep radioactive matter under strict control so that no bomb-makers can get their hands on it.

Although not one of his best stories, it sold well and did nothing to harm his literary standing. Many critics admired his still-soaring imagination, though not his balsa-wood characters or the rickety plot. In one of the most complimentary reviews, the *New York Times* saw beyond the limitations of *The World Set Free* and glimpsed why historians, if not literary scholars, would study this 'magnificent' book a century later: 'It is the development of the control and employment of radioactivity that lies at the root of the changes prophesied . . .'[31] Wells did not fully deserve this praise. In the next three decades, he did next to nothing to promote his notion that atomic energy could be important in war and peace. When he wrote a new introduction to the story in 1921, he scarcely mentioned the nuclear science underpinning it.[32]

In the year after *The World Set Free* first appeared, Wells was praised not so much for his scientific vision as for his prediction of the outbreak of what would become the First World War. A few days after the conflict began, his American publisher took out an advertisement boasting that 'the European conflict now in progress' had been 'foreseen and described' by 'the world's greatest imaginator'.[33] During the early stages of the conflict, Wells watched Churchill burnishing his own reputation as an orator of singular power and wit, running the Admiralty and eager – too eager for some – to learn the art of war.

Churchill was well qualified to play a leading role during the conflict. At the Royal Military College at Sandhurst, he had been trained in fortification and other military tactics, though he was never taught anything about bombs, he later wrote, as 'these weapons were known to be long obsolete'.[34] He had then served in the army, killed in battle and demonstrated a strong grasp of both geopolitics and military strategy. In Asquith's government, he had played a role in the founding of the British intelligence services MI5 and MI6 and repeatedly stressed the importance of equipping the country's fighters with the latest technology.[35] He had encouraged officials to get in touch with the Wright brothers in February 1909 to explore the military potential of their invention, the aeroplane – this was four months after Wells published *The War in the Air*, which Churchill read with 'astonishment and delight'.[36]

According to Prime Minister Lloyd George, Churchill did more than anyone else in the Cabinet to promote Wells's idea of 'land ironclads', subsequently known as tanks.[37] Churchill invited Wells to see prototypes in action and helped to ensure that the vehicles became standard equipment for the army. Wells gave him great credit for this and Churchill later repaid it, testifying in court that the tank was solely his friend's idea.[38] Although the jury accepted the case, the truth was that several other inventors had independently hatched the concept.

Churchill's judgement at the top table in wartime proved to be erratic. Within nine months of the start of hostilities, after the disastrous campaign in the Dardanelles, he was obliged to resign his post in the Admiralty. He became so deflated and depressed that his wife thought he might die of grief, but he picked himself up, reported for duty in the army on the Western Front and developed his new hobby of painting, later his favourite pastime. Back in the government, as Minister of Munitions, in less than two years, he supplied the army with increasing quantities of guns, shells and tanks. His return to office in mid-1917 coincided with the first bombing raids on London by Gotha aeroplanes, when Wells stood defiantly on a balcony to witness the beginning of the aerial bombardment of cities that he and others had foreseen.[39] He had long been critical of the government's wartime deployment of scientists and new inventions, especially the aircraft.[40]

It was their views on the Soviet Union that first led Wells and Churchill to fall out, publicly and spectacularly. Wells had welcomed the Bolshevik Revolution of 1917 and supported Lenin's vision of an organised, godless society that embraced science and technology. Always fiercely anti-Communist, Churchill was the British government's most outspoken critic of the Bolshevik regime. It was a 'cancer', he said – a 'monstrous growth swelling and thriving upon the emaciated body of its victim' – and must be eradicated.[41] After he was appointed Secretary of State for War in January 1919, Churchill was fixated on the Soviet threat, hoping that the Allies would 'declare war on the Bolsheviks' and 'send huge forces there'.[42] His words, and the limited British Expeditionary Force sent to Russia, would return to haunt him some two decades later, when he had to work with Soviet leaders who remembered his vilification and his attempts to smother their regime before it could mature into an international force.

Wells took a very different view of the Bolsheviks. Although more critical than many British Socialists of the new Soviet government, he was prepared to excuse some of its failings as unfortunate consequences of a development that was for the best in the long term. Wells defended Lenin's administration as the only possible Russian government, and even defended the murderous Red Terror that accompanied the civil war.[43] In the autumn of 1920, he toured a number of Russian cities and described his experiences in a series of articles that called on other powers to help the Soviets create 'a new social order'.[44] Churchill snapped, attacking him for his naivety and for giving solace to evil fanatics. Wells's reply was weak, but he made one astute point:[45]

> [Churchill] believes quite naively that he belongs to a peculiarly gifted and privileged class of beings to whom the lives and affairs of common men are given over, the raw material for brilliant careers . . .

Churchill was a menace to world peace, Wells harrumphed – he should retire from public life and concentrate on his painting.[46] The two men, professional writers with skins of titanium, quickly put this spat behind them, neither bearing a grudge.

Afterwards their relationship was friendly, intermittently hostile but never poisonous – even after January 1923, when Wells published his political satire *Men Like Gods*, which featured a thinly disguised version of Churchill in the character Rupert Catskill, an Empire-obsessed warmonger, though 'fundamentally a civilised man'.[47]

In November 1922, Churchill lost his seat in the Commons. During his time away from Parliament, he edged back towards the Conservative Party and developed his parallel career as a writer, by far his main source of income. He had already published the first volume of his insider's account of the First World War, *The World Crisis*, described by former Prime Minister Lord Balfour as 'a brilliant autobiography, disguised as a history of the universe'.[48]

At the same time, Churchill wrote dozens of articles and regarded most of them as potboilers. He was, however, especially proud of one, which focused on the future of warfare.[49] This was his first attempt at Wellsian prognostication and it was here that he first alluded to his sometime friend's 'atomic bombs'.

1924–1932

Churchill glimpses a nuclear future

> 'It would be much better to call a halt in material progress and [scientific] discovery rather than to be mastered by our own apparatus and the forces which it directs.'
>
> WINSTON CHURCHILL, November 1931[1]

When Churchill wrote his first predictions for the long-term future of warfare, he was assisted by Professor Frederick Lindemann, who by then had become the main influence on his thinking about science. The men had first met privately in August 1921 at a special dinner arranged by the Duke of Westminster, a mutual acquaintance.[2] Lindemann was keen to befriend Churchill, but their relationship was slow to gel, perhaps in part because the two men, though both aristocrats, were quite different.

Lindemann was a man of complex lineage. His father Adolf was a German and had emigrated to Britain around 1870, later becoming a wealthy business executive. In an observatory and laboratory built in the garden of the family's palatial home in Devon, Adolf spent most of his leisure hours pursuing his hobby of astronomy, the young Frederick often at his side, learning fast.[3] Lindemann's mother Olga was, like Churchill's, American. The Prof always resented that she had given birth to him in Germany, during a visit – he regarded himself as English to the core and strongly denied that, despite his surname, he had any Jewish blood.[4]

Lindemann was a bachelor, teetotal and a vegetarian, much more Conservative in his political outlook than Churchill and with none of his generosity of spirit. Twelve years younger than Churchill, the Prof was untiring in his enmities, notably of Socialists, Jews and colleagues who – in his view – overvalued the arts compared with the sciences. Nor did he much like the company of those with a skin colour different from his own, or even people he judged to be ugly.

Nine months after the two men met, the Prof wrote his father a newsy letter, tinged with excitement and mentioning that he had just received a cable from Clementine Churchill, inviting him to lunch. In the letter, he also commented that he had recently met H. G. Wells ('of all people') at Blenheim Palace, commenting that the writer was 'very second rate as regards brains' and had been put in his place by someone who was 'not considered clever at all'.[5] Besides, the Prof noted, Wells was a comically bad dancer. This note is classic Lindemann – written in his neat hand, it was crammed with obsessive high-society gossip and references to lords, ladies, dukes and duchesses, whose company he adored and whose approval he craved. In his description of Wells – a Socialist and self-evidently not to be trusted – he vents his feelings about someone he plainly regarded as an interloper in his circle. Lindemann may also have been concerned that he had a rival for the ear of Churchill in matters of science.

It took Lindemann almost five years to win Churchill's friendship, but then he never lost it. The Prof proved himself to be as loyal as a lapdog, a charming companion at the dinner table and good company for Clemmie on the tennis court – he was a player of international quality, once progressing to the second round in the men's doubles at Wimbledon.[6] Most important, Churchill was dazzled by the Prof's ability to analyse and solve technical problems, by his skill as a writer of jargon-free summaries on difficult topics, and by his gift for précis.

No one could deny Lindemann's scientific credentials: Oxford University appointed him as a professor of physics in

1919 and he was soon afterwards elected a Fellow of the Royal Society, Britain's academy of science. In the world of physics, he seemed to know everyone worth knowing, including Einstein, who had recently become a global celebrity almost overnight.

Churchill had been used to a steady flow of technical advice from politically neutral civil servants and scientists working for the government. What he wanted was the private and confidential counsel of a tame scientist, whenever he needed it, partly to give him the edge over other politicians. In the early spring of 1924, when Churchill agreed to write about the future of warfare, he had 'a good many ideas' but asked to meet Lindemann to talk about the topic, and apparently did not consult Wells. The Prof responded immediately. He was rightly sceptical of a sensational report of a new 'death ray' that Churchill had read about, and sent him a copy of 'Daedalus', a spirited essay on the future of biology published a few months before by the Marxist geneticist J. B. S. Haldane.[7]

Although the resulting article, 'Shall We All Commit Suicide?', lacks Churchill's usual wit, it is in many ways typical of his prose – easy to read, self-assured and informative but with no pretensions to original scholarship.[8] In one passage, he noted how Germany's sudden surrender in the First World War meant that the world only narrowly avoided an 'immense accession to the power of destruction'. In particular, he feared an escalation in the use of poison gases 'of incredible malignity', not mentioning that, five years before, he had approved their use against the Bolsheviks and, in a note about the strategy in the war against Mesopotamia, had declared himself 'strongly in favour of using poisoned gas against uncivilised tribes'.[9] This would have been easy to arrange – the British government had set up a laboratory facility to develop and produce these weapons in March 1915, at Porton Down in Wiltshire. In the field of chemical warfare, he wrote, 'Only the first chapter has been written of a terrible book,' and he looked ahead to biological weapons, including 'Anthrax to slay horses and cattle [and] Plague to poison not armies only but whole districts'.

Towards the end of 'Shall We All Commit Suicide?', Churchill alluded to the type of weapon that Wells foresaw in *The World Set Free* – 'a bomb no bigger than an orange' possessing 'a secret power . . . to concentrate the force of a thousand tons of cordite and blast a township at a stroke'. He also suggested that conventional explosives could be delivered more effectively, foreseeing vehicles we now know as drones: 'Could not explosives . . . be guided automatically in flying machines by wireless and other rays, without a human pilot . . .?'

In what would become a familiar theme in his ruminations on the new weaponry, Churchill worried that politicians would not be able to handle the terrible devices scientists were about to put at their disposal. Also, these weapons might well get into the wrong hands: 'A base, degenerate, immoral race [could subjugate a more virtuous enemy simply by possessing] some new death-dealing or terror-working process [if they] were ruthless in its employment.' This purple passage made prescient reading fifteen years later, when it seemed that Hitler's Germany might well beat its enemies to the acquisition of nuclear weapons.

Churchill believed that humanity's best hope of avoiding 'what may well be a general doom' was to support the League of Nations. This may have been a nod to international attempts to prohibit the first use of chemical and biological weapons – initiatives that led Britain, the United States, Germany and other countries in June 1925 to sign the Geneva Protocol. The document was registered in the League of Nations Treaty Series four years later.

Of all the essays Churchill wrote around this time, 'Shall We

All Commit Suicide?' made the biggest splash.[10] The article first appeared in Britain in *Nash's Pall Mall Magazine* on 24 September 1924 and was equally successful in North America two months later. The well-regarded former president of Harvard, Charles W. Eliot, praised the piece warmly: 'This statement . . . should be placed forthwith in every American household.' By the time Churchill read this endorsement, shortly after his fiftieth birthday, his political career was again back on track. Having been out of Parliament for almost two years, he won the safe seat of Epping and was appointed Chancellor of the Exchequer by Prime Minister Stanley Baldwin. Churchill carried out his duties in rumbustious style, dominating the Commons with his oratory, calming the House to a contemplative hush moments before summoning roars of laughter. In his first Budget, however, he made what he later called the biggest blunder of his life by accepting the advice of his officials to take the British currency back onto the Gold Standard, at a rate that made British exports disastrously uncompetitive. Among his many enemies in Parliament, many of them Tories, this was yet more evidence of his dubious judgement. Never entirely comfortable at the Treasury, Churchill was easily distracted, even – as Lindemann saw in the spring of 1926 – by atomic physics.

It was a wonder the adult Churchill had any appetite at all for science after the training he had received as a boy. Unlike Lindemann, whose governesses and tutors had given him a first-class education, Churchill had been miserable in his early years at school. He had seen little of his parents, who packed him off as a boy to a disciplinarian school in Ascot, which he hated – he was much happier at home, playing with his collection of some fifteen hundred toy soldiers, most of them from Napoleonic regiments.[11] He fared better under the more kindly regime of his next school, in Brighton, although he appears to have been one of the naughtiest boys in his class. He hated rote learning, especially of Latin and other subjects that did not capture his interest. Yet he was already a precocious reader.

For the young Churchill, mathematics was a trial. Labouring in its 'Alice in Wonderland world', he ground his way through exercises on square roots and the propositions of Euclid, unconvinced that such things were of much use in the real world. After he entered Harrow as a thirteen-year-old, he struggled with the subject, although one of his teachers, a Mr Mayo, was able to convince him that it 'was not a hopeless bog of nonsense, and that there were meanings and rhythms behind the comical hieroglyphics'.[12] Churchill was more inclined to the arts, especially literature. He won a school prize for reciting from memory a 1,200-line poem by the nineteenth-century essayist and historian Thomas Macaulay, whose thinking exerted a powerful influence on him. Macaulay, laureate of British imperialism, argued passionately that it was right for his country to colonise less developed ones and impose 'civilised' values on 'savage' cultures, notably in India.[13]

Believing that Churchill was not bright enough to go to university, his father successfully encouraged him to embark on a military career. At the age of fifteen, the young Winston began to prepare for the entrance examination to Sandhurst, which proved to be a challenge for him. Among the subjects he had to master was mathematics, obliging him to toil in its 'dim chambers lighted by sullen, sulphurous fires . . . reputed to contain a dragon called the "Differential Calculus"'. When he passed the examination at the third attempt, he was relieved to have seen the back of mathematics for ever: 'I am very glad that there are quite a number of people born with a gift and liking for [it],' he later recalled. 'Serve them right!'

As he later wrote, after he 'passed out of Sandhurst into the world' in December 1894, he was perpetually busy: he saw himself as 'an actor' in the 'endless moving picture' of life.[14] Some of his pursuits were cerebral rather than physical – at the end of August the following year, when he was a twenty-year-old cavalry subaltern in Aldershot, he decided to read *The History of the Decline and Fall of the Roman Empire* by Edward Gibbon, who soon became one of his heroes.[15] A year later, in Bangalore, southern India, Churchill was still reading Gibbon's volumes, while lying on his charpoy in the punishing afternoon heat, waited on hand and foot by servants. It was not until the winter of 1896 that 'the desire for learning' belatedly came up on him, he later recalled.[16] In a six-month feast of reading, he gave himself something like the liberal education he wished he had received at university. While improving his polo skills, his intellectual focus was on the dozens of challenging books he approached 'with an empty, hungry mind with fairly strong jaws'.[17] At the end of March 1897, he wrote to his mother listing the books he had gulped down, including a translation of Plato's *Republic*, twelve volumes of Macaulay and four thousand pages of Gibbon.[18] Later, Macaulay and Gibbon were the most important influences on his literary style, their perspectives, rhythms and mannerisms imprinted on virtually every paragraph of his mature writings and speeches.

During his time in Bangalore, Churchill also read at least two science books: Darwin's *On the Origin of Species* and the more superficial *Modern Science and Modern Thought*, a popular introduction by Samuel Laing, a politician and railway administrator.[19] Laing's colloquially written but dense text enabled Churchill to build on his schoolboy science without the pain of complicated mathematics. The author concentrated mainly on the life sciences, but did include an account of the fundamental contents of the material universe – 'ether, matter and energy' – including reverential passages on the progress scientists had made in probing the sizes and masses of atoms. The second part of the book, endeavouring 'to show how much of religion can be saved from the shipwreck of theology', may well have reinforced Churchill's coolness about religious faith. In a letter to his mother, he looked forward to a time when 'the great laws of Nature [are] understood', when 'the cold bright light of science & reason will shine through the cathedral windows' so that 'we can dispense with the religious toys that agreeably fostered the development of mankind'.[20]

Although enriched by this reading, he did not allow scientific methods to intrude on his personal life. He later wrote:[21]

> I therefore adopted quite early in life a system of believing whatever I wanted to believe, while at the same time leaving reason to pursue unfettered whatever paths she was capable of treading.

Almost everyone knew that Churchill was at heart a politician and a man of letters, not an academic and certainly not a scientist. Lindemann, however, claimed to regard him as 'a scientist who missed his vocation':[22]

> All the qualities . . . of the scientist are manifest in him. The readiness to face realities, even though they contradict a favourite hypothesis; the recognition that theories are made to fit facts, not facts to fit the theories; the interest in phenomena and the desire to explore them, and above all the underlying conviction that the world is not just a jumble of events but that there must be some higher unity . . .

This was idiosyncratic to the point of perversity: Churchill was much more skilful as a rhetorician than as an analyst. Yet the Prof's comment is perceptive. Churchill had the combination of imagination and – when it suited him – scepticism that characterises all good scientists. Even in his most partisan speeches, there is a sense that he knew he was advancing a theory of events that might well have to be revised in the light of evidence. 'I have often had to eat my words,' he once said, 'and I must confess that I have always found it a wholesome diet.'[23] Churchill consistently demonstrated that he wanted to keep one step ahead of orthodox thinking about the impact of new science and technology on the human race, especially when it was at war.

The Prof also cultivated Churchill's interest in curiosity-driven science, and found him notably receptive in the spring of 1926 when he gave him a book on the new quantum theory of the atom. The account – whose title is not known – grabbed Churchill's imagination and distracted him from the Budget he was to deliver a few weeks later.[24] On the first Sunday of April, Churchill was at home with his family at Chartwell, his country seat, and spent much of that morning thinking about the book. He was working on the first floor in his spacious study, whose recent refurbishments included a moulded architrave installed in the Tudor doorway. The room – its walls lined with paintings and books – was dominated by a mahogany table, on which rested a porcelain bust of Admiral Nelson, and another of Napoleon.[25]

Chartwell is high on the North Downs in Kent, barely half an hour's drive from Westminster but as quiet as a forest. Set in a little over eighty acres of wooded grounds, the rambling redbrick house then had five reception rooms, nineteen bedrooms and dressing rooms, and a dining room with a gorgeous view across the Weald.[26] When Churchill was not in London, he lived there with his wife Clemmie and their four children, attended by a cadre of some eighteen servants, including a butler, a footman, a chauffeur, a chef and a few gardeners. He relaxed by painting in his studio, by building brick walls on the site, and by supervising the gardening or planning the construction of a new water feature. Yet on that unseasonably warm morning, the grounds of Chartwell coming into bloom, atomic science took precedence over the welcome of spring.

Wanting to make sure that he had correctly grasped the book's gist, Churchill dictated a summary to an assistant and arranged for a typed transcript to be sent to Lindemann for checking. The Prof did not have to make many annotations, as Churchill's distillation was accurate enough to do credit to a scholarship student. In his summary, he pictured electrons in a typical atom moving rapidly around, some of them able to make quantum jumps from one allowed orbit to another – an idea introduced a decade earlier by the Danish physicist Niels Bohr. Churchill's summary includes his first written reference to the atomic 'nucleus', whose existence had been deduced in 1911 by the experimenter Ernest Rutherford, Bohr's mentor. This atomic kernel, as Churchill had read, is tiny and extremely dense, occupying typically only about a billionth of the atom's volume, and almost all its mass. It is the energy stored there that is released in radioactivity and that, in H. G. Wells's imagination at least, might conceivably be used to make weapons.

Most atoms of the known chemical elements are completely stable and hold together for ever, Churchill noted. But some of the elements, such as uranium and radium, undergo radioactive decay, their nuclei transmuting into other varieties and ejecting a few other smaller particles, while releasing huge amounts of energy by atomic standards. At the end of his notes, when he comes to the subject of nuclear disintegrations, he lets loose his imagination and draws an analogy with geopolitics: '[The process of radioactivity] . . . constitutes a liberation of energy at the expense of structure. It suggests the breakup of Empires into independent States, and the breakup of these again into village communities.'

'There are a great many points I want to ask you about,' Churchill wrote in a covering letter to Lindemann, mentioning one that especially preoccupied him. 'Have the relations between music and mathematics been examined in the same way as those between mathematics and physics? If so, there will be a correspondence between music and physics other than mere sound wave [sic].'[27] He had grasped that great advances in theoretical physics sometimes invite an aesthetic response, as Einstein had shown when he commented that Bohr's atomic theory exemplified 'the highest form of musicality in the sphere of thought'.[28]

This summary of Churchill's reading in atomic theory is one of the best testimonies to his scientific curiosity. Although his detractors could reasonably complain about his impetuosity and egocentricity, his intellectual energy was undeniable. Lindemann was one of the few who knew that this vitality extended to science.

After the General Election of May 1929, Churchill's fortunes took a sharp turn for the worse – the electors ejected the Conservatives, and Ramsay MacDonald, no admirer of Churchill, again became Prime Minister. Churchill left 11 Downing Street and would be out of ministerial office for over a decade.

He had starred in Parliament for twenty-seven years, and played ten leading roles on the Commons stage, but now he had only a walk-on part. A small consolation was that he now had much more time for lucrative journalism. One of his most widely reprinted articles was an appreciation of George Bernard Shaw, whom Churchill regarded as 'the greatest living master of letters in the English-speaking world' but also as 'the world's greatest intellectual clown'.[29] Churchill predictably took the opportunity to take a swipe at the government of Communist Russia, Shaw's 'spiritual home'.

After quarrelling with Stanley Baldwin over economic policy, Churchill took a break from politics in the summer of 1929 and went on a three-month lecture tour of North America. He was usually well disposed to the United States – his mother had been born there, and used to unfurl the Stars and Stripes and wave it in front of her two sons every Fourth of July.[30] American culture had long been one of his interests: at twelve, he had seen his hero Buffalo Bill perform in London, and thirteen years later he had been introduced to an audience in New York by Mark Twain.[31]

Churchill's wife once remarked with a smile, 'Winston is half-American and all English.'[32] In the 1920s, however, he – like many of his Commons colleagues – sometimes took a dim view of American foreign policy. As a new Chancellor of the Exchequer in 1924, he rejected the American claim for a share of Germany's war reparations, on the grounds that the US had not signed the Treaty of Versailles.[33] More seriously, he was angry with President Coolidge's administration for having the temerity to seek parity with Britain's naval power, as he told the Cabinet in July 1927:[34]

> No doubt it is quite right in the interests of peace to go on talking about war with the United States being 'unthinkable'. Everyone knows that this is not true . . . We do not wish to put ourselves in the power of the United States. We cannot tell what they might do if at some future date they were in a position to give us orders about our policy, say, in India, or Egypt, or Canada . . .

In the early summer of 1929, he was nonetheless looking forward to being among Americans, whom he believed to be 'a frailer race' with 'more hopes and more illusions'.[35] During his tour, he took full advantage of the money-making opportunities the United States afforded and also experienced the nadir of the Crash, when he walked down Wall Street aware that he was about to make huge losses. At today's prices, they were about half a million pounds.[36]

In late 1929, when he returned to the UK, which was also plunging into a depression, he found himself excluded from the Conservative Party's inner quorum. There was even talk that he might quit Parliament. But plans discussed in Parliament to give Dominion status to parts of India outraged his imperialist sensibilities. 'I shall certainly not retire from politics', he said, 'while the question of our retention of India is still to be decided.'[37]

In the Commons tea room, his critics muttered that he came to the House only to hear the sound of his own voice.[38] It was pointless to spend much time there, he decided – he would be better off developing his career outside Westminster. Having become a 'fan of the wireless' in 1927, his authoritative and witty delivery made him a popular attraction on radio – one producer called him 'the perfect broadcaster' – though he complained that the monolithic British Broadcasting Corporation was muzzling him.[39] During these 'locust years',[40] he continued to live well, dining high on the hog at the Savoy, drinking the finest champagne and running Chartwell in style. By all accounts it was a happy home, though he had his share of sadness, having to cope with the antics of his bumptious son Randolph and the frequent absences of his wife. While he worked long hours in his study and fraternised with his associates, she often took off on vacations alone.

Most of Churchill's time in the decade from 1929 onwards was consumed with literary projects. He wrote a charming memoir, *My Early Life*, a multi-volume biography of his ancestor the first Duke of Marlborough, and began what would become his elephantine *History of the English-Speaking Peoples*. Another money-spinner was *Great Contemporaries*, a collection of essays he had written over the previous few years on 'Great Men of our age' including Bernard Shaw, Hitler and Franklin Roosevelt.[41] In addition to the million or so words contained in these books, he contributed hundreds of articles on a wide variety of topics, many of them spun to bolster his nationalistic creed: 'I am all for old England going on, year after year, century after century, building up each generation, and losing nothing.'[42] Editors got used to him barking down the phone with both a proposal for a new series of articles and a statement of his eye-watering fee.[43] Clemmie knew that these assignments were completed mainly to fund his extravagant lifestyle – most of these journalistic pieces, she complained, were beneath him.

Shortly before he fell out of political favour, he published the penultimate volume of his account of the First World War, *The World Crisis: The Aftermath*, which includes some uneasy reflections on the future of human conflict. In the opening chapter, he dreams of what might have been, given that 'science had produced weapons destructive of . . . whole cities and populations, weapons whose actions were restricted by no frontiers'.[44] He imagines the victorious leaders meeting promptly and resolving that 'the new instrument of world order should be armed with the new weapons of science'. Under the auspices of the League of Nations, an 'International Air Force' is set up, its pilots dedicated to maintaining peace, like latter-day knights of the air. The difficult question of chemical warfare is handled through 'a universal decree forbidding any nation to practise it'.

These ideas were faithfully echoed sixteen years later, when the victors of the Second World War had to consider the most destructive of the next generation of weapons, nuclear bombs. Churchill considered them explicitly for the first time in his article 'Fifty Years Hence', published towards the end of 1931. This was a year of international upheaval, when the world's banking system almost fell apart and Japan seized Manchuria, the first act of what would become a long conflict in the Far

East. In January, the commissioning editor of the *Strand Magazine* Reeves Shaw had written to Churchill suggesting a series of articles including 'Fifty Years Hence . . . a forecast of the state of affairs all over the world – Britain, America, India'.[45] Churchill accepted immediately but changed the focus to 'science, morals and politics'.[46] He was taking a shot at a subject recently covered by his best friend, the rapier-tongued Earl of Birkenhead, who had just completed *The World in 2030 AD*, a book Churchill almost certainly read and discussed with him over dinner at the Savoy Hotel during the fortnightly meetings of The Other Club, which they had co-founded two decades before.[47] If so, Churchill will have read about the changes that might come if nuclear energy were to supersede fossil fuels, making electrical power much cheaper. He will also have seen that Birkenhead's opening chapter ended with a warning about the coming of nuclear weapons: 'As you are reading these words, some disinterested researcher may detonate an atomic explosion which will involve the world, and reduce it to a flaring vortex of incandescent gas.'[48]

To help with the writing of 'Fifty Years Hence', Lindemann sent Churchill an eleven-page draft.[49] The Prof's ideas ranged from mobile phones to the rudiments of a new biotechnology, and he looked forward to a time when 'we shall escape the absurdity of growing whole chicken in order to eat [just] the breast or the wing'. Most important, 'New sources of power, vastly more important than any we yet know, will be discovered,' enabling human beings to have unprecedented control over their environment – he was thinking of nuclear energy. It will not have taken long to adapt the draft – it needed only an inviting introduction, a satisfying conclusion and a few editing tweaks to give the piece a Churchillian sheen.[50]

Among the cultural references in the article was the recent London premiere of the play *RUR* by the Czech writer Karel

Čapek, who popularised the word 'robot' coined by his brother, and Olaf Stapledon's new science-fiction novel *Last and First Men*, which explored the development of the human race over the next two billion years. The book left Churchill cold. He was more taken with the melancholic wisdom of the prophetic poem 'Locksley Hall', written in 1835 by Alfred Tennyson.[51] Churchill quoted six of its couplets and praised the accuracy of its predictions, including the conquest of the air for commerce and war: 'the heavens fill with commerce, argosies of magic sails'.

Strand Magazine published 'Fifty Years Hence' in its bumper, December edition, with a Christmassy cover advertising its most prominent contributors, including Churchill and P. G. Wodehouse.[52] Illustrated with suitably apocalyptic drawings, the article included all of Lindemann's ideas, and highlighted the importance of nuclear energy, which is 'incomparably greater' than the familiar types of energy in use today and by no means a distant prospect:

> There is no question among scientists that this gigantic source of energy exists. What is lacking is the match to set the bonfire alight . . .

Scientists were looking for this match, Churchill wrote, and if they found it, the human race would have in its inexperienced hands 'tremendous and awful powers . . . explosive forces, energy, materials and machinery . . . upon a scale which can annihilate whole nations'. Possession of such powers put human life in jeopardy unless *Homo sapiens* could develop morally and spiritually, he argued. Now that 'the busy hands of the scientists are already fumbling with the keys of all the chambers hitherto forbidden to mankind', he warned that 'without an equal growth of mercy, pity, peace and love, science herself may destroy all that makes human life majestic and tolerable'. This was not the most cheerful Christmas fare for the hundreds of thousands of readers who pondered 'Fifty Years Hence' over their sherries and mince pies. Nor would they have felt much better had they known that a scientist in Britain was, within the next eleven weeks, to discover the particle that would enable the release of the 'gigantic source of energy' that Churchill had mentioned, finally making nuclear weapons possible.

During the following summer, Churchill put together *Thoughts and Adventures*, a collection of some of his best essays, including 'Shall We All Commit Suicide?' and 'Fifty Years Hence'.[53] In his preface, he drew attention to the two pieces and underlined his hope that these 'two nightmares' would be read as more than 'the amusing speculations of a dilettante Cassandra'. He had written the essays 'in deadly earnest as a warning of what may easily come to pass if Civilisation cannot take itself in hand . . .'

By 1932, Lindemann had become one of the most frequent guests at Chartwell. Even his arrival was an event. Emerging from his limousine, attended by his liveried chauffeur and valet, he looked less like a scientist than an investment banker, complete with a velvet-collared Melton overcoat and bowler hat, even during the dog days of summer.[54] Churchill's chef had to make special provision for the Prof, whose vegetarian diet featured an exceptionally narrow range of meals, including dishes made from egg whites (not the yolks), skinned tomatoes, waxy potatoes and only the highest-quality fresh mayonnaise. No one seemed to mind the inconvenience of catering to his tastes. Clementine was extremely fond of him, and the Churchill children treated him like a favourite uncle – he always remembered their birthdays and never left them before pressing a banknote into their grateful hands.

No longer in the limelight and shunned by many colleagues, Churchill put great store by his conversations with Lindemann and a few other acolytes, especially Brendan Bracken, his principal business adviser. Bracken was an Irish-born Tory MP and publisher, unique in his ability to sate even Churchill's appetite for flattery. Bracken, Lindemann and Churchill believed they were living in dangerous times – all the signs were that the 1930s were going to be turbulent. The unfinished business of the Great War made Europe a breeding ground for aggressive dictatorships and conflict.

Over dozens of meals at Chartwell, Churchill and his colleagues reflected on the parlous state of the British economy, on the rise of totalitarianism in Europe, on Germany's rearmament and, very probably, on Stanley Baldwin's belief that it was impossible to resist an attack by enemy aircraft: 'The bomber will always get through.'[55] Yet politics was by no means Churchill's only interest. He spent much of his time writing and was still widening his horizons, reading novels such as Tolstoy's *Anna Karenina*, though he was 'not much attracted by these thin-skinned, self-disturbing Russian boobs'.[56]

The gatherings at Chartwell were grand affairs.[57] The Churchill family – lively, affectionate and sometimes boisterous – would assemble round the vast oak table in the dining room, the head of the household almost always arriving late. The huge windows flooded the room with light and offered a splendid view far out across the Weald, the kind of vista that could lift the most jaded spirits. Churchill – wide-shouldered and, at 210 pounds, considerably overweight – sat at the head of the table like a feudal lord, able to dominate the dining room even when he was not speaking. After one of his occasional monologues, he would relight his cigar in the flame of a candle standing in the silver Georgian holder on the table. The main course, perhaps beef with Yorkshire pudding, was incomplete without a glass or two of wine, only a small part of Churchill's daily consumption of alcohol. He had a well-developed ability to hold his drink and – despite many reports to the contrary – was apparently never drunk, a state he had been brought up to abhor.[58]

Later, Churchill's daughter Sarah remembered one particular lunch in early 1932, soon after Lindemann had published his book *The Physical Significance of the Quantum Theory*.[59] The servants were pouring the coffees and the after-dinner drinks. Lindemann may well have had a brandy, having long before been persuaded by Churchill occasionally to abandon strict teetotalism.[60] Having decided that it was time to display Lindemann's talent for synopsis and simplification, Churchill placed his gold watch on the table and asked Lindemann to summarise quantum theory in no more than five minutes, using words of one syllable. Sarah recalled: 'Without any hesitation, like quicksilver, he explained the principle and held us all spellbound. When he had finished we all spontaneously burst into applause.'

Performances like this impressed Churchill. He appeared to be unaware that his science adviser had a reputation among experts for misunderstanding new and fundamental ideas in theoretical physics, and was increasingly becoming alienated from his peers, among them the undisputed doyen of the nuclear community.

1932

Rutherford: nuclear sceptic

> 'In a recent book, H. G. Wells has discussed in an interesting way some of the future possibilities if this great reservoir of [nuclear energy] were made available for the use of man . . . The possibility . . . does not at present seem at all promising.'
>
> SIR ERNEST RUTHERFORD, Washington DC, 21 April 1914[1]

Rutherford, Lindemann's opposite number in Cambridge, was the Christopher Columbus of the atomic nucleus. He had discovered it, explored it, helped to clarify its strange behaviour and shown that it stores comparatively huge amounts of energy. Quite apart from his pre-eminence as a scientist, he was an accomplished operator in Whitehall, a leading adviser to the British government on the application of science to military problems. By the time it became possible to use nuclear energy to make explosives, he was dead, and it was Lindemann who, through his closeness to Churchill, became by far the most influential British scientist on the early development of the new weapons. Many of Lindemann's peers saw this as a tragedy – Rutherford believed him to be 'a scientist *manqué*', a physicist who had failed to live up to expectations, while many of his colleagues damned the Prof as a scientific amateur.[2]

The most graphic proof of Lindemann's weak grasp of modern fundamental physics arrived in 1932. Early in the year, Lindemann published *The Physical Significance of the Quantum Theory*, his unconventional perspective on the most revolutionary theory physicists had produced for centuries, about the behaviour of matter on the smallest scale. In effect, he tried to explain why the world's leading quantum physicists were wrong about the new theory and why he was right. The book arrived on Rutherford's desk in the Cavendish Laboratory in early January along with a brief note: 'I trust you will not be displeased by the endeavour to base our concepts upon observation rather than adopt the mystical outlook . . . so fashionable of late.'[3] The Prof was appealing to Rutherford's distaste for abstract theory and to his insistence that by far the best way of uncovering nature's secrets was to come up with well-chosen experiments.[4]

Rutherford does not appear to have replied to the note, but it is all but certain that he discussed the book with the quantum theorists in Cambridge. One was his son-in-law Ralph Fowler, who was offended that Lindemann had even considered writing on a subject he knew so little about: 'It is an impertinence of Lindemann to write a book on quantum theory,' he snorted.[5] In agreement with Fowler was his former student Paul Dirac, a co-discoverer of the new theory, whose belief in the power of mathematics in fundamental physics some regarded as mystical.[6] After one of the Prof's lectures, when one audience member despaired of its wrong-headedness and utter lack of originality, Dirac disagreed, making one of his rare interjections: 'No. Only Lindemann could have made those mistakes.'[7]

In the coming months, the Prof's book – in particular, his confused thinking about space and time in quantum theory and the role of mathematics in fundamental physics – lost him the respect of many of the ablest theoreticians.[8] The experience will have been painful for him as it coincided with an especially glorious period for Rutherford and his young researchers.

By 1932, the science establishment had run out of garlands to lay over Rutherford's shoulders. Then sixty, and recently ennobled, he was President of the Royal Society, had a Nobel Prize, and was internationally recognised as 'the generalissimo of the atom-smashing artillery', as the *New York Times* later described him.[9] It was difficult to tell his status from his demeanour and appearance. Tall and thickset, with a bay window of a belly, he had a large drooping moustache and more often than not had a cigarette or cigar hanging from his bottom lip.[10] His tongue was not refined – he was given to swearing like a trooper at apparatus that failed to behave itself – and he had no taste for either fine art or great literature or demanding classical music, much preferring military bands to string quartets. If conversation in the Trinity Common Room after dinner became too highfalutin, he was known to shout in his shrill yet booming voice – still with clear traces of the accent of his native New Zealand, 'Anyone for the Marx Brothers?', before heading off to the movies.[11]

He had largely given up working at the laboratory bench, having proved himself a great experimenter with an unrivalled nose for productive lines of research. It was at McGill University in Montreal, Canada, in his late twenties and early thirties, that he had done much of his pioneering work on radioactivity, mainly in collaboration with Frederick Soddy. They demonstrated that the process – initially a complete mystery – usually involved one chemical element transmuting into another. By 1904, Rutherford had understood from these experiments that 'an enormous amount of energy could be obtained from a small quantity of matter',[12] foreshadowing the energy–mass equation Einstein set out a year later, $E=mc^2$, c denoting the speed of light in a vacuum. This simple relationship implied that even the mass of a penny was equivalent to a huge amount of energy – enough, in principle, to run a small city for hours and wipe it out in seconds.

After he moved to the University of Manchester, where he

discovered the atomic nucleus, the First World War slammed the brakes on his career as a nuclear physicist. He switched his focus from nuclei to submarines and devoted most of his phenomenal energy to finding new ways of using sound waves to locate enemy vessels underwater. Although he had no reservations about working on military projects such as submarine detection, he drew the line when it came to developing gruesome weapons. Max Born suggested to him in 1933 that he meet the chemist Fritz Haber, who played a prominent role in developing and deploying chemical weapons during the war, but, as Born later remembered, Rutherford 'declined violently'.[13]

Rutherford was a prominent Whitehall adviser during the global conflict and served on several government committees, including the new Department of Scientific and Industrial Research, set up to encourage private and public investment in university work. Soon after America entered the war in April 1917, he was a joint leader of the Anglo-French Mission to the United States, sent to brief the Americans on everything the Allies had learned about the application of science to war. It was all 'somewhat one-sided', he and a military colleague reported soon after they returned, though they believed that the advice they had given would soon produce results 'of great value, not only to America, but to the Allied cause in general'.[14] Sharing British secrets had been entirely acceptable to him – it brought to the Alliance, after all, the internationalism he practised in science – as it would be to his researchers and associates in the next global conflict, especially when they came to develop the Bomb.

Within a year of the end of the First World War, Rutherford moved to Cambridge to run the Cavendish Laboratory, and brought with him a list of contacts in the Department of Scientific and Industrial Research, which would later become one of his most munificent funders. As he had done at Manchester, Rutherford regenerated his laboratory, running it like a benevolent dictator, focusing the teaching and research mainly on 'fundamental physics', rather than other physics, such as crystallography, which he dismissed as 'stamp collecting'.[15] Much of Rutherford's greatness lay in his skills as a leader. Soon after he arrived, he recruited several young researchers, many of them foreign, wanting to put the war behind them, and trained as engineers rather than physicists.[16] To most of his 'boys', as he called them, he became a role model – although sometimes domineering and unreasonable, he supervised his researchers with a light touch provided they appeared to be on a productive track, and he would encourage scientists who wanted to take risks on what seemed impossibly long shots.

When Rutherford became director of the Cavendish Laboratory, he was an admirer of Lindemann and gave him a reference for the Oxford chair, a contribution the Prof regarded as crucial.[17] Lindemann told him: 'I am most anxious to work in the closest collaboration with Cambridge and schools of physics.' For several years, the two men exchanged friendly and sympathetic correspondence, though their characters and styles were quite different. Lindemann seemed to be less at home in his laboratory than in stately homes, his head turned by anyone who was rich or who had a sufficiently impressive title, preferably both. One joke doing the rounds went: 'Why is Lindemann like a coastal steamer?' 'Because he runs from peer to peer.'[18] Whereas Rutherford focused strongly on fundamental physics and successfully sought funding from industrial partners to support it, Lindemann supported a wide variety of research topics and set up profitable collaborations with industry.

Like Lindemann, Rutherford was a Conservative, though much closer to the middle of the road and more tolerant of colleagues of different political persuasions.[19] Several of the

Cavendish researchers – especially Patrick Blackett and Peter Kapitza, a committed Soviet citizen who returned to Russia in 1934 – spent evenings and weekends engaged in left-wing political debates, but Rutherford remained unconcerned, provided they left their views at the door when they entered the laboratory.[20] Rutherford was a tough-minded advocate for the Cavendish within Cambridge University, but appears to have been a good deal more popular with his colleagues than Lindemann was at Oxford, where his name was a byword for authoritarianism and aggressive empire-building. J. J. Thomson, Rutherford's friend and predecessor at the Cavendish, fumed when he heard reports of the Prof's management style: 'He seems to think he can run his Laboratory by the methods of a Prussian dictator.'[21]

In Whitehall, Rutherford argued strongly that it was the State's role to support curiosity-driven research, even if it had no obvious commercial or military applications, a view that often irritated the politicians who wanted State-funded science to bear fruit quickly.[22] Nuclear energy, however, was not among the findings of basic research that were likely to prove useful in the foreseeable future, as he noted in a lecture in Washington DC a few weeks after *The World Set Free* was published. Four months later, during a visit to New Zealand, when a reporter asked him if 'atomic bombs' were likely to be made soon, Rutherford's tone was even more dismissive: 'The suggestion of Mr Wells must be considered as a dream of the future,' as up to the present there was 'not the slightest evidence' that radioactive energy could be released quickly enough to make explosives.[23]

By the beginning of 1932, Rutherford's career as a productive scientist was showing signs of tapering off. His hands were so unsteady that he could no longer perform experiments, and his laboratory's results over the previous few years had been thin by his standards.[24] His fortunes changed one morning a few days after Lindemann's book on quantum theory arrived at the Cavendish, during one of the 11 a.m. meetings Rutherford had every day with his deputy, James Chadwick. That morning, 'Jimmy' told him that it may be possible to prove the existence of the neutron, a sub-nuclear particle that Rutherford had hypothesised twelve years before, but that had not yet been observed. Hardly anyone else except Chadwick had taken the idea seriously.[25]

Chadwick raised his already-accomplished game to an even higher level: after scattering helium nuclei by beryllium nuclei on his laboratory bench, and then carefully interpreting the results, he nailed the neutron. It was now becoming clear that a typical atomic nucleus is built not only from protons – each with a positive charge equal and opposite to that of each orbiting electron – but also from Chadwick's electrically neutral neutrons. The discovery was especially exciting as the new particle promised to be a useful probe: a beam of neutrons would not be deflected by atomic nuclei and so should be able to penetrate deep inside the nucleus, though no one knew what would happen after the disruption. Fundamental science was about to be transformed again.

For most people, it was hard to get excited about this – the new particle was billions of times smaller than anything human beings can see and had no obvious uses. The *Manchester Guardian* announced the discovery in a scoop on 27 February, its correspondent James Crowther – Rutherford's press officer in all but name – assuring his readers that the new particle's practical applications 'will doubtless be discovered before long'.[26] Newspapers all over the world reported the story, genuflecting at what scientists assured them was a great discovery. Only *The Times* in London introduced a note of caution, commenting that even if the existence of the neutron were confirmed, for 'humanity in general' the results of this and other nuclear experiments 'would make no difference'.[27] Within thirteen years, the world would see the mushroom-cloud image that confirmed that the neutron was anything but an irrelevance – it was the particle that would trigger the first nuclear explosions.

Chadwick's discovery had given new impetus to the public interest in sub-atomic science, which now expanded along with Rutherford's public profile. A poll of Britain's best brains conducted by *The Spectator* magazine in 1930 had ranked him bottom of the seven scientists on the list, and well below H. G. Wells, Churchill and the runaway winner, the playwright and critic George Bernard Shaw.[28] After 1932, Rutherford could have expected to fare rather better, especially as the discovery of the neutron turned out to be only the beginning of his Indian summer.

Late in the afternoon of Wednesday 28 April 1932, Lindemann was on his way to a special meeting where his nose would be rubbed into his professional inferiority. Rutherford had invited him to be a platform speaker at a gathering of British atomic physicists held in the Royal Society's headquarters in Piccadilly. Billed as a review of the latest findings about atomic nuclei, the meeting was sure to be a regatta of Cavendish showboating, though Lindemann and most of the other participants had no idea that Rutherford was planning another spectacular scoop.

From the beginning of the proceedings Rutherford was in his pomp, delivering a bravura *tour d'horizon* of nuclear physics. The audience probably then expected him to hand over to Chadwick, the man of the moment, but Rutherford surprised them when he cast aside his pre-circulated script and announced that two of his 'boys' had taken one of his own experiments a step further. In 1917, in breaks between his war work, he had become the first successful alchemist when he converted nitrogen into oxygen by bombarding nitrogen nuclei with helium nuclei emitted from a radioactive source. In common parlance, he had split the atom – a finding so surprising that he spent over a year checking it.[29] Pleased as punch, Rutherford reported that two of his young colleagues, the Englishman John Cockcroft and the Irishman Ernest Walton, had split the atom with *artificially* produced beams of particles.

Using a proton beam shot from a purpose-built particle accelerator, Cockcroft and Walton had bombarded lithium nuclei and transmuted some of them into two helium nuclei. The two virtually unknown young physicists had begun one of the most productive techniques of modern particle physics, probing deep into the heart of matter by bombarding sub-atomic particles with as much energy as could be mustered.[30] As a bonus, Cockcroft and Walton's results gave the first direct confirmation of Einstein's equation $E=mc^2$, which enabled the energy of the process to be fully accounted for. A few days later, they demonstrated this to Einstein himself, who soon after wrote of his visit to the experiment and his 'astonishment and admiration'.[31]

Rutherford could not talk in detail about the discovery as it was under wraps until *Nature*, the house journal of British science, published it. But he said enough to draw applause when, with a sweeping hand gesture, he asked Cockcroft and Walton to stand up. Lindemann knew that this latest discovery underlined once again the superiority of the Cambridge physics department over his own, which was so backward when he arrived that it lacked even a proper supply of electricity.[32] Nuclear physics was not one of his strong suits, so it was perhaps cruel of Rutherford to invite him to speak, and perhaps foolhardy of Lindemann to accept and make a contribution that was only a notch above footling.[33]

Two days later, Rutherford was back in Piccadilly, giving an after-dinner speech at the Royal Academy of Arts' annual banquet, attended by royalty, foreign ambassadors and ministers, and hundreds of dignitaries, including the elderly composer Edward Elgar.[34] Rutherford, elated after *Nature*'s publication of Cockcroft and Walton's paper that day, sat at the top table close to Winston Churchill. It is no surprise that they got on well:[35] each had a penchant for independent-minded high achievers, and they were both unabashed supporters of the British Empire, with a Falstaffian presence at the dinner table.[36] Churchill made 'a strong impression' on Rutherford, telling him that Hitler was a man riding a tiger.[37] The acquaintance between the men, however, went no further – it would not have been easy to yoke together in friendship two such rampant egos.

The first speakers at the event – including Prince George, a son of the King – fretted about Britain's broken economy and suggested feebly how the arts might help. It was left to Churchill to lighten the evening. In a perfectly judged speech comparing the Academy with Parliament, he soon had the banquet hall in gusts of laughter. He spoke with his usual blend of pride and self-deprecation about his differences with the current Tory leadership, likening himself to an art teacher who, having fallen out with the Academy's organising committee, was 'not exhibiting this season'.[38]

After the gathering toasted 'Science', Rutherford responded with his usual spirit, taking an ill-informed swipe at the new 'metaphysical' thinking in theoretical physics and risking one speculation that probably went down well with his audience: 'A strong claim could be made that the process of scientific discovery might be regarded as a form of art.'[39] He was probably pleased to read these words quoted in the *Observer* the following morning, though he was in a grumpy mood. Despite a press embargo, the story of Cockcroft and Walton's triumph had been broken by the populist *Reynolds's Illustrated News*, which announced it in the lead article on its front page, riddled with errors and hyperbole.[40] A posse of journalists and photographers was standing expectantly outside the Cavendish early the next morning. Rutherford refused to oblige them with an interview or permission to enter the building, but eventually consented to pose briefly with his protégés outside the laboratory. The result was a classic image: two tired but thrilled young physicists alongside that monster, spruce in his Homburg hat and loose-fitting three-piece suit, proud as a new grandfather.

The discovery was reported in all the leading British newspapers, most of them unsure what it presaged – the *Daily Mirror* commented, 'Let [the atom] be split, so long as it does not explode.'[41] Most of the articles on the discovery reported correctly that when Cockcroft and Walton split a nucleus, they got out more energy than they put in. Yet almost all the reports ignored or glossed over the crucial point that their protons hit the target only once in every ten million shots – much more energy was wasted than released.[42] 'Surely I have explained often enough that the nucleus is a sink, not a source of energy!' Rutherford roared.[43] He filed most of the speculative articles and even kept a few of them folded in a trouser pocket, telling the journalist Ritchie Calder that they were all 'drivel' and 'rot'.[44]

Even before Cockcroft and Walton had published their experiment, its consequences were being talked about by actors on the London stage. At the West End's Globe Theatre, the creaking melodrama *Wings Over Europe* considered how a British government might deal with scientists flaunting the power to use nuclear weapons. The play had opened on 27 April, the day before the Royal Society meeting on nuclear physics. Written by the American Robert Nichols and the

English Maurice Browne, *Wings Over Europe* had premiered successfully on Broadway some three years before and had then been performed across America.[45] The action focuses on a hyperactive young scientist who, having discovered how to unleash nuclear energy, attempts to dictate policy about the new resource to the Cabinet. When he fails, he threatens blackmail but ends up shot in the heart by the Secretary of State for War. In the closing seconds, the politicians receive a message from the 'United Scientists of the World', declaring that they too know the scientist's secret and that, unless world leaders can agree on the wise use of the new energy source, atomic bombs will be dropped all over the world. The London critics were sniffy, though several pointed out the play's prescience. Desmond MacCarthy, critic for the *New Statesman and Nation*, went so far as to say that society was now at a turning point:[46]

> The destiny of mankind has slipped from the hands of politicians (we are all aware of it) to the hands of scientists, who know not what they do, but pass responsibility for results on to those whose sense of proportion and knowledge are inadequate to the situations created by science.

When journalists tried to link Cockcroft and Walton's discovery with the moral questions raised in *Wings Over Europe*, Rutherford and his colleagues found the debate distasteful and declined to join it, underlining MacCarthy's point.

The fears were fanned again a few months later by the publication of the satirical novel *Public Faces*, about a future British government's handling of newly available 'atomic bombs'. One of the characters was Winston Churchill. The book's author Harold Nicolson, a diplomat and a former member of Oswald Mosley's New Party, had begun to write it shortly before *Wings Over Europe* opened in London. Nicolson's story takes place during a June weekend in 1939, about eighteen months after the 'disastrous' adventurist government run by Churchill and Mosley has been ousted. The new, brittle British government learns that it is possible to make atomic bombs, each no bigger than an inkstand, using metal available only in one of its colonies in the Middle East. When France, Germany, Russia and the US oppose the monopoly, panicked officials worry whether Churchill and 'his crowd' know about the atomic bomb and will criticise their timidity. In the farcical climax, a foretaste of *Dr Strangelove*, the Cabinet hears that a nuclear bomb has been accidentally dropped three miles east of the Carolinas, killing thousands of Americans. The British government soon agrees to destroy its atomic bombs and to stop manufacturing them.

Public Faces was a hit with critics in Britain and the United States, and had soon sold tens of thousands of copies. Nicolson had helped to bring the possibility of 'atomic bombs' to public attention, though it was still not a popular talking point, not least because Rutherford and other experts declined to give the theme the slightest encouragement.

At the annual Cavendish Dinner shortly before Christmas, Rutherford and his 'boys' always let their hair down – in 1932 they had especially good reason to celebrate. After a splendid meal in Trinity College of filet mignon, roast goose and cognac-laced mince pies, topped off with a *canapé marinière* in case anyone still had an appetite, Rutherford stood to give his usual toast to 'the Laboratory'.[47] He and his 'boys' had seen some lean years, but the Cavendish was completing perhaps the greatest year in its history. Fortified by fine wines and spirits, he may have struggled to hold back the news that yet another banner-headline discovery was ready to be announced.[48]

Rutherford was now the most admired scientist in the

country. He was, in Einstein's words, 'one of the greatest experimental physicists of all time, and in the same class as Faraday'.[49] No less impressive than Rutherford's strength as a scientist was his ability to cultivate the talent of his 'boys', most of whom regarded him as a hero and tried in some ways to emulate him. Within a few decades, several of them would play central roles in the story of how nuclear energy was used, militarily and commercially.

Although the New Zealander had his shortcomings as a friend and colleague, he got on well with all his associates and peers. Except one: in the next few years, he came to loathe Frederick Lindemann.[50]

MARCH 1933 TO DECEMBER 1934

The Prof advises 'a scientist who missed his vocation'

'[Lindemann] is a genuinely horrible figure . . . He is the only person, I think, whom I have ardently wished to murder.'

ISAIAH BERLIN, 1936[1]

'In his peculiar and indefinable way, Old Prof was a good sport.'

JAMES TUCK, 1961[2]

In the late afternoon of Wednesday 15 March 1933, Winston Churchill was preparing to take a few hours away from Parliament, to chair a talk on nuclear physics. The event, in aid of charity, was to take place in Mulberry House, a Westminster mansion where Frederick Lindemann was to lecture on 'Some Recent Discoveries in Science'. Lady Forbes-Robertson, Lord Ratendone, the Dowager Lady Swaythling and Lady Cynthia Mosley, wife of Britain's most prominent Fascist, were among the blue bloods who had bought tickets for what promised to be a singular event in the season's calendar. By five-thirty, most of the guests were sipping their pre-talk cocktails and getting ready to take their seats.

On an occasion like this, Churchill was careful to look his best – an extra shave, a clean shirt, a splash of lavender water.[3] He was probably feeling cheerful that afternoon as his name was all over the newspapers, following a well-received speech he had given on the state of Britain's air defences. A few weeks before, he had been buoyed by the arrival in the White House of Franklin Roosevelt, who began to implement his New Deal with an expeditiousness that Churchill found deeply impressive.

The owner of Mulberry House, Lord Melchett, then in his mid-thirties, was a veteran of the Great War. A writer and poet whose aspirations were rather greater than his talent, he was also a director and part owner of Imperial Chemical Industries. The wealth he had recently inherited from his father was reflected in the opulence of his home's new refurbishments, much admired in London society: Greek marbles and vases, monumental images from ancient legends carved into walls of travertine stone, bronze doors with polished marble architraves.[4] Melchett knew Lindemann well and had appointed him to a lucrative membership of ICI's Research Council, a luncheon club that brought together some of the company's scientists with senior academics.[5]

Lindemann was a popular figure at this type of event.[6] It was rare for the nobility to have among them such a distinguished figure from the scientific elite, especially one who so relished being in their company and who conformed so willingly to their conventions. He was well known for his fascination with society news, his quiet effervescence, and for the high-pitched grunt that always followed the punchlines of his jokes.

When Churchill introduced the talk, he probably praised the Prof's skills as an expositor. But Lindemann's ability as a speaker was limited. His soft drone often barely carried beyond the front few rows, leaving audience members at the back struggling to keep their eyes open.[7] That evening, however, he appears to have been on his best form. He began with a stimulating, two-minute tribute to Churchill, including this encomium to his intellect:[8]

> [Churchill] has pre-eminently the synthetic mind which makes every new piece of knowledge interlock with previous knowledge:

where the more ordinary brain puts each new experience to the scrap heap, he insists on fitting it into the structure of the cantilever jutting out over the abyss of ignorance.

When Lindemann turned to nuclear physics, he stressed that he would avoid 'mystical questions'. It was rare for him to speak for long without flaying other people's sloppy thinking, so it was no surprise that he quickly found fault in most reports of Cockcroft and Walton's experiment, 'which the newspapers so inaccurately describe as "Splitting the Atom"' (he was right – the duo had split atomic nuclei, not atoms). Aware that most of his listeners barely knew what atoms were, he went back to basics, setting out the types of particle from which atoms were then believed to be made – the electrically charged electron and proton, and Chadwick's electrically neutral neutron, which Lindemann dubbed 'the eunuch'. Like most other scientists presenting to non-specialists, he went out of his way to quash fears that the atom-splitting experiments would set off some sort of global conflagration. The important point was that they clarified our understanding of matter – this was why he added nuclear physics to his department's already wide portfolio of interests.[9]

He moved on to the scientific topic of the day, the amazing photographs of showers of cosmic rays raining down on Earth, taken by two of Rutherford's 'boys', the Englishman Patrick Blackett and his Italian colleague 'Beppo' Occhialini.[10] This last of the hat-trick of great discoveries at the Cavendish may have been painful for Lindemann – the two experimenters had designed and built just the kind of device the Prof admired for its cunning design and brilliant efficacy. Their technique of making charged cosmic rays photograph themselves as they passed through a detector had rendered the competition obsolete and made it relatively easy to detect antimatter, until then extremely elusive. It probably did not make things any easier for Lindemann that he had met Blackett, an outspoken Socialist, and taken an instant dislike to him.[11]

When Lindemann sat down, Churchill rose to give the vote of thanks. The audience was soon laughing, though his jokes about splitting the atom and Bohr's atomic model were uncharacteristically lame. He nonetheless successfully drove home the essential point: although the new discoveries looked complicated, in the main the march of scientific thought and discovery was not towards an ever greater heap of ideas, but towards simplification.

Among the press representatives in the audience, the *Daily Telegraph*'s reporter was impressed by Churchill's skill in his 'new role' of reducing abstruse science to its simplest terms. The next morning, the *Telegraph* featured an article entitled 'Splitting the Atom', with the sub-headline 'Mr Churchill's aid to science', concentrating on his light-hearted summary.[12] The reporter had been struck by Churchill's comment about how marvellous it was that the new particles could be manipulated 'as if they were animals in the Zoological Gardens'. Apparently echoing Churchill's conclusion, the article ended by positing that the results of the new research 'at any moment might break upon the world and almost certainly revolutionise the life of man'. One possible revolution would involve the use of nuclear energy to make bombs. Although Lindemann had dismissed chatter about the imminent arrival of nuclear weapons, in the next few years he would see that such notions were not so far-fetched. As Germany and other European countries became more repressive and racist, he helped to provide sanctuary for refugee scientists who later played crucial roles in raising awareness that nuclear weapons might soon be built and be in Hitler's hands.

*

Lindemann's audience at Mulberry House had witnessed his most appealing trait: the unswerving loyalty he showed to his friends. He had also demonstrated his skill in the underrated art of rendering a jumble of scientific ideas, statistics and opinion into a lucid conspectus. If, however, the conversation that evening ventured into territory outside science – as it probably did – the audience may also have been given a taste of his philistinism. 'He had stone-blind spots outside his own territory,' in the opinion of Churchill's friend Violet Bonham Carter, who later recalled Lindemann's sniffing at a portrait by Rembrandt or Velázquez, commenting that such pictures were a waste of money – much better likenesses were achieved with a camera.[13] On another occasion, he chided a friend for revering a poet who, Lindemann said, was merely a 'feller [who] writes quite decent verses'.[14]

Lindemann had arrived in Oxford in 1919, a professor at the age of thirty-three, knowing that the university authorities wanted him to breathe new life into their moribund Clarendon Laboratory and put it on a par with the Cavendish.[15] The university appeared to have found the ideal person to deliver its dream, as he had already proved himself a first-rate researcher and was held in high regard by the leading scientists of the day. In Berlin before the war, he had flourished at the court of Einstein. The Prof was an admired experimentalist and theoretician, working successfully on a wide range of topics, from the way matter behaves at ultra-low temperatures to astrophysics. He was no mean inventor, too, and took out several patents.[16] As soon as the war broke out, he returned to England and, after applying unsuccessfully for a commission to fight, joined the Royal Aircraft Factory at Farnborough, the UK's principal centre of aviation research. There, he famously worked out the principles involved in getting aircraft out of a tailspin, and courageously showed that his ideas were correct by taking the controls of a demonstration flight.[17]

Lindemann seized the opportunity Oxford had given him, reinvigorating its physics department and winning resources for new staff, equipment and buildings.[18] In getting things done so quickly, however, he made more than his share of enemies, some from his attacks on the university's bias towards the arts, some from his pungent sarcasm, some from a combination of the two. No sooner had he put down his roots as a leader of scientists than he had gone to seed, repeatedly demonstrating that he did not fully understand the revolutionary import of Einstein's theory of relativity.[19] It was not long before there were mutterings in the Common Rooms that he was not quite the scientist he had been cracked up to be.[20]

Lindemann had withdrawn from front-line research in physics, knowing he was never going to be able to compete with scientists of the calibre of Einstein and Rutherford. During a soirée at the house of the Fabian economists Beatrice and Sidney Webb, Lindemann was asked why he no longer did research, and replied candidly: 'I can understand and criticise anything, but I have not got the creative power to do it myself.' Deep down, he was aware that he was Salieri to Rutherford's Mozart.[21] Although he affected modesty when talking with his friends, they knew he believed himself to be nothing less than a great man, accepting that only a handful of others, including Einstein and Churchill, were above him.[22] Although he was never going to be a great physicist, he could – if he played his cards right – use his knowledge as a scientist to inveigle his way into the highest echelons of the Conservative Party, and put his brilliance on display. As a result, he chose to spend much of his time and energy becoming one of Churchill's most attentive courtiers, a choice that cost him some of his dignity but which certainly paid off.

On the day Lindemann gave his talk at Mulberry House, he read in *The Times* that Hitler's seizure of power in Germany was almost complete.[23] In a series of maniacal speeches, the Führer had continued to exploit widespread resentment over the terms forced on Germany by the Versailles Treaty and had whipped up anti-Semitism into a frenzy, now backed by the full resources of the State. Gangs of brown-shirted storm troopers were rampaging across Germany, intimidating, beating and sometimes murdering Jews and Communists.[24] The report in *The Times* began: 'During the past week the Nazi steam roller has passed over every one of the seventeen Federal States of the Reich and has left a Brown uniformity behind it.'

In Britain, Hitler was not yet perceived to be an international threat. Even Churchill was initially inclined to accept the Foreign Office's view that the German leader was either a harmless lunatic or a gallant ex-corporal out to restore the morale of his country.[25] Less than a month after the Mulberry House talk, however, Churchill told the Commons that, with an aggrieved Germany rapidly acquiring military parity with its neighbours, 'We should see ourselves within a measurable distance of the renewal of general European war,' words that then had no resonance.[26] He worried, too, about the governance of India, a subject not likely to win him a significant popular base of national support.[27] Out of tune with the times, he was now easily put off his stride in Parliament. One commentator thought Churchill's oratory in the Commons was on the wane and that he was now like a 'great romantic actor trying to become a heavy tragedian'.[28]

Lindemann had many Jewish friends and associates in the German physics community and he knew they were being harassed, beaten up or worse, and that many excellent Jewish scientists were soon likely to be thrown out of their jobs.

The dam was about to burst: in January 1933, Germany was home for about 525,000 Jews, about half of whom would soon emigrate.[29] Lindemann's attitude to Jewry was contradictory: he was often gratuitously anti-Semitic but was appalled by Nazism and its treatment of Jewish scientists.[30] Yet he was the first senior British scientist to help his displaced colleagues in Germany. He knew that if he could recruit some of the best of them, the academic standard and status of his department would rise sharply, so he sought funding from his old friend Harry McGowan, the dictatorial chairman of ICI, who readily agreed despite his company's relatively poor financial health.

On 14 April 1933, when the first reports of the arrival in London of Jewish refugees were published, Lindemann was being chauffeured to Germany in his Rolls-Royce, sometimes taking a nap in the bed installed in the rear compartment.[31] That Easter, he spent several days meeting with Jewish scientists and offered some of them an opportunity to move to England by taking up the temporary posts he had established in his laboratory.[32] The result was an influx of top-quality talent into Oxford University's physics department – Lindemann had seized an opportunity to do good, and to do well.

Six weeks after Lindemann's trip to Berlin, British academics began to set up an organisation to help the growing number of refugees from German universities and other scholarly institutions. The prime mover was William Beveridge, director of the London School of Economics, who had recently seen something of the panic among persecuted Jewish academics in Germany. Within a few weeks, British newspapers published a letter announcing the formation of the Academic Assistance Council, with Rutherford as its president. The letter was signed by forty-one academics and leading public figures, including J. B. S. Haldane, the poet and classicist A. E. Housman, the physiologist A. V. Hill and the economist John

Maynard Keynes.[33] Lindemann was not among them. Behind the scenes, the Council hoped that the refugee crisis would be short-lived, so it awarded its first grants for only a year and on the condition that beneficiaries could not apply for permanent academic posts, which were assumed to be reserved for British applicants.[34] This did not prevent the circulation of leaflets in Cambridge urging that university students should be taught only by Britons, not by 'politically biased aliens'.[35]

Donors had contributed almost ten thousand pounds to the Council's coffers by the end of July, but Lindemann attracted more for his private initiative, especially from his friends in ICI.[36] The Council appears to have regarded Lindemann's work as complementary to its own, although his policy of enticing the best scholars to Oxford by paying double the top grant given by the Council was bound to be divisive. Lindemann played the game skilfully: he promoted his own Oxford-focused interests, while offering the Council's broader project enough support to deflect charges of selfishness. To anyone of Rutherford's shrewdness, however, Lindemann's tactics were transparent.

One of the scientists the Prof assisted was Einstein. In early May 1933, Lindemann received a letter from Einstein, asking if 'a small room' in Christ Church could be made available for a short visit.[37] He was writing from his 'very pleasant exile', billeted with his wife in a villa on the Belgian coast near Ostend, protected twenty-four hours a day by two armed guards. 'I shall never see the land of my birth again,' he predicted correctly. A few days later he wrote again, confirming Lindemann and Churchill's view of what Hitler was up to: 'I am reliably informed that [the Nazis] are collecting war material and in particular aeroplanes in a great hurry. If they are given another year or two the world will have another fine experience at the hands of the Germans.'[38]

Einstein had visited Britain in the previous two summers, co-hosted by Lindemann, and was planning another stay before taking up a permanent position in the US, at the Institute for Advanced Study in Princeton. Before setting sail for the US, Einstein visited England at short notice, going to the House of Commons and meeting several politicians, including Churchill. The two men – each in exile in their own way – met at Chartwell for lunch on a stifling Saturday afternoon in late July, with Lindemann in attendance.[39] Churchill wore no tie and sported a Stetson-like hat, while Einstein was draped in a suit of white linen that looked as if he had slept in it. A few hours after the meeting, Einstein wrote to his wife that Churchill 'is an eminently wise man . . . It became clear to me that these people have taken precautions and will act resolutely and soon.'[40] Perhaps Einstein had misunderstood his host's influence, for Churchill was in no position to speak for his country.

Lindemann was one of Einstein's hosts during his visits to Britain, enabling him to spend weeks in Christ Church, away from the clamour of the press. The Prof disliked publicity and wanted nothing to do with humanitarian appeals,[41] so he played no part in an evening event organised by the Academic Assistance Council and other groups to highlight the need to support refugees. Einstein was the star attraction. On 3 October, he appeared with Rutherford on the stage of the Royal Albert Hall in front of ten thousand people. Rutherford gave him a messiah's welcome, inviting him to address the crowd by thrusting out a comradely arm. The evening gave a boost to the coffers of the hard-pressed Council, which was straining to support the dozens of desperate refugees who were arriving every week. By the outbreak of the war, some two thousand academics had registered with the Council, most of them receiving financial aid.[42]

Four days after the Albert Hall event, Einstein set sail for America, never to return to Europe.[43] Two months later, a week before Christmas, Einstein wrote to congratulate Lindemann on his 'wonderful work for exiled scientists'.[44]

The sage of Princeton now seldom ventured outside his new hometown but did accept an invitation to speak in December 1934 at a meeting in Pittsburgh, where he talked about his equation $E=mc^2$.[45] Before the event, an interviewer asked him whether it would ever be possible to harness energy from atomic nuclei, as dozens of popular articles and books had speculated. Pointing out that he was not a scientific prophet, he made what amounted to a prophecy: 'I feel absolutely sure, nearly sure, that it will not be possible . . . it will be like shooting birds in the dark, in a country where there are few birds.' Quoting his opinion, the *New York Times* assured its readers that even if there were another terrible war, there was virtually no chance that either side would be able to use bombs of the type Wells had envisaged. Over in London, however, one scientist was telling anyone who would listen that Einstein was wrong – and that nuclear weapons were a distinct possibility.

SEPTEMBER 1933 TO FEBRUARY 1935

Szilárd's nuclear epiphany

> 'Even in 1925, Szilárd felt he was already someone important; so, he reasoned, all scientists would benefit from his acquaintance.'
>
> EUGENE WIGNER, 1992[1]

To tell Leó Szilárd that something cannot be done was to spur him on to a crisp refutation. He liked nothing better than to disprove dogmas and to shock everyone with his own boldness and ingenuity.[2] In this way, he came to be the first to outline how nuclear energy might be harnessed, possibly to make bombs, and to warn that such weapons might soon be in the hands of tyrants. In the mid-1930s, he was obsessed with bringing the danger to the attention of the highest authorities in politics and science – a cosmopolitan pied piper, albeit one who struggled to attract followers.

One of Szilárd's strengths was that he was extremely well connected – among his many friends and collaborators was Einstein. The two had met in Berlin in late 1920, when Szilárd was a twenty-two-year-old postgraduate student with prematurely receding hair and always in need of a comb.[3] He was already in a hurry to make his mark in physics, having recently arrived in Berlin, fleeing the political upheavals and anti-Semitism of his native Hungary. Within a few months of the young man's arrival at the city's university, he walked up to the world's most famous scientist and requested private tutorials for himself and a few colleagues. Einstein agreed. This was typical both of Einstein's generosity and Szilárd's directness – the young Hungarian's self-confidence was always on display, like plumage.

Among Szilárd's friends in Berlin were his fellow countrymen Eugene Wigner, a theoretical physicist, and the mathematician John von Neumann, fellow Jews educated in the same school district and destined to become leading lights in the American effort to build nuclear weapons. Of the three, Szilárd was not the brightest but was by far the most self-promoting, always teeming with imaginative ideas that he would press on everyone he deemed worthy of his attention.

The 1920s were Szilárd's best years as a scientist. He proved himself as a physicist, hobnobbed with the subject's royalty and made his name. Although his assertiveness sometimes bordered on impertinence, he could not be ignored. More quickly than many of his colleagues, Szilárd saw that Germany's seething discontent with rampant inflation, rising unemployment and the Versailles Treaty was festering into a dangerous compound of militarism and xenophobia. Yet, as he studied H. G. Wells's *The Open Conspiracy*, admiring its statement of the problems facing the world in the late 1920s, he believed that democracy might survive in Germany for a generation or two. Hitler soon disabused him and, by early 1933, life had become intolerable for him and his fellow Jews. The last straw was Goebbels's crackdown in mid-March 1933 on the employment of Jewish academics – universities could now employ Jews only in the same proportion as their presence in the population, about one per cent. On the afternoon of 30 March, two weeks before Lindemann was chauffeured into Berlin, Szilárd packed two suitcases and left his lodgings. He headed for the railway station and took a train to Vienna, about to embark on his career as an itinerant nuclear ambassador.

Out of the blue in the early autumn of 1933, while crossing a London street, he hatched the idea for harnessing nuclear energy. At that time, he was living near the British Museum at the ostentatiously ornate Imperial Hotel, its redbrick frontage

– complete with corbels and gargoyles – overlooking the small park on Russell Square.[4] His savings, some of them generated by profits from patents, enabled him to live there for a few weeks. Such comfortable accommodation was well beyond the means of most newly arrived refugees, the majority of them almost destitute and forced to come to terms with an alien culture, a foreign language and the possibility that they would not see their homeland for years. Though proudly unsentimental, Szilárd was as homesick as any of them.

He had a desk at the nearby London School of Economics, where he worked for the Academic Assistance Council, which – at least in his own mind – he had conceived shortly after he left Germany, during a conversation in Vienna with William Beveridge.[5] Szilárd was not the easiest of colleagues at the Council – he was never going to be a popular figure in the English establishment, its officials more accustomed to cringing obeisance than to his brand of unflinching directness. Szilárd approached everyone in the same way, firing words at them like bullets.[6]

He followed international politics closely, reading *The Times* in the hotel lobby every morning, before returning to his room for a few hours' soak in the bath. The reports from Germany painted a picture of violent racism and increasing brutality, fostered by Hitler, who for many people in Britain was still a richly comic figure, almost Chaplinesque, surely soon to be found out and pelted in the stocks with custard pies.[7] Szilárd knew that life for many in Germany, especially Jews, was much worse than most people in Britain appreciated.

He planned to change his routine on Monday 11 September by travelling to 'atom smashing day' at the annual meeting of the British Association for the Advancement of Science in Leicester, where Rutherford and a few of his 'boys' were to talk. But that morning Szilárd woke up with a cold and decided to stay in bed. The next day, he read about the session in *The Times*, its long report featuring an eye-catching comment from Rutherford: 'Anyone who looked for a source of power in the transformation of . . . atoms was talking moonshine.'[8] Those words made the front page of the *New York Times*[9] and ricocheted around the British press for days, with some reports giving the impression that Rutherford had pronounced the pursuit of nuclear energy to be futile.[10] Szilárd immediately rose to the challenge. Perhaps Rutherford would soon be writhing in posterity's elephant trap, with all the other mavens rash enough to pontificate on the future of science?

In the next few days, Szilárd thought about Rutherford's remarks as he was walking around London. Crossing the busy thoroughfare in front of his hotel, he stopped for a traffic light and, just as it turned green, he had his epiphany:[11]

> It suddenly occurred to me that if we could find an element which is split by neutrons and which would emit *two* neutrons when it absorbed one neutron, such an element, if assembled in sufficiently large mass, could sustain a nuclear chain reaction.

He had envisaged a runaway release of energy: one neutron would beget two, which would beget four more, which would beget another eight, and so on, releasing more and more energy at each stage. If something like this happened in nature, then it might be possible to capture energy on an industrial scale and build nuclear bombs. Szilárd quickly realised that he had been introduced to such a possibility the year before, when he read H. G. Wells's *The World Set Free*, which, as Szilárd later wrote, 'made a very great impression on me', although he 'didn't regard it as anything but fiction'.[12] He now saw that chain reactions had the potential to turn Wells's Alpine whimsy into a reality that could be catastrophic if Hitler's scientists were first off the mark with nuclear technology.

Szilárd was now a man possessed, on a mission to save the world, albeit with limited resources.[13] During the winter, he was feeling the pinch financially and moved several times to save money on accommodation, once renting a room that used to be a maid's closet.[14] Having concluded that it would be easiest to set up a nuclear chain reaction using the chemical element beryllium, he began his campaign to persuade other scientists to back the idea. Szilárd knew that an ill-prepared charge into the Cavendish could well be disastrous, so he talked first with some of Rutherford's former colleagues, including Patrick Blackett and G. P. Thomson, now both running their own shows in London.[15] Neither was interested, and nor was Sir Hugo Hirst, founder of the General Electric Company, despite Szilárd's attempt to stimulate his interest by sending him a copy of *The World Set Free*.

By early spring, Szilárd had worked out his chain-reaction idea in detail and he decided to take the unconventional step of patenting it. He filed his twelve-page application on 12 March, and then went for broke, seeking an audience with Rutherford, who agreed to meet briefly on the first Monday in June.[16] By then, Rutherford's glory days were over – the most exciting nuclear science was being done elsewhere, notably in Paris and Rome, where physicists had demonstrated that perfectly stable nuclei could be forced to become radioactive by bombarding them with other nuclear particles, including neutrons.[17] The atomic nucleus was behaving oddly, but this was nothing compared with what was about to be revealed.

Shortly before noon, Szilárd arrived at Rutherford's office. It was bestrewn with papers, with dormant apparatus scattered and one item of high-tech equipment perched on his desk – a telephone, one of only two in the entire building.[18] According to Chadwick's recollections much later, the idea of the chain reaction was not new to Rutherford.[19] Szilárd nevertheless ploughed on, explaining the idea in terms he thought Rutherford would find especially congenial, in the process making errors that the world's most accomplished nuclear physicist was not slow to point out.[20] The last straw was Szilárd's remark that he had patented his thinking on the subject, flouting the convention that no idea in basic science should ever be anyone's preserve, still less a source of income.[21] Szilárd later recalled how the meeting ended: 'I was thrown out of Rutherford's office.'[22]

Licking his wounds, Szilárd decided that he had no option but to behave out of character and grovel. He sent Rutherford what amounted to a begging letter. When this failed,[23] Szilárd tried to mend fences with the great man, asking him to regard the patent as the property of the entire community of physicists, an argument unlikely to move Rutherford. Later, Szilárd signed over the chain-reaction patents to the British Admiralty, but only after the War Office had turned him down, saying they saw 'no reason to keep the specification secret'.[24] Rutherford still offered him nothing more than a cold shoulder.

Szilárd spent the summer and autumn lobbying companies for funds and doing nuclear research at St Bartholomew's Hospital.[25] There, he and his colleague Thomas Chalmers invented a neat way of making chemical elements radioactive and medically useful, by bombarding them with neutrons. Still used today, this work was widely admired by his peers and did much to establish him as a respected nuclear physicist. He was, however, finding it hard to make progress with his ideas on releasing nuclear energy, except with the physicist G. P. Thomson, who told the Academic Assistance Council that the Hungarian gadfly might be on to something:[26]

The chance of atomic disintegration becoming of commercial importance in the future is very real, and the type of experiment which Dr Szilárd proposes seems as likely to lead to it as any other.

There was, however, no room for Szilárd in Thomson's laboratory at Imperial College or, for that matter, any of the other university departments he approached. Szilárd seemed to have been blackballed, especially by physicists within Rutherford's circle.[27] So it made sense for him to approach Oxford University's Clarendon Laboratory, an especially welcoming harbour for refugees, growing in status and now aspiring to compete with the Cavendish. By the end of 1934, Lindemann's project to rejuvenate his department was beginning to take flight.[28]

Despite the harsh economic climate, funds were coming in steadily and he was beginning to make plans for new premises that would give his physicists much-needed additional facilities. Using mainly funds he had won from ICI, the Prof had recruited a strong group of refugee scientists specialising in cryogenics, the science of how matter behaves at ultra-low temperatures. Foremost in the group was one of Lindemann's friends from his Berlin days, Franz Simon, another fugitive from the Third Reich and a world-class scientist, universally popular for his warmth and approachability (he later anglicised his first name to Francis).[29] His wife swore she would never return to Germany, even if it meant scrubbing floors in Britain for the rest of her life.[30]

The Simons' cottage home, a short walk from the Clarendon, was a welcoming pied-à-terre for refugees when they first arrived in England. The family was always ready to put an extra seat or two at the dinner table and find their guests a bed for the night. So Szilárd was wise, when he arrived in the city in January 1935, to knock on their door and announce that he wanted to work at the university. Franz Simon fixed him up with modest accommodation nearby and an opportunity to meet Lindemann to make his case for a post at the Clarendon. But the Prof was able to offer Szilárd nothing more than a desk.

Most émigré physicists who had recently arrived in Britain shared the same plight. The authorities treated them decently and enabled them to make a living, but made it clear that they were only visitors and should seek permanent sanctuary elsewhere. Of the sixty-seven physicists known to have arrived in the UK before the war, only three secured a permanent academic job, while almost half (thirty-two) re-emigrated, mostly to the US.[31] Szilárd was one of those who took that path, sailing to New York in February 1935. He had a brief change of heart, returning in the early summer to London, where he wrote a flattering letter to Lindemann, once again appealing for help.[32] This time it did the trick: the Prof secured Szilárd a Fellowship, enabling him to return to the UK on a comfortable salary. With a chutzpah that will have surprised no one who knew him, Szilárd took up the position but then travelled to America whenever it suited him, spending only part of his time in Oxford as a strolling player in the field of nuclear theory. Szilárd had grievously abused Lindemann's patience and generosity, but showed no remorse.[33]

With the Prof's laboratory now flourishing, his task of establishing Oxford as a leading research centre was almost complete, and he turned his attention to a larger canvas – national politics. At the same time as he was trying to secure financial support for Szilárd, Lindemann was beginning his campaign to be the next Conservative MP for Oxford University, though his unpopularity among his colleagues was his undoing, and he lost the election.[34] He also wanted to be involved in setting the direction of British military research, which he and Churchill believed was being run so incompetently that it put the country's security at serious risk.

Lindemann probably first heard about the possibility of

building nuclear bombs from Szilárd.[35] Yet the Hungarian's ill-substantiated warnings did not register with the Prof, who was much more concerned with pressing matters of national defence. As Szilárd and everyone else who spent any time in the Oxford physics department knew, Lindemann's focus had now shifted to London. The Prof was about to take his place alongside Churchill in Whitehall in their first major political battle.

FEBRUARY 1934 TO OCTOBER 1938

Churchill fears war – and that nuclear energy will soon be harnessed

> 'In the fires of science, burning with increasing heat every year, all the most dearly loved conventions are being melted down; and this is a process which is going continually to increase.'
>
> CHURCHILL to the Commons, 21 March 1934[1]

The coming of the aeroplane – 'this cursed, hellish invention' – had revolutionised the position of Britain, Churchill believed.[2] Only twenty years before, his country knew it was able to defend its islands with its mighty navy, but no longer: 'That is the thing that is borne in upon me more than anything else.' His nagging worry was that the Luftwaffe would launch a devastating attack on a scale hundreds of times worse than in the First World War, which had left Britain – especially military experts in Whitehall – fearful of a Wellsian blitz. For Churchill, London was 'the greatest target in the world', as he told Parliament in July 1934: the city was like a 'fat cow tied up to attract the beasts of prey'.[3] George Bernard Shaw agreed, but construed the implications differently – he proposed in a radio broadcast that the best response to an aerial bombardment of London would be to surrender, as that is what the enemy would also do when the British retaliated by bombing their capital.[4] Aerial bombers were, in Shaw's dream world, 'angels of peace'.

In an unusually well-received Commons speech in November 1934, Churchill predicted what would happen if London were subjected to such a bombardment – at least 'thirty thousand or forty thousand people would be killed or maimed', with some 'three million or four million people' driven out of the city.[5] The government should increase its expenditure on Britain's air force, substantially and without a moment's delay, he said, adding that it would be a great mistake to neglect the scientific research into preventing attacks by aerial bombers:

> Certainly nothing is more necessary, not only to this country but to all peace-loving and peace-interested powers in the world and to world civilisation, than that the good old earth should acquire some means or methods of destroying sky marauders.

Churchill pestered the government to invest in military defence, berating ministers for their complacency and ineptitude. Among leading British politicians at the time, Churchill did more than anyone else to draw attention to the German threat, apart from his more conciliatory friend Austen Chamberlain (now often forgotten, perhaps mainly because he died in 1937, before events proved him right). It was Churchill, however, who spoke most vigorously in support of military science, though many in the Commons thought he was being alarmist and – with his India cause petering out – looking for a new hobby horse to ride back into the spotlight.

In his speeches and writings, he seemed to hold the work of scientists in awe. When he spoke of these 'high authorities', he gave his reflex romanticism full vent, regarding them as an alien tribe of super-humans 'gathering knowledge and power with ever increasing and measureless speed', as he had written in 'Fifty Years Hence'. He spoke in the same vein in the Commons three years later: 'Science and invention are sweeping all before them,' and it was vital to support them so that their results could be applied in every branch of the military. Pure science, pursued for the sake of curiosity and without heed to practical benefits, was of relatively little interest to him. He had no more time for abstractions, whether in Einstein's science or in Picasso's art.[6] The greatest achievement of modern science for him was not relativity or quantum theory but the Wright brothers' aeroplane, even though it had brought humanity not only an ability to fly, but also a vicious new form of warfare.[7] Yet he was prepared to bet that scientists were going to find a way of coping with this aerial menace – whenever a military problem had been put before them, they had delivered like fairy godmothers:[8]

> My experience, and it is somewhat considerable, is that in these matters when the need is clearly explained by military and political authorities, science is always able to provide something.

Churchill, with Lindemann at his side like a stony-faced bodyguard, made advancing the science of air defence his cause célèbre. In early 1935, Churchill was invited to join the government's Air Defence Research Committee, a useful platform from which he could harass the government over its indolence, though he was sworn to keep its proceedings secret.

As everyone in Parliament knew, the Prof regarded Churchill as the man who should be leading the country, while Churchill regarded the Prof as the only source of advice on military science worth listening to. For several years, the two men were one side of a venomous political battle against some senior government officials and several leading academics, including one of Britain's finest nuclear scientists. Four years later, when many of the combatants first started to think about developing nuclear bombs, the pool of poison was still bubbling.

After evidence of the Luftwaffe's rapid expansion became undeniable, Ramsay MacDonald's government finally responded. In one initiative, taken in the closing weeks of 1934, the British Air Ministry set up a committee to consider new technologies that might help defend the islands against aerial attacks. The members of the committee were all from the heights of the science establishment, and all highly regarded by Rutherford: the physicist Patrick Blackett, Nobel Prize-winning physiologist and expert on anti-aircraft gunnery A. V. Hill, and the Ministry's director of scientific research Harry Wimperis. The chairman was Henry Tizard, rector of Imperial College and generally considered to be the country's leading research administrator. After Lindemann heard he had been excluded, he fired off a note to Churchill, complaining that although Tizard was 'a good man', neither Hill nor Blackett 'have ever had anything to do with aeroplanes'. Besides, Blackett held 'himself out as a Communist'.[9] Angry and disappointed, Churchill was not going to stand aside and do nothing.[10]

A clash between Lindemann and Tizard had been a long time coming – both wanted to be top dog among the academic scientists advising the military. Besides, the two men had form. Thirty years before, as young researchers in Berlin, they had been close colleagues, but their friendship had cooled, apparently after they fought a friendly boxing match that left Lindemann decidedly worse off. Later, though Lindemann proved himself to be the better academic scientist, Tizard's administrative career took off, leaving the Prof envious and resentful. After he was passed over for an appointment to a committee, Lindemann believed Tizard was responsible and marked his card. When applying science to challenges posed by the military, Lindemann had no time for Tizard's communitarian approach, which involved painstakingly fostering collaboration between scientists and the combat troops who would use the weapons. Lindemann's top-down style was all of a piece with his self-belief, which contrasted sharply with the anxiety that always seemed to shadow Tizard's face.[11]

The Tizard Committee first met at the end of January and got off to a strong start. By happenstance, only a few days earlier the Air Ministry had received Robert Watson-Watt's proposal for the technology that would later be called 'radar'.[12] Soon, however, the Committee was under intense pressure from Churchill, who agreed with Lindemann that its terms of reference were 'totally inadequate', lacking status, power and resources. By June, there was bad blood among the committee members after Lindemann had been foisted on them, following a back-room campaign by Churchill. In an eight-page memorandum, Lindemann enumerated the ideas he wanted government scientists to develop immediately. Lukewarm about radar, by far his top priority was the development of aerial mines, explosive devices attached to wires that could be dropped in the paths of enemy bombers. He argued sensibly that it would be unwise to neglect ideas apart from radar, which the enemy might be able to foil and which would not be useful at night.[13]

When his fellow Committee members dismissed his ideas as impracticable or old hat, he was furious. He complained of their foot-dragging and lack of imagination to Churchill, who backed the Prof's judgement to the hilt, but struggled to understand the underlying science. After he asked the Committee secretary Albert Rowe to explain the difference between radio and sound, Rowe wrote a 'tiny-tot note on the subject' – as he later recalled – but Churchill said 'it was beyond him'.[14]

Less than a month after the Prof joined the Committee, its atmosphere was septicaemic. After one tempestuous meeting, Blackett and Hill despaired of working constructively with Lindemann and submitted their resignations.[15] Ministry officials wound up the Committee but reconstituted it a few months later with the same members, apart from Lindemann, whose place was taken by the more collegiate Edward Appleton, another accomplished physicist. Churchill complained to the Secretary of State for Air that the Tizard Committee's progress over the past year had been like watching a film in slow motion.[16] He was, however, ignored. For now, his campaign to have the Prof at the top table of air-defence research was over.

Until Friday 13 March 1936, Churchill kept his powder dry, moderating his public criticisms of the Conservative government in the hope of securing a Cabinet post. But in vain. On that day, the new Prime Minister Stanley Baldwin created the post of Minister for the Coordination of Defence and gave it not to Churchill but to the conciliatory Attorney General Sir Thomas Inskip. Churchill now knew for sure that there was no place for him in any government run by his Conservative peers, who regarded him as too self-obsessed, too wayward, too much of a blow-hard and insufficiently committed to their party.

Lindemann described Baldwin's action as 'the most cynical thing that has been done since Caligula appointed his horse as consul'.[17] Still fuming at what he regarded as the pathetically inadequate progress on air defence, the Prof decided to try to join Churchill in the Commons, and in the autumn attempted again to become an MP for Oxford University. In what would turn out to be the last time Lindemann stood for public office, he was humiliated once again, finishing bottom of the poll.[18]

This election campaign took place during the abdication crisis, sparked by King Edward's relationship with his American lover Mrs Wallis Simpson. Baldwin handled 'the King's matter' with much-admired tact, easing His Majesty into a position where he had to choose between the throne and marriage to the divorcee. Churchill, a friend of the King, regarded this as unnecessary and made his case in an unconvincing speech in the Commons, where he was shouted down. The debacle underlined once again his reputation as unreliable, disloyal to his party colleagues and a man of poor judgement.[19] His parliamentary career had hit rock bottom.

*

Churchill's literary career, however, was flourishing as never before, enabling him to live the life of ease, like a member of the landed gentry. Between 1935 and 1936, his income peaked at about sixteen thousand pounds, more than thirty times his parliamentary salary.[20] Many of his most lucrative publishing contracts were struck by his Hungarian literary agent Emery Reves, an international syndication specialist who had settled in Paris after leaving Germany in April 1933, when he had been flung out of his Berlin office by Nazi storm troopers.[21] Assisted by researchers and secretaries who struggled to keep up with him, Churchill toiled until late into the night on a stream of popular articles and books, striving to meet the targets he had set for his daily word count. The quality of his prose was good enough to win praise from Rudyard Kipling, who wrote to Churchill 'craftsman to craftsman'.[22]

Any hope Churchill may have had of a return to the Cabinet was scotched in May 1937, when Baldwin was succeeded by the drab Neville Chamberlain, who had even less taste for Churchillian fireworks. The disappointment was only a brief distraction for Churchill. That year, in addition to hundreds of thousands of words he wrote on the life of Marlborough, he dashed off sixty-four newspaper articles, about half of them for London's *Evening Standard*, owned by his friend Lord Beaverbrook. Another of Churchill's favourite platforms was the *News of the World*, a Sunday newspaper that boasted 'the largest circulation in the world',[23] secured by providing its sixteen million readers with a fruity diet of titillation and scandal. In October 1937, he submitted an article that drew 'hearty congratulations' from the newspaper's principal proprietor, Sir Emsley Carr. The article concluded by focusing on a subject rarely seen in the newspaper's columns, nuclear physics. Carr believed the piece 'one of the most interesting so far as our general readers are concerned',[24] apparently not realising – or perhaps not caring – that Churchill had played one of his favourite tricks of reusing old material – the article was almost identical to the first half of 'Fifty Years Hence', published six years earlier. Sir Emsley had enabled him to increase the essay's readership a hundredfold, and paid him four hundred pounds for the privilege of recycling it.

The *News of the World* published the piece 'Vision of the Future Through the Eyes of Science' on 31 October, flagging it prominently on the front page. Taking up most of page twelve of the newspaper, the article was accompanied by a photograph of the author, in a flattering pose, looking thoughtfully downwards, at beatific peace.[25] The climactic line, about the possibility of releasing nuclear energy, was set in bold type: 'The new fire is laid, but the particular kind of match is missing.' This time Churchill did not go on to say, as he had in 1931, that the result might be an explosion.[26]

This article was well timed, as it was published a few days after the funeral of Lord Rutherford, who had died following a bungled operation when he was only sixty-six, four years before he intended to retire.[27] The government did him proud, making him the first scientist born 'in the overseas dominions' to be laid to rest in Westminster Abbey, next to the remains of Newton and Darwin.[28] Among the thousands of mourners were Tizard and also Lindemann, who may have given a thought to the implications of the event for his own career – he had lost his most influential detractor, while Tizard had lost his most powerful supporter. The funeral seems to have made the Prof reflect on the future of nuclear science and its possible consequences for warfare – on that day, he drafted another popular article on the subject for Churchill, who did not use it immediately but put it in his files.

A week after the 'Vision' article appeared, the *News of the World* featured the second half of 'Fifty Years Hence'.[29] This time, Churchill's tone was ominous as he fretted about the monsters that scientists might unleash, knowingly or otherwise, particularly by tinkering with atomic nuclei. Could a few nuclear scientists be the successors to Mary Shelley's Frankenstein? The article's baleful headline 'Life in a World Controlled by the Scientists' reflected a deeper worry about how the country might change if, as H. G. Wells had hoped, scientists and engineers were soon playing an active role in government. Churchill believed that unelected scientists had no more right to a say in government than, say, bankers or dentists. Again, he highlighted the impact that nuclear science might one day make on everyday life. In the second paragraph, printed in bold, he reiterated a point he had first made six years before, that nuclear energy could be released not only by splitting atomic nuclei but also by combining them:

> . . . if one could induce the atoms of hydrogen in the Serpentine to combine to form helium they would produce enough heat to change the whole climate of England for a year.

He was alluding to nuclear fusion, whose power would nevertheless amaze him fifteen years later, when a hydrogen bomb was first detonated.

In his article in the following week's *News of the World*, he focused on how science was about to change the practice of war, a subject he may well have discussed with H. G. Wells, who in mid-August had been a weekend guest at Chartwell, much to Clemmie's delight.[30] In public, the two men still scrapped like kittens, but a bond of affection remained, and Wells had 'a perfect time', having basked 'in the glow of Winston's approval'.[31] A few months before, on a whim, Wells had dedicated his new novel *Star Begotten* to Churchill, who responded with a special grace: 'It gives me real pleasure to feel that my early admiration of thirty-five years ago for [your] wonderful books should have come to rest in our later times in a harbour of personal friendship.'[32] In his newspaper article, Churchill foresaw 'the obliteration of the personal factor in war'; the outcome of the conflict depended less on the charismatic generals and their heroic soldiers than the faceless puppet masters drinking coffee at their desks. He imagined a future battle in which 'some spectacled "brass hat" . . . extinguished some London or Paris, some Tokyo or San Francisco, by pressing a button'.[33] There was an encouraging consequence, he wrote: 'The idea of war will become loathsome to humanity. The military leader will cease to be a figure of romance and fame . . . It may well be that the chemists will carry off what credit can be found.' This was pure Wells, though shorn of his political agenda.

On 11 March 1938, Nazi troops marched into Austria and occupied it without firing a shot. Churchill was on his feet in a subdued House of Commons three days later, warning that Europe was 'confronted with a programme of aggression . . . unfolding stage by stage'. It was a great speech, all the more powerful for the absence of self-congratulation.[34]

In his public utterances, by turns fiery and cautious, Churchill repeatedly drew attention to the threat posed by Hitler, but argued that Britain should keep its distance from the Fascist rebellion in Spain and Japanese aggression in China. He was, however, consistent in courting the friendship of the United States. In radio broadcasts and articles published in the US, he praised 'the majestic edifice' of Anglo-American friendship, and usually avoided commenting on its internal quarrels.[35] At home in the UK, his message was clear and far-sighted: 'We must not ask too much of the United

States . . . we may find them with us at the end of the road.'[36]

Apart from his far-receded hairline and his bulging belly, Churchill did not look like a man in his mid-sixties. He was still as lively as ever, the disappointments of the past decade having etched surprisingly few lines into his face.[37] The upper half of his body somehow spoke of his determination: the fat cigar jutting from his mouth, his heavy shoulders carried a little forward and the dome of his forehead – they all gave him the air of a bull about to charge.

He was on the rampage again in June 1938, condemning what he believed to be the feeble progress of air-defence research and pressing harder than ever for Lindemann to be given a seat on Tizard's Air Defence Research Committee.[38] Officials compiled a rebuttal, refuting each of his accusations and alleging – with a hint of menace – that his own record of implementing new technological ideas before the First World War was far from perfect. According to Tizard, Churchill had failed adequately to prepare for the threat of the U-boat and displayed 'his total lack of real scientific imagination and foresight'.[39] Churchill responded with more sadness than aggression, perhaps partly because he did not want such allegations aired. During the summer, there was an unsatisfactory stand-off, but in the end resistance to Churchill's onslaughts was futile. By November, Tizard reluctantly allowed Lindemann to join the Committee, on the condition that A. V. Hill also became a member.

This clash of wills between Tizard and Lindemann had damaged them both. Even sympathetic colleagues muttered that Tizard treated Lindemann with too little respect and too much scepticism,[40] while in Whitehall, Lindemann was now widely regarded as an impossible colleague. As the distinguished government scientist Frederick Brundrett later commented on this ugly rivalry: 'For two extremely intelligent grown-up people, their attitude to each other was singularly childish.'[41]

As Hitler prepared to invade Czechoslovakia at the end of September 1938, it seemed that war with Germany was all but certain. Churchill wanted the Prime Minister Neville Chamberlain to warn Hitler that Britain would go to war if his soldiers so much as set foot in Czechoslovakia.[42] This was an unpopular view. Was the freedom of the people in a small, distant foreign country worth a repeat of the carnage of the First World War, this time likely to begin with a Luftwaffe attack that would lead to the deaths of hundreds of thousands in British cities?[43] The sense of dread was later well caught by Harold Macmillan: 'We thought of air warfare in 1938 rather as people think of nuclear warfare today.'[44]

Chamberlain was cheered when he returned in late September from his negotiations in Munich with Hitler and other leaders, waving the 'peace for our time' agreement. In Lindemann's favourite science journal, *Nature*, a gushing editorial praised the 'beautiful solution' the Prime Minister had helped to negotiate and urged that he be awarded the Nobel Prize for peace (the Prof's snort would have been worth hearing).[45] After Churchill opposed the agreement in the Commons,[46] Tories in his constituency threatened to dislodge him as their MP, though he soon saw off the challenge.

Unlike the public, and most of his colleagues in Parliament, Churchill knew that Britain was developing radar defences and, crucially, that they were not yet fully operational. He was also more aware than most of his colleagues of other technologies that might soon be unleashed in wartime. In late October 1938, he returned to this theme in another article for the *News of the World*, based on the piece written by Lindemann a year before, on the day of Rutherford's funeral.[47]

This article, 'What Other Secrets Does the Inventor Hold?', published on 23 October,[48] pays less attention to physics than to biology: 'It is in this field that the impact of invention and discovery is likely to be most formidable,' Churchill wrote. In a section on eugenics, he bewails the invention of contraceptives, which he complains have jeopardised the chances of survival of 'more civilised peoples' in a world where 'the barbarian is breeding against them'. He predicts a forthcoming age of genetic engineering, wondering what will happen if 'the very make-up of mankind becomes the plaything of the bureaucrat'. Churchill had changed scarcely a word of Lindemann's draft. In the only striking departure from the original, Churchill took a passage midway through Lindemann's text and placed it at the top, making it the article's hook. The passage he chose was about nuclear science.

'Almost every day, the scientists tell us,' Churchill begins, 'discoveries are being made about the artificial building-up or breaking-down of the nuclei of the atoms.' After six short paragraphs about nuclei and the energy locked up inside them, he concluded:

> With these immense resources of power available, it seems likely that means will ultimately be found to tap them. If this were achieved, man's control over nature would take a step forward greater than any since in Palaeolithic times when he discovered how to make fire.

Again, this does not go so far as to suggest that the release of nuclear energy might lead to a new type of weapon, perhaps to avoid more accusations of alarmism. As usual with his articles about the military applications of future science, this one is pessimistic. He doubts whether the human race would be able to handle the scientists' latest inventions, adding that it was impossible to say 'whether they will lead to a Utopia or to the extinction of the human race'. The conclusion is one of unmitigated gloom – the new devices 'may spell not only the ruin of the civilization we know, but the end of human dominance of this planet'. The article may have had a special resonance with the hundreds of thousands who had read or heard about J. B. Priestley's light-hearted novel *The Doomsday Men*,[49] an unlikely summer hit that ended with physicists detonating a nuclear bomb in a South Californian desert. The yarn was also popular in the US, where the *New York Times* reviewer concluded: 'we might do well to keep an eye on those atomic physicists. If *they* go crazy, something might happen.'[50]

Eight weeks after millions in Britain read Churchill's thoughts about a possible nuclear future, a single experiment made it more plausible to believe that nuclear energy might soon be released on a large scale, perhaps to make weapons. This experiment took virtually all scientists by surprise, including even the world's most accomplished nuclear theoretician, later to be regarded by Churchill as probably the most annoying physicist he ever met.

NOVEMBER 1938 TO SEPTEMBER 1939

Bohr thinks the Bomb is 'inconceivable'

> '[Niels Bohr] utters his opinions like one perpetually groping and never like one who believes himself to be in possession of definite truth.'
>
> ALBERT EINSTEIN[1]

With Rutherford dead, no one had a surer intuition of how nuclei behave than his only protégé theoretician, Niels Bohr. Yet even he was amazed when experimenters discovered that a uranium nucleus could be split in half when it was struck by a neutron, with a few other high-energy particles emitted at the same time. It was as if a granite monument could be cleaved by throwing a pebble at it. Bohr was in good company – no one had clearly foreseen this discovery, which soon commanded worldwide attention. Few were surprised when Bohr was among the first to shed a bright light on the fission process.

Bohr had first learned about nuclear science from Rutherford, who had almost been 'a second father' to him.[2] They met shortly before the First World War, when the twenty-six-year-old Dane won a scholarship from the Carlsberg Foundation to spend time in the UK. At Manchester University, using mainly high-school mathematics, Bohr developed the idea that electrons in a typical atom move in quantised orbits around the nucleus, an insight that enabled scientists to understand swathes of experimental data that had previously appeared unconnected. Rutherford considered Bohr's work as 'one of the greatest triumphs of the human mind'.[3]

In 1921, the year before he won the Nobel Prize for physics,

Bohr became director of the new Institute for Theoretical Physics in Copenhagen, accommodating about twenty scientists. Within a few years of opening its doors, it became the maternity ward of the nascent quantum theory, with Bohr its most attentive obstetrician. On first acquaintance, he did not appear to be greatly distinguished, though he was an imposing presence, with his huge head and hands, bushy eyebrows and a pipe that for him was virtually a prosthesis. He ran the Institute with a uniquely cerebral brand of avuncularity, discoursing in the corridors, running up its stairs two at a time, spending hours in the lunch-room chatting with his young colleagues about the latest paradox to catch his eye. After becoming the Institute's director, he wrote relatively few original scientific papers, though his colleagues admired him – to the point of idolatry – for his intellectual depth and his generosity of spirit.

No one counted communicative skills among Bohr's strengths, however. His writing was as opaque and tortuous as his speech, so hushed and garbled that he sometimes left his audience unclear about which language he had been speaking. Yet he loved to talk, not only about physics, but also about the many subjects that fascinated him, including cubism, economics, genetics and cowboy movies.[4] Once his interest was captured by what he believed to be an important cause, he would pursue it politely but with the fearlessness of a terrier.

By the early 1930s, the atmosphere at the Institute had darkened. Its glory days were coming to an end and conversations were turning ever more insistently to international politics. The dinnertime banter, for years relaxed and free-wheeling, was becoming tense and stilted – German nationalists such as Bohr's young friend Werner Heisenberg had to respond to tales of Nazi anti-Semitism circulating round the table. More than any other leader, Hitler had made a mockery of the internationalist ideal of science that Bohr and his colleagues cherished. From the mid-nineteenth century, Berlin could have claimed to be the capital of world science but, by late 1938, the city was home to only a few eminent research scientists, most of them fearful, demoralised and feeling cut off from their international community of peers. Over a year earlier, *Nature* had been banned by the German Education Ministry.[5] Hitler boasted that the Third Reich would last a thousand years, but he seemed indifferent to the destruction of German scientific culture, until then one of the most powerful and prolific the world had ever seen.[6] Yet out of this husk came the sensational discovery of nuclear fission.

By November 1938, Germany was in the grip of an anti-Semitic hysteria. In the national pogrom that became known as Kristallnacht, there were coordinated attacks on Jews, their businesses and places of worship, while tens of thousands of them were dispatched to concentration camps. In Berlin, festooned with swastikas, assaults were a spectator sport, respectable middle-class mothers on street corners holding up their babies to see the 'fun'.[7]

On Monday 19 December, the city was in the grip of an Arctic cold snap. As midnight approached, the packed beer halls resounding with Christmas merrymaking, Otto Hahn was alone at his laboratory desk in the Kaiser Wilhelm Institute for Chemistry, contemplating a discovery that would soon shock scientists all over the world. Awaiting the return of his collaborator Fritz Strassmann, Hahn was describing his results in a letter to the Austrian-Jewish physicist Lise Meitner, a close colleague until only five months before, now exiled in Sweden. Two weeks later, Meitner was spending the Christmas vacation on the Swedish coast with her nephew Otto Frisch – a refugee from Nazi Germany working at Bohr's Institute in

Copenhagen – and she talked in detail about the new results described in Hahn's letter. They were the first to understand that Hahn and Strassmann had observed the process that Frisch termed 'nuclear fission'. According to Einstein's equation $E=mc^2$, the energy released by each cleaving of a nucleus would not deflect the path of a mosquito, but on the atomic scale the energy was huge. Every one of Hahn and Strassmann's neutrons had less energy than a particle of visible light, but the energy released when one of them split a uranium nucleus was a billion times greater. With this level of return on energy investment, it might be possible – contrary to Rutherford's dictum – for a nucleus to be a source of energy, rather than a sink.

Early in the New Year, Frisch returned to Copenhagen desperate to share his news with Bohr, who struck his forehead in disbelief: 'Oh, what idiots we have been!' he said. 'Oh, but this is wonderful.' Within a few days, Frisch had tested the explanation he and his aunt had devised and had found experimental evidence for it. A simple experiment showed that the energy released in the process really was as large as their calculations suggested. Everything added up, though for many scientists in the coming weeks the explanation seemed too far-fetched to be true.

A few days after speaking with Frisch about the new nuclear process, Bohr set sail for the United States, where *Time* magazine had recently named Hitler its Man of the Year. Bohr had arranged to work for a few months in Princeton, an hour's train ride from New York, at the Institute for Advanced Study, where colleagues hoped he might collaborate with Einstein on the challenges of interpreting quantum theory.[8] But those plans came to nothing: Bohr's top priority now was to understand the fission process by thinking of nuclei as drops of liquid, an approach he and others had earlier pioneered. By virtue of his brains and charisma, he would soon become a leading player in the drama of fission, whose next act was to be set in the United States, which had now eclipsed the centuries-old dominance of Europe in physics.[9]

Bohr was too absent-minded to be a reliable keeper of secrets. In Copenhagen, he had promised not to divulge the Frisch–Meitner theory of nuclear fission until *Nature* published it, but he discussed it in detail with his close colleague Léon Rosenfeld during their voyage to New York. Within a few hours of arriving in freezing Princeton on 16 January, Rosenfeld shared the news at an evening gathering of scientists, who could scarcely contain their excitement.[10] By the next morning, the news had spread via phone to hundreds of confidants.[11] After Bohr had seen a copy of the fission paper in print, he agreed to talk about the theory ten days later at a meeting in Washington DC in an unscheduled presentation with Enrico Fermi, who had arrived from Europe shortly before, having fled Fascist Italy with his family.

This first public announcement of the discovery of fission was anything but polished – Fermi and Bohr scribbled hieroglyphics on a blackboard and struggled to explain the concept clearly in English. Afterwards, the *New York Times* reporter William Laurence rushed up to the speakers and blurted out: 'Does this add up to an atomic bomb?' Startled, Bohr looked up to the ceiling. Fermi, no less surprised, looked quizzical and after a long pause conceded that the Bomb may be twenty-five or even fifty years away.[12] The physicists in the audience left the lecture theatre excited and ready to spread the word, but Laurence – believing it a 'foregone conclusion' that the Nazis would build the Bomb first – went home a 'frightened man'. He might well have been gratified had he known that, within a week, a schematic diagram of a nuclear bomb would be on the blackboard of one of America's leading theoreticians, Robert Oppenheimer.[13]

Since Bohr heard about the discovery of X-rays as a ten-year-old, he had hardly ever seen an experiment cause this much excitement. The American press did it full justice. Newspapers covered the initial announcement and, within a week, the *New York Times* pointed out that Bohr's mentor Lord Rutherford may have been wrong to write off the idea of harnessing nuclear energy. 'Romancers have a legitimate excuse', the editorial declared, 'for returning to Wellsian utopias.'[14] It was, however, easy to miss such topics among the saturation coverage of America's battered economy. Roosevelt had been re-elected by a landslide two years before but, struggling to secure a return to national prosperity, was now losing popularity.[15]

Nuclear fission also featured on the radio, which carried a special interview with Fermi, billed as the most recent Nobel Prize-winning physicist and a welcome new immigrant.[16] In the Groundhog Day edition of CBS's popular series *Adventures in Science*, the presenter Watson Davies could barely contain his excitement in his introduction: 'The world may be on the brink of the release of atomic power.' Fermi refused to play along – he spent the interview carefully considering his host's optimism before finally deflating it.[17] Inevitably, the press soon pointed out that nuclear bombs were no longer quite as unlikely as they had been led to believe. In the US, the suggestion first appeared on 11 February 1939 in the article 'Atomic Bombardiers' in the weekly *Science News Letter*.[18] It warned that 'scientists are anxious that there be no public alarm' – it was best to ignore the 'wild speculations' of H. G. Wells, J. B. Priestley and the writers of the play *Wings Over Europe*: the world was not about to be 'blown to bits'.[19]

The first new experimental insight into what was going on in nuclear fission came not from Fermi and his colleagues, but from a laboratory in Paris. Frédéric Joliot-Curie and his colleagues demonstrated that when a uranium nucleus fissions, a few neutrons are typically released at the same time, like a coconut splitting cleanly in two, but shedding a few splinters. No one understood the implications of this better than Leó Szilárd. He had finally settled in the United States and talked his way into a position at Columbia University alongside the steadier, more focused, more practical Fermi.[20] Szilárd later said that when he heard of Joliot-Curie's results, he saw straight away that a chain reaction might well be possible in uranium, with the neutrons from the fissioned nucleus possibly going on to split another one, releasing more neutrons that could split even more nuclei, and so on, rather like breeding rabbits. Huge amounts of nuclear energy would then be released. Szilárd later recalled: 'All the things which H. G. Wells predicted appeared suddenly real to me.'[21]

Fermi began trying to set up a nuclear chain reaction, with the often unwelcome participation of Szilárd, who declared that experiments were not for him and hired a young scientist to do the work for him. It turned out that a chain reaction was most likely if uranium nuclei were bombarded by slow neutrons, much slower than the ones initially produced by fission. Fermi and his group – like other colleagues in Europe investigating these reactions – looked into ways of decelerating the neutrons, for example by making them collide with 'moderators' such as heavy water or graphite, whose molecules helped to slow the particles down. Heavy water was in extremely short supply in the US, so Fermi concentrated on the other option. Unlike his backseat driver, he was sceptical that the experiments would soon yield useful results. Scared that the Nazis would now be able to develop a nuclear weapon, Szilárd began to campaign for nuclear fission researchers to keep their work under wraps ('I invented secrecy,' he later claimed[22]). He and two colleagues pleaded with physicists in Europe, urging them to self-censor their work on fission, but they were given a dusty response. Nor was Fermi going to abandon any time soon the centuries-old scientific tradition of publishing results at the first opportunity. He replied to Szilárd: 'Nuts!'[23]

One morning in early February, during a five-minute walk across Princeton's snow-carpeted campus, Bohr had one of his brainwaves.[24] The upshot of the idea, worked out in detail with the American theorist John Wheeler, was that Hahn and Strassmann had engineered the fission not of all uranium nuclei but only of very rare ones containing 235 nuclear particles (each denoted ^{235}U). Almost all of the nuclei in their uranium target – well over ninety-nine per cent – contained 238 particles (^{238}U) and underwent no fission at all. Though widely disputed, if this reasoning was correct and a nuclear bomb were built from ordinary pure uranium, very few of its nuclei would fission and release energy, making it the dampest of squibs.

That winter, Bohr and Szilárd chewed over the possibilities of setting up nuclear chain reactions.[25] The sight of these two loquacious men, both poor listeners, trying to converse in a foreign language, must have been richly comic: the ruminative Dane weighing up the dozens of possibilities; the Hungarian wanting to drive home only one. For both men, Wednesday 15 March was a red-letter day. The news that morning was terrible: the Nazi and Hungarian armies had marched into Czechoslovakia and 'snuffed out' its government, as the *New York Times* reported on its front page.[26] Hitler pronounced the Munich Agreement dead, ending virtually all hopes of appeasing him, so it now required no imagination at all to visualise Fascist armies controlling every city on the European continent.

Soon after Bohr and Szilárd heard the reports, they met in Einstein's office with John Wheeler and the theoretical physicist Eugene Wigner, a refugee from Hungary (Einstein himself was out of town). For hours, they talked over the possibility that nuclear fission could lead to the manufacture of enormously destructive bombs.[27] Bohr and Wheeler explained that, if their theory was correct and only ^{235}U nuclei undergo fission, nuclear bombs were simply not viable. 'Then separate out the ^{235}U', Szilárd shot back, 'and use it to make atomic bombs.' Bohr told him he had nothing to worry about. To isolate enough of the extremely rare ^{235}U nuclei would be a colossal industrial undertaking, requiring 'the efforts of an entire nation'. Such an outcome was, Bohr assured him, 'inconceivable'.[28]

When Bohr left the United States in early May and sailed for Europe, the ideas he and Wheeler had set out were still not widely accepted, though the consensus was heading in their direction. Bohr stood by his theory and was convinced that nuclear-fission bombs were not feasible. That stance was under fire again in early June, when Otto Hahn's assistant Siegfried Flügge published the article 'Can Nuclear Energy be Utilised for Practical Purposes?' in the German journal *Die Naturwissenschaften*, the first of a few papers on the subject.[29] He wrote that 'the energy liberation should . . . assume the form of an exceedingly violent explosion'. Scientists soon began to take seriously the remote possibility that a fission bomb could be built and, later that month, virtually all scientists outside Germany and Italy accepted an unwritten voluntary agreement, unprecedented in fundamental physics, to self-censor new research on nuclear fission.[30]

Later that month, when Bohr visited physicists in Britain – including his friends at the Cavendish and G. P. Thomson at Imperial College – they almost certainly talked over the possibility of nuclear weapons.[31] The implications were clear from a glance at the newspapers, full of anxious reports on the increasing political tensions in Europe.[32] The bookshops were now selling Churchill's new book *Step by Step*, a collection of his speeches, many of them warning of the dangers of appeasement and spelling out the need for Britain to rearm swiftly and improve its air defences. Now apparently vindicated, Churchill had won admiring reviews, *The Times* commenting that he 'is clearly entitled to his triumph' after his stance on rearmament.[33]

In the United States, concerns over the possibility of nuclear weapons were growing among the cognoscenti. An editorial in the July edition of *Scientific American*, entitled 'Incomparable Promise or Awful Threat?', pondered the question currently 'worrying the physicists'.[34] Should they terminate their fission experiments before the science is used to destructive ends? That would be 'merely to abandon them to . . . war makers and conquerors'. The article ends by pointing out that the physicist is only a minor player in the great scheme of human events:

> He cannot control his own discoveries, once they are given out, for he is far outnumbered. And if the human race won't leave its new playthings alone, and gets badly hurt, that's its own funeral. In a few years we may have the answer.

Predictably, Szilárd was now taking matters into his own hands, and considering how to take his warning to the White House. He was right to be alarmed. The Third Reich had become the first government in the world to think seriously about harnessing nuclear energy. In Berlin, several of its top military officials and a few civilian scientists, including Otto Hahn, met in secret to consider whether fission might have opened the way to nuclear bombs.[35] Bohr knew nothing about the Nazis' initiative and, even if he had, he would probably have been unperturbed, as he was still convinced it would take many years before nuclear weapons could be built. As he said in a public lecture soon afterwards, there was 'no cause for alarm'.[36]

By mid-July, when Bohr and his family began their vacation at their summer home at Tisvilde on the Danish coast, Europe was on the edge of catastrophe. While the Bohrs were away, the news became even more alarming, culminating in Hitler and Stalin's non-aggression pact, which also delineated their spheres of influence in north-eastern Europe. War now seemed inevitable.

Back in Copenhagen on 1 September, Bohr heard that war had begun after the German army had crossed the border into Poland. It was now only a matter of time before Britain and its Empire entered the conflict. A day later, in the Danish capital, the newspaper *Politiken* reported that Neville Chamberlain was taking steps to include in his War Cabinet a politician many believed had been out of government for too long: Winston Churchill.[37]

2

WORLD WAR II

AUGUST TO DECEMBER 1939

Churchill – nuclear weapons will not be ready for the war

'I expect Lindemann's view is right, i.e. that there is no immediate danger [that nuclear weapons will be developed], although undoubtedly the human race is crawling nearer to the point when it will be able to destroy itself completely.'

WINSTON CHURCHILL, 13 August 1939[1]

On the first day of the war, the opening words of the editorial published that morning in *Discovery* magazine did not make reassuring reading: 'Some physicists think that, within a few months, science will have produced for military use an explosive a million times more violent than dynamite.'[2] The editor, C. P. Snow, was reflecting on the magazine's article 'Energy from Matter' by the science-fiction specialist Douglas Mayer, about the latest nuclear-fission research. Snow concluded with a question: 'Shall we have a Wellsian chaos with each nation dropping bouquets of uranium bombs in a policy of encirclement?'

Snow warned that 'the power of most scientific weapons has been consistently exaggerated but it would be difficult to exaggerate this'. Laboratories in the United States, Germany, France and England 'have been working on it feverishly since the spring', he wrote, and President Roosevelt was well aware of the new developments. Snow's conclusion was ominous:[3]

If it is not made in America this year, it may be next year in Germany. There is no ethical problem; if the invention is not prevented by physical laws, it will certainly be carried out somewhere in the world. It is better, at any rate, that America should have six months' start . . . Such an invention will never be kept secret . . . within a year every big laboratory on earth would have come to the same result. For a short time, perhaps, the US government may have this power entrusted to it; but soon after it will be in less civilised hands.

Such a conclusion would normally trigger a full-blown panic in the press, but it made no impact at all – on that day the country's eyes were focused solely on the unfolding crisis in Europe. On the morning of 1 September, Churchill was chomping at the bit, refreshed after a working vacation in France.[4] He did not have long to wait to get involved in the war effort – that afternoon he was summoned to 10 Downing Street, and a few hours later Neville Chamberlain offered him a place in the War Cabinet he was assembling, expecting hostilities to begin at any moment.[5] London was on a war footing – sandbags on the streets, children evacuated, houses and streets blacked out in expectation of an aerial onslaught that the Home Office had predicted would lead to 2.5 million casualties in ten weeks.[6] At 11 a.m. the following Sunday, Britain declared war on Germany, Chamberlain announcing his decision in an uninspiring radio broadcast. After the Commons met later that afternoon, Chamberlain summoned Churchill again and this time offered him the Admiralty, a job he had held during the First World War. 'That's a lot better than I thought,' Churchill told Clemmie.[7]

In spite of Hitler's preoccupation with crushing Poland, he quickly showed Chamberlain that he meant business, sinking the British ship SS *Athenia* a few hours later. Yet no bombs fell on London, and there was no Wellsian cataclysm,[8] only the beginning of 'the Sinister Trance', as Churchill put it, as Britain braced itself for aerial attack and, perhaps, invasion.[9]

Although the opening months of the war were an anticlimax, the British navy went on the offensive.

Britain was well placed to fight.[10] It had a strong army, by far the largest economy in Europe, the continent's most powerful navy, and was about to retake from Germany the world record for annual aircraft manufacture. Britain also had the crucial support of its dominions and colonies, along with a global trading network second only to that of the United States. Hitler, however, had a much greater momentum and self-belief, with a well-drilled military machine backed by an industrial base working as if its destiny depended on breaking every productivity record. No one was quite sure of the contents of Hitler's arsenal: perhaps he had bacterial weapons, gliding bombs, pilotless aircraft or even new self-guiding torpedoes. Any one of these might leave his enemies dangerously exposed. The nervous British press jumped on a cryptic phrase in the speech he made in Danzig (now Gdańsk) on 19 September, several newspapers mistranslating the words as 'a secret weapon against which there was no defence'.[11] Whitehall officials wondered whether this 'secret weapon' might be nuclear – not crop-eating locusts or death rays as others suggested – and sought advice from the Department of Scientific and Industrial Research.[12] When Churchill heard the story, he may well have surmised that Hitler was bluffing, as Lindemann had predicted a few weeks before.

For several months, the Prof had been keeping a sardonic eye on press reports of the imminent advent of nuclear weapons. After Bohr and Fermi's announcement of the discovery of fission in January, he may have seen that a gossip columnist in the *Observer* was quick off the mark with a warning: 'Unless we learn the lesson, it is, in all probability, only a matter of time before atomic energy becomes a form of warfare.'[13] Those words were easily missed. Not so the article 'Scientists Make an Amazing Discovery' splashed across almost an entire page of the *Sunday Express* several weeks later, on 30 April.[14] The article's apparently pseudonymous author, C. A. Lyon, writing from 'somewhere in England', gave millions of readers a well-informed if slightly hysterical account of fission research, concluding that 'a nation at war might be able to wipe another nation right off the face of the earth in a second'. The story 'reads like H. G. Wells, but it is strictly authentic', Lyon insisted.

Overblown stories like this in the newspapers, which Lindemann read in bed over breakfast, were enough to make him choke on his truffled scrambled egg whites.[15] He knew that several government officials were carefully weighing up the options for research into nuclear weapons. Tizard had received the support of his Air Defence Committee after he hurriedly dictated them a note estimating that the chance of the release of energy having military applications was 100,000 to 1.[16] This was a spuriously precise number, though he was careful to add that even this low probability should not be ignored.

Lindemann was worried that press articles about the atomic bomb could lead the public to believe that the Nazis had the weapon and might use it to intimidate the British government. In early August, he decided to make his concerns public, by drafting a letter to the *Daily Telegraph*. Beginning with an oblique reference to the *Sunday Express* article, Lindemann homed in on the latest reports of the possibility of producing nuclear chain reactions (he wrote inaccurately that scientists had already observed them).[17] The essential point, in his view, was to realise that 'there is no danger that this discovery, however great its scientific interest, and perhaps ultimately its practical importance, will lead to results capable of being put into operation on a large scale for several years'.

Lindemann predicted that the press would be as gullible as ever: 'Attempts will no doubt be made by the Fifth Column', he wrote, 'to induce us by means of this threat to accept another surrender.'[18] There was no substance to the fears that the Nazis had such a weapon, he insisted, concluding with the sneer he reserved for his intellectual opponents: 'Dark hints will no doubt be dropped and terrifying whispers will be assiduously circulated, but it is to be hoped that nobody will be taken in by them.' Churchill read the draft during a mid-August weekend at Chartwell and forwarded it privately to the Air Ministry, asking if there was any objection to Lindemann's sending the letter to the *Daily Telegraph*.[19] Although officials raised no objections to the letter – one of them commenting that Lindemann's views were 'consonant with the best scientific opinion in this country' – the letter was never published.[20] With the country about to go to war, someone apparently decided that it would not be good for national morale to raise even the thought that a single one of the bombs expected to rain down on Britain might be nuclear.

When Churchill returned to the Cabinet, he was three months short of sixty-five and eligible for a pension. After almost eleven years in the wilderness, he was thrilled to be back in government. Based in Admiralty House, an eighteenth-century redbrick building in Whitehall, he worked hyperactively, as if to make up for lost time.[21] Arriving at his desk by seven in the morning, he began a day of cross-examining his admirals, badgering civil servants, dictating memoranda (known as 'minutes') to a fleet of stenographers, telephoning colleagues and poring over the colourful charts and maps often covered with black cloths to hide them from unauthorised personnel.[22]

To stay fresh, he usually took a long nap in the late afternoon, bringing to his harassed department a welcome calm that ended abruptly after he awoke. Later, he relaxed again for a couple of hours over a good dinner, washed down with champagne, wine and brandy. The working day ended for him only in the small hours of the morning, when he would retire to the suite of rooms he and his wife shared on the top two floors of Admiralty House.

Six days after he returned to office, the Admiralty notice-board displayed an announcement that surprised no one:[23]

Professor F. A. Lindemann FRS to be personal adviser to the First Lord on scientific development. The appointment will be temporary and unpaid. It will take place from September 9th.

Lindemann had left Oxford for London as soon as he heard that Britain was at war, having only just moved into his office in the new Clarendon Laboratory building.[24] He now had an office next to the war room, with another for his valet Harvey, who was always on hand to launder his clothes, prepare his meals and chauffeur him between meetings.[25]

Two days after Lindemann officially took up his post, he received an amiable note from Tizard, scientific adviser to the Chief of Air Staff.[26] Relations between the Prof and Tizard were still tetchy, if not as poisonous as they had been a few years before. Lindemann believed that not nearly enough had been done to develop his ideas and that even the top-priority radar programme had been handled poorly. In a party of officials who witnessed a demonstration of the new radar technology two months before, he alone had been unimpressed and filed a critical report.[27] On the day after the official announcement of Lindemann's appointment as science adviser in the Admiralty, Tizard wrote to him: 'I am sure that you will agree with me that any remnant of a private hatchet should be buried . . . we should remember old friendship and cooperate as much as we can.'[28] Lindemann replied immediately from London's Carlton

Hotel, where he lived during the working week: 'Hatchets are made to be buried, above all when so many trenches are available,' adding that all he wanted was 'to cooperate as much as we can in the common cause' and concluding with a warm invitation to lunch.[29] Tizard then heard nothing more from him.

Lindemann's appointment put several noses out of joint. Churchill was signalling that he did not wholly trust the opinions of the scientists already employed by the Admiralty and that he wanted the Prof to subject every statistic, every opinion, every recommendation to his unforgiving scrutiny. This undermined the confidence of Churchill's colleagues, who knew their advice would always be less important than the assessment given by a disdainful outsider. Contrary to the image of Olympian detachment Lindemann cultivated, he had a rich palette of prejudices, especially against government officials, whom he regarded as feckless buck-passers – a view summarised in a mock prayer that he kept in his files:[30]

> O Lord grant that this day we may come to no decisions, neither run into any kind of responsibility, but that all our doings may be so ordered to establish new and unwarranted departments for ever and ever. Amen.

In many ways, Lindemann's skills complemented Churchill's, but they shared a weakness for any wheeze and gadget that might give even the slightest advantage over the enemy. Churchill promoted new ideas he had hatched with Lindemann, including the trench-cutting tank, but it seems that none of them proved especially useful. The admirals, many of them exhausted by Churchill, resented these time-wasting suggestions and came to despise Lindemann for his arrogance, his interferences and his second-guessing of the ideas they put in front of their leader. Yet they knew that Lindemann had a unique ability to calm their boss down. Most evenings, he would arrive in Churchill's private office around midnight, after the Admiralty evening conference and another round of dictations ('Are you ready?' Churchill once said to a typist, 'I'm feeling very fertile tonight').[31] The two men would then settle down on the sofa by the fire and talk, Churchill thoughtfully sipping his whisky.

One of the good ideas Churchill came up with, within a month of taking office, was the need for a continuous stream of high-quality statistical advice and for the staff to research and collate it. The information from the various departments in the Admiralty was unreliable and incoherent; besides, as Minister for the Navy, he needed to be kept well informed about all other matters relevant to his brief, in other words almost everything to do with domestic economics and the other armed forces. Using this empire-builder's logic, and in the teeth of opposition from his senior Admiralty staff, he issued a minute to announce the formation of 'The First Lord's Statistical Branch', to be run by Lindemann and with a modest number of staff. When a colleague recommended to the Prof that he employ more people, he replied disbelievingly, 'Who would join me?', and did not raise the matter again.[32] Like Churchill, he placed a high premium on personal fealty. He recruited a handful of bright assistants – all of them from Oxford University – distinguished by their ironclad loyalty, and also by their willingness to ask awkward questions. Two of the first recruits were the economist Roy Harrod, who had worked closely with Maynard Keynes, and the resourceful physicist James Tuck, who later made crucial contributions to the design of the first nuclear weapons.

At the War Cabinet table, Churchill was a bold, imaginative risk-taker, just as he had been at the opening of the First World War, and he was confident that the war was going well for Britain. Even in those hectic days, he found time to relax, to meet with friends and to continue working on his *History of the English-Speaking Peoples*. Soon, the pressure of work forced him to shelve the project until after the war, but he did not give up the pleasure of dining with his friends, particularly the members of the Other Club.[33] Eight days after Britain entered the war, he sent a telegram to H. G. Wells inviting him to one of the Club's meetings, probably at its usual venue of the Savoy Hotel.[34]

The war seemed to have given the ailing Wells a new lease on life. He proffered advice and comment to the government in a fusillade of public letters to newspapers and magazines, impressing the poet T. S. Eliot so much that he wrote an article on Wells's sudden return to prominence, comparing it to Churchill's, both men having apparently been 'slowly and unwillingly retiring from public life'.[35] Eliot particularly welcomed their bluntness, 'all too rare among the loudspeaker voices of our time'. Wells and Churchill stayed in contact during the war by letter and telegram, but no record remains of their conversations at the Other Club, where they will have had their most candid exchanges. Wells had long since lost interest in making technological predictions and had not publicly mentioned 'atomic bombs' for almost two decades.

Churchill took some time off over Christmas, celebrating at Admiralty House with his wife and a few 'padlock' friends, including Lindemann and other Chartwell regulars.[36] Impatient for action, he had begun to develop plans to mine Norwegian waters in order to choke off the supplies of iron ore Germany needed to sustain its war machine. Most of the War Cabinet dithered, not wanting to provoke Germany or the neutral countries, but Churchill championed the Norwegian offensive, taking care to stay on the right side of the Prime Minister. Chamberlain was looking increasingly sick, disillusioned and out of his depth, while Churchill appeared ever more assured.

Among the hundreds of files, reports and minutes that Churchill and his colleagues waded through in those early months of the war, a few mused on the possibility of nuclear weapons. As one of dozens of long-shot threats that the politicians had to contend with, it did not command much attention. Gradually, however, a handful of nuclear physicists, working outside the huge government-run science laboratories, began to take the idea seriously. Among the experts assessing the danger was the recent winner of the first Nobel Prize for nuclear physics, awarded for discovering the match that could light a nuclear bonfire.

SEPTEMBER 1939 TO FEBRUARY 1940

Chadwick doubts that the Bomb is viable

> '[James Chadwick] put his duty, whether to his country, university or college, before everything, so he never did what he really wanted to do with his neutron. That is, apply it to the medical aspect for the cure of cancer.'
>
> AILEEN CHADWICK, 1974[1]

James Chadwick did more than any other British scientist to give Churchill the Bomb, though the two men never met. In the course of the war, as nuclear weapons became a reality, Chadwick was transformed from an introvert experimentalist into the first and most influential of the new breed of nuclear physicist-diplomats, and was tested almost to breaking point. He knew better than anyone that he would be judged by how well he filled the outsize shoes of the scientist who would almost certainly have done the job, had he lived – his former mentor, Lord Rutherford.

Chadwick had left the Cavendish to take up a professorship at the University of Liverpool in 1936, a few weeks short of his forty-fifth birthday. Within a few months, he had demonstrated that he was not only a brilliant experimenter but that he could also lead his own department effectively, winning new resources, chopping down the dead wood and cultivating new young talent. He had left Cambridge to avoid an altercation with Lord Rutherford, who had spent years resisting his younger colleagues' pleas to invest in some of the new and extremely compact particle accelerators that could probe deep inside nuclei[2] – 'I won't have a cyclotron in my laboratory,'[3] he told his boys, brooking no further discussion.[4]

When Chadwick was offered the Liverpool professorship, he jumped at the chance to run his own show, mix more freely with industrialists and live in the gritty native city of his wife Aileen, daughter of a prosperous local family. The Chadwicks and their eight-year-old twin daughters moved to the affluent suburb of Aigburth Vale, where they had a grand home with a tennis court in its grounds. Aileen was content with domestic life, while her husband worked in his department on Brownlow Hill, flanked on a road leading up from Lime Street railway station by some of the worst slums in Britain.

Soon after he arrived, Chadwick won the first Nobel Prize to be awarded for nuclear physics. The prize must have been all the sweeter for being unshared, though it was not enough to win him the job of Rutherford's successor as director of the Cavendish Laboratory. To his well-hidden disappointment, Cambridge University authorities gave the post to the less distinguished Lawrence Bragg – rumour had it that Chadwick's gauche manners ruled him out of contention.[5] Soon, the Cavendish broadened its interests, but it was no longer a world centre of nuclear physics.

Chadwick soon realised the task he faced in revitalising a department that had been running down for thirty years.[6] Its annual budget was 'less than some men spend on tobacco', he later recalled, and some of the laboratories did not even have alternating current.[7] None of this deterred him: during the First World War, he had managed to do research even as an internee at the liberally run German prisoner-of-war camp at Ruhleben. There, he gave fellow inmates lectures on nuclear physics and even set up a basic laboratory in his barracks, where he did worthwhile experiments using radioactive material in toothpaste.[8] He walked out of the camp in November 1918, his digestive system ruined, and soon returned to Manchester, where Rutherford gave him his first big break by securing him a scholarship.[9] Chadwick's career quietly flourished until his discovery of the neutron enabled him to step out of his boss's shadow.

At first it seemed that he was going to have a quiet war. Six days after Britain entered it, *Nature* declared uncontroversially that 'the interests of pure science as an intellectual pursuit and discipline must remain in abeyance'[10] – nuclear physics and other curiosity-driven research would be put on the back burner so that scientific talent could be redirected to projects most likely to help win the war. In 1939, Chadwick had begun the new term in early October with the usual complement of students but only half its roster of lecturers – the rest had left to work on radar, a technology that never seemed to interest him. It was going to be an uphill struggle to maintain the growth of his department.

Chadwick had not expected war to be declared. When proved wrong, he kept his head down and made the best of things, privately grousing that his younger colleagues had lost their grip, 'dithering about and feeding on their imaginations, instead of getting on with the job'.[11] It was when his department was in this febrile state that a letter arrived in his office indicating for the first time that his nuclear expertise might be useful to the government. The note was from his former Cavendish colleague Edward Appleton, then running the Department of Scientific and Industrial Research, which was being inundated with queries about a possibile uranium bomb. Chadwick was the ideal scientist to assess the worth of what appeared to be nuclear scare stories, as he was famous for the sobriety of his judgement.

When Appleton first enquired whether this possibility was anything to be worried about, Chadwick played it straight:[12]

It is not easy to say anything definite about it. It is certainly a possibility that under suitable conditions the uranium fission process might develop explosively. There is one point which must have occurred to you at once and that is, how to prepare a uranium bomb that will not blow up immediately. 'Hoist with his own petard' will almost certainly be the fate of the man first successful with this process. But I do not think this difficulty is insuperable.

He concluded by saying that he would 'look into it carefully and write again'. This was typical of Chadwick: asked a question, he could be reliably expected to give a sensible, cautious reply off the top of his head and then undertake to think about it more deeply.

Chadwick ran his physics department in Liverpool with the same authority as Rutherford had led the Cavendish, without his charm though with rather more tolerance. Lean, dark and brooding, Chadwick would have passed as an ambitious bank manager, his manner brisk, his attire spruce, his hair combed neatly over the sides of his shining forehead. His health had been poor since he left the detention camp in Germany. In his correspondence, he comes across as a man who, if not a hypochondriac, spent most of his time either sick, going down with an illness or recovering from one. Although universally admired for his scientific achievements and for his industriousness, he was a forbidding figure to his younger colleagues – his 'boys', as he called them.[13] When one of his students knocked on his door in the laboratory, Chadwick would not answer, obliging his visitor to peer round the door, often to find him sitting with his head in his hands. 'What do you want?' he would ask, looking over the glasses perched on his nose, before standing up and pressing his hands into his back and complaining about his lumbago ('My God, my back'). Once he had warmed up, however, he was courteous, helpful and inspirational, in his downbeat way.

Chadwick's connections with local industry ensured that funds flowed copiously into his department's coffers and that he was not short of visitors, many of them seeking his advice. The most able scientist to beat a path to his door was Joseph Rotblat, a bright-eyed Polish physicist in his early thirties. He had arrived in Liverpool in the spring of 1939 without his wife – he was too poor to support her – and knew so little English that he had to steel himself before beginning a conversation. Homesick and miserable, his mood was not improved a few weeks later when he read Siegfried Flügge's paper on the possibility of building an explosive nuclear device. If Flügge knew this much, there was a good chance that at least some of his colleagues were actually working on the idea. After weeks of hand-wringing, Rotblat agreed with most of his colleagues that 'the only way to stop the Germans from using it against us would be if we too had the bomb and threatened to retaliate'.[14] Sensing that war was imminent, he returned to Warsaw to bring his wife back to England, but appendicitis made it impossible for her to make the journey. He left her behind, intending to pick her up later.[15] When he read soon afterwards of Hitler's savaging of Poland he felt, as he later wrote, that 'the might of Germany stood revealed, and the whole of civilisation was in mortal peril'. Hearing nothing from his wife, he was bereft.

At Chadwick's suggestion, Rotblat started to work on the nuclear bomb. Chadwick had filled pages of his notebooks with calculations on the putative weapon's viability, and after five weeks came to a disappointingly vague conclusion, which he sent to Appleton: 'I can give no definite answer to this question.'[16] He promised, however, to pursue the problem experimentally and delegated it to the 'very able and very quick' Rotblat, assuming they were able to get hold of enough uranium oxide.

Appleton, though ignorant of the underlying nuclear physics, kept the Cabinet informed and reassured.[17] G. P. Thomson, at Imperial College London, and Mark Oliphant, another of Rutherford's former 'boys', at Birmingham University, had long been independently working on chain reactions. Eight months before, Thomson had felt like a character in a pulp thriller when he sheepishly requested a ton of uranium oxide from an Air Ministry official, for purposes he was not at liberty to explain.[18] Thomson's status as a recent winner of the Nobel Prize for physics would not have harmed his case. He was duly supplied with the ore and became the first scientist in Britain to secure government resources to investigate nuclear chain reactions. Appleton visited the Air Ministry to get a proper brief, after one of their officials had described the state of affairs as an 'illuminating example of the scientists' right hand not knowing what their left hand was doing'.[19] Ministry staff had been even less amused when Tizard put forward an idea – conceived independently by him and Thomson – that the government should lay a trap for the Germans by issuing a spoof report stating that Britain had made a nuclear bomb, in effect replying to Hitler's 'secret weapon' speech.[20] As Tizard recalled later, the authorities were 'horrified'.

The idea of a nuclear weapon was now being taken seriously in Whitehall. Government officials, many of them worked off their feet, now found themselves looking into what they thought was arcane science, boning up on geochemistry and locating the world's most plentiful supplies of uranium. It turned out that uranium ore was relatively cheap (uranium-based chemicals then cost about two dollars a pound) and was being mined in the Congo – ruled by the friendly Belgian government – and the Great Bear Lake region of Canada, a British ally.[21] The bad news was that the most abundant sources of high-quality uranium ore were to be found in Joachimsthal, in Nazi-occupied Czechoslovakia, and were being refined in Berlin.

Chadwick focused on the key question: if a nuclear chain reaction could take place, how could it be used to make a bomb? He and his Liverpool colleagues were using their brand-new cyclotron to investigate whether a nuclear chain reaction might be possible, squeezing in precious hours of research between teaching and administration.[22] Yet these and similar experiments were only desultory, not part of a coherent programme. In January, Chadwick wrote to his former colleague John Cockcroft, then working on radar: 'Our laboratories seem to be so disorganised that very little useful work of any kind – peace or war work – is going on.' If the chaos lasted much longer, Chadwick fretted, 'we shall be hopelessly behind the Americans'.[23]

He was right to be concerned. Although the Great Depression had left many physicists in the United States short of resources, the country's nuclear physics had been given a substantial boost by the arrival of hundreds of first-class refugee scientists. Several of them had landed first in Britain but were then redirected to America, the British authorities waving them on like mindless policemen on a traffic island, eager to keep the vehicles moving. One of those who had been waved on was the Austrian refugee Viki Weisskopf, who emerged as a leader in the nuclear community. He later said that he would have happily stayed in Britain if the authorities had shown any interest in keeping him: 'The English had a very short-sighted policy toward refugees,' he said in 1966. 'Gosh, what they could have gotten at that time for nothing.'[24]

Europe, the crucible of both quantum theory and relativity, was losing its lead over the US in the field of physics. American physicists, so far free of the pressures of war, could scarcely be better placed to pick up and run with ideas hatched in a devastated Europe. And if such an initiative could win the support of the American government, there would be no stopping them.

Unknown to Chadwick and his colleagues, the idea of developing a nuclear bomb was being pondered not only in American laboratories but also in the White House.

OCTOBER 1939 TO JULY 1940

FDR receives a nuclear warning

> QUESTION: Why are Franklin Roosevelt and Columbus alike?
> ANSWER: Like Columbus, Roosevelt didn't know where he was going or where he was when he got there, nor where he had been when he got back.
>
> Joke told shortly before the war in the presence of a stony-faced Churchill.[1]

When the war broke out, Franklin Roosevelt was approaching the end of his second Presidential term and, with the economy still in the doldrums, another period in office looked – to use one of his favourite words – iffy. His focus over the past seven years had been on his homeland, but Hitler and his allies put paid to that: by early October 1939, the President was walking on eggshells,[2] trying to persuade sceptical lawmakers to repeal the 1937 Neutrality Act, which had attempted to make it impossible for the United States to become embroiled in foreign conflicts. He faced opposition both from adamant isolationists, who wanted their country to stay out of what they saw as an imperialist power game, and from those who thought it the duty of America to stop the spread of Fascism.[3]

On the day Britain entered the war, Roosevelt gave one of his fireside chats on the radio, reassuring his listeners that their country 'will remain a neutral nation', though adding, 'Even a neutral cannot be asked to close his mind or close his conscience.'[4] The prevailing view in the country was that it was best not to get involved: for every American voter who wanted to send the US army to Europe, nineteen did not.[5]

The possibility of nuclear weapons was first drawn to his attention at a meeting in the White House on Wednesday 11 October 1939. A few days before, he had spoken to Churchill on the phone for the first time, a month after sending him a telegram to congratulate him on his appointment to the War Cabinet.[6] The two men had not been in touch since October 1933, when Churchill sent a copy of the first volume of his biography of the Duke of Marlborough to the new President, inscribing the book: 'With earnest best wishes for the success of the greatest crusade of modern times.'[7] In his congratulatory telegram, the President underlined his preference for direct personal diplomacy: 'What I want you and the Prime Minister to know is that I shall at all times welcome it if you will keep me in touch personally with anything you want me to know about.' Thus began Churchill and Roosevelt's wartime correspondence, which eventually ran to some two thousand telegrams and letters.

The 11 October meeting took place after another day of hard lobbying on Capitol Hill. Between the President's first appointment at 11 a.m. and his pre-dinner dip in the White House pool, his diary was packed with fourteen engagements. Somehow, his secretaries had to squeeze in time for a meeting with his occasionally obtrusive acquaintance Alexander Sachs, who had been pressing for weeks to bring the Bomb to the President's attention.[8] Sachs had been one of the President's economic advisers since 1932, a director of Lehman Brothers and an avid reader of popular articles on modern nuclear physics.[9] Though a quick learner, he had no talent for précis and was well known for writing documents that had to be fought through rather than read. During an earlier meeting with Roosevelt, Sachs had mentioned that nuclear weapons might be possible, but Enrico Fermi had as usual smothered the idea in scepticism.[10] Leó Szilárd, having sniffed out Sachs as a willing messenger to the White House, sought to trump Fermi's wariness with a supportive message from Albert Einstein, who had visited the White House in January 1934 and chatted with the President about their shared interest in sailing.[11] Assisted by his fellow Hungarian refugees Edward Teller and Eugene Wigner, Szilárd had visited Einstein in August at his vacation home on Long Island to explain to him that it might be possible to build nuclear weapons, and to press him to alert Roosevelt.[12] The idea of a nuclear chain reaction was a revelation to the world's most famous scientist, who did not take much persuading to write to the President: on 19 August he signed a letter, drafted by Szilárd, warning the President that it might be possible to construct new and 'extremely powerful bombs', based on nuclear chain reactions in uranium. He concluded by pointing out that the Nazis might be working on this project and now had access to uranium ore.

When Sachs walked into the Oval Office that autumn afternoon, the President greeted him with his customary geniality: 'Alex, what are you up to?'[13] Among the clutch of papers Sachs was cradling in his arms were Einstein's signed letter, Szilárd's ramblings and his own eight-hundred-word summary of nuclear energy and 'bombs of hitherto unenvisaged potency and scope'. Sachs read his words to the President, who got straight to the point: 'What you are after is to see the Nazis don't blow us up.' After Sachs agreed, Roosevelt – sitting at his huge desk – told an aide, 'This requires action.' Within minutes, the government's Advisory Committee on Uranium was gestating. It was suggested that a good chairman would be the veteran Lyman Briggs, head of the National Bureau of Standards (America's national physics laboratory), once a soil scientist, now a competent bureaucrat, but never anyone's idea of a visionary. The Committee's first meeting on 21 October was attended by military personnel, Sachs and the Hungarian scientists Szilárd, Teller and Wigner, but not Einstein, who was already backing away from front-line involvement.[14] The American nuclear bomb project was now under way.

Roosevelt had behaved true to form. If a trusted colleague presented him with a well-argued idea, backed up by expert opinion, he would usually go along with it. If, however, he had already made up his mind about the topic under discussion, he could be extremely tricky – he liked to tell officials and other visitors what they wanted to hear, even if he later had to spend a little time backtracking or soothing wounded egos. For him, consistency was overrated as a virtue – what mattered was that events went in broadly his direction, ideally in *precisely* his direction. Beneath the veneer of bonhomie was a resolution so strong it could be chilling. His successor Harry Truman later remarked that he was 'the coldest man I ever met'.[15]

Roosevelt entered politics when he was twenty-eight, in 1910. Like Churchill, he had been born into a wealthy family and was blessed with so many political gifts that his rise to power was all but inevitable. Not only was he bright, charming and energetic, he even looked the part – broad-shouldered, six feet two inches tall and weighing 190 pounds. Former President Woodrow Wilson called him 'the handsomest young giant I have ever seen'.[16] Roosevelt's political career had been blessed and without serious misfortune until, at the age of thirty-nine, he contracted polio and had to spend most of the rest of his life in a wheelchair. Never quite giving up hope that he would walk again, he crawled from room to room, submitted to being carried around like a child. The press took care to photograph him only from the waist up – usually capturing his toothy, confident grin – so few Americans knew of his disability.

Unlike Churchill, Roosevelt had few interests outside politics, and they did not include modern science and technology.

At school in Groton, he had studied basic science and in his first two years at Harvard had taken courses in geology and palaeontology. He left full-time education, however, with a Churchillian indifference to abstraction, but with little of Churchill's concern for using the products of scientific thinking to help the military.[17] For Roosevelt, science was something that for the most part went on outside politics – he had no Lindemann and never showed much interest in acquiring one. The idea of Roosevelt taking time off from preparing an important speech to review his understanding of atomic physics is unimaginable – his idea of a quiet morning's pleasure was to spend a few hours annotating his collection of postage stamps.

Roosevelt's New Deal policies of high taxes and aggressive government intervention had made him many enemies, who regarded him, in his own words, as an 'ogre – a consorter with Communists, a destroyer of the rich, a breaker of our ancient traditions'.[18] Even Churchill, an admirer of the President's boldness and spirit, openly criticised his record. In an article published in late 1937, Churchill complained that 'the Washington administration has waged so ruthless a war on private enterprise' that it was 'leading the world back into the trough of depression'.[19]

The President was unpopular among some of the leaders of the cash-strapped science community who had long been pressing for government money to fund basic research.[20] Even though Roosevelt had little money for them, he always had plenty of warm words, assuring physicists in 1935 that he was 'wholly in sympathy' with a government programme to support them. But his heart was not in curiosity-driven science that was of no immediate benefit to the ravaged economy, and none of the physicists' federal funding initiatives made much progress.

The President was more interested in the role science could

play in improving society. As he remarked in 1937, during his second inaugural address, government husbandry was essential to 'create those moral controls over the services of science which are necessary to make science a useful servant instead of a ruthless master of mankind'.[21] By the time nuclear fission had been discovered in late 1938, American physics was in good health, largely because of the munificence of private funders, such as the Rockefeller Foundation. With physicists of the calibre of the cyclotron inventor Ernest Lawrence and the theoretician Robert Oppenheimer, both of them at Berkeley and running excellent research programmes, the community was well placed to answer any call on its services the President might care to make.

Within two weeks of the inaugural meeting of Lyman Briggs's Uranium Committee, its first report was delivered to the White House. The problem was that the document was too dull to stand any chance of inspiring the President, whose heart must have sunk when he read the yawningly predictable recommendation that the government should fund more research into nuclear chain reactions. More likely to have caught his eye was the statement that if the reactions were explosive, then they 'would provide a possible source of bombs with a destructiveness vastly greater than anything now known'.[22] Yet even that phrase did not seize Roosevelt's attention, assuming he read it. Without a Lindemann to keep a scientific eye on incoming technical papers, to translate them into readable prose and draw out the key points, he gave a pat response. He merely asked an aide to ensure that copies were sent to the army and navy but otherwise to 'keep it on file for reference'. For the next few months, the file gathered dust. Szilárd, Fermi and their colleagues in New York received the funding and made progress, but heard nothing from Washington, where Briggs's dozy leadership had ensured that the idea of investigating the possibility of a nuclear bomb lay deep in Capitol Hill's longest grass.

The project might have stayed there had it not been for the shake-up initiated by a spry newcomer in Washington DC, Vannevar Bush, known to his friends and colleagues as 'Van', recently appointed President of the Carnegie Institution. He had arrived in Washington eighteen months before, determined to make his mark on the national administration of science by joining several influential committees. Born to a family of New England seafarers, he rose to a professorship of electrical engineering at the Massachusetts Institute of Technology, where his intellect and technical ingenuity marked him out, his lucrative patents affording him a measure of financial independence. His appointment to the vice-presidency of MIT did not sate his ambition – which was to be the chief science policy-maker in Washington, and to leave an indelible mark on his country's history. Just turned fifty, he had the energy, ability and will to shake things up in the capital: determined and business-like, his self-assurance was obvious from every word he spoke.[23] When upset, he didn't talk, he growled.

A little less than six feet tall and slightly stooped, he was always smartly dressed, wearing wire-rimmed spectacles and habitually puffing on a pipe he had carved in his home workshop. Although he looked like a beardless Uncle Sam and had the folksy charm of the cowboy entertainer Will Rogers, Bush had the hard-headed pragmatism of a seasoned political operator. A registered Republican and a scabrous critic of the New Deal, he was nonetheless happy to do business with Roosevelt: 'I knew you couldn't get anything done in that damn town unless you organized under the wing of the President.'[24]

Bush was aware of the discontent among scientists on the spavined Uranium Committee – a symptom, he knew, of a much wider malaise: there was no mechanism to bring America's scientists together with the military and to ensure constructive collaboration. He achieved this on a June afternoon in 1940 in a power grab that had the audacity of a heist. A few weeks before, Bush had inveigled Roosevelt's uncle Frederic Delano into persuading the President's friend and colleague Harry Hopkins to give him a coveted slot in the White House diary.[25] In the Oval Office, with Hopkins looking on, Bush introduced himself and set out his ideas for mobilising America's military technology by coordinating and supervising scientific research on military equipment and weapons. Theatrically pulling out a single sheet of paper summarising his plans, Bush handed it to Roosevelt, who promptly instructed Hopkins to write on the paper 'OK – FDR'. The meeting had lasted about ten minutes.

Roosevelt's intuition was sound – and his appointment of Bush was a masterstroke. He had sanctioned what became known as the National Defense Research Committee, soon to give new energy and focus to the application of science to warfare by American academics and military personnel. It went without saying that Bush would chair the Committee. Working with Hopkins, he quickly appointed its members and recruited dozens of scientists, giving them accommodation in his own Carnegie Institution and the National Academy of Sciences; soon his staff was spilling over into other buildings. Delegation was the key to his success: rather than setting up laboratories and institutes to do the Committee's work, he devolved projects to thousands of contracts with universities and industrial organisations all over the United States. He was building up what one American reporter aptly described two years later as 'a scientific organization such as the country has never seen before'.[26]

At first, the putative nuclear weapons barely figured in the Bush Committee's deliberations. A year before, when the media chatter about the possibility of these bombs and of cheap power was at its most feverish, Bush ridiculed the journalists' most outlandish predictions and went out of his way to debunk them.[27] At one point, he feared a repeat of the panic that followed Orson Welles's radio drama based on H. G. Wells's *War of the Worlds*, which persuaded thousands of Americans that Martians had landed in New Jersey. Soon, however, Bush accepted that the press speculation was harmless and that the 'great impracticability' of the supposed bombs meant that it was wise to continue 'soft pedaling a bit'.

Van was right to give the President cautious advice as no one had any idea how to build such a bomb. Szilárd and his colleagues were in some ways 'all talk' – they were terrified of something that appeared unfeasible to everyone who had an informed opinion on the matter. Soon, however, Van heard that it may well be possible to make the weapon, following a brilliantly simple idea conceived in England by two scientists classified there as 'enemy aliens'.

MARCH TO JUNE 1940

Frisch and Peierls discover how to make the Bomb

> 'As a weapon the super-bomb would be practically irresistible. There is no material or structure that could be expected to resist the force of the explosion.'
>
> OTTO FRISCH and RUDOLF PEIERLS, March 1940[1]

The timing was, perhaps, not ideal. In March 1940, when the world was about to plunge into the most destructive global conflict in its history, two Jewish refugee physicists – the Austrian Otto Frisch and the German Rudolf ('Rudi') Peierls – showed that nuclear bombs could, in principle, be built.

A memorandum written by the two physicists in just a few days provided a basic blueprint of a nuclear bomb to the British government. The document obliged officials to consider the possibility that the Nazis might also be developing such a weapon, and the unpalatable truth that Britain was poorly placed to build one of its own. The consequences for Britain and the United States were momentous – not for nothing has the Frisch–Peierls document been called 'a memorandum that changed the world'.[2] Yet, for decades, many experts on nuclear weapons – scientists, politicians and military strategists – were unaware that this was the memo that kick-started the development of the Bomb. This was partly because it went missing after the war. A copy of the main part of the text turned up two decades later in the UK Atomic Energy Authority's strongroom – according to legend, the document was stuffed inside a Corn Flakes packet.

Frisch and Peierls, both in their thirties, were not working in the academic powerhouses of Cambridge and Oxford but in the relatively modest ambience of Birmingham, capital of the West Midlands. Apart from Hitler, the person most responsible for bringing them together was Mark Oliphant, since October 1937 one of Birmingham University's most energetic professors, and yet another of Rutherford's 'boys'. Just as Rutherford had done in Manchester and Cambridge, Oliphant arrived in his new department like a buccaneer storming a galleon. He reshaped Birmingham's courses and research programme, made adventurous new appointments, and aggressively sought out research funds, stressing that nuclear experiments might one day benefit industry and have medical applications.[3] He persuaded the car manufacturer and philanthropist Lord Nuffield to buy for the department 'the world's largest cyclotron', reassuring locals that the new 'atom-splitting machine' was nothing to worry about.[4] Oliphant become well known in Birmingham for promoting nuclear research to journalists and for persuading them of its potential usefulness. After an interview with him about his cyclotron, one of them wrote: 'The purpose of the machine is not to release energy for destruction. It is for humanitarian work.'[5]

Lord Nuffield also paid for the handsome new building where Frisch and Peierls were working in March 1940, a few yards apart, in a single-storey redbrick extension to the main block. The whiff of fresh paint, wood shavings and glue still lingered in the offices. Frisch, then thirty-four, had arrived in Birmingham during the previous summer, after working in Bohr's Institute in Copenhagen for the past five years. Fearful of a Nazi occupation of Denmark, Frisch had taken soundings in Britain about short-term posts he might apply for, and accepted an invitation from Oliphant to stay for a couple of months to work on nuclear fission.[6] He was homesick, did little work and spent most of his time worrying about the fate of his wider family, including Lise Meitner, downhearted in

Sweden after having turned down a post in Cambridge.[7] An only child, Frisch felt a special responsibility to take care of his parents, who had also settled in Sweden after his father had spent months in Dachau, the Nazis' first concentration camp.[8] In the autumn, he began a long and wearying campaign to try to persuade the British Home Office to allow his parents to join him.[9]

By the time war broke out Frisch felt at home in Birmingham, which was just as well, as there was no possibility that he would return to mainland Europe. After Oliphant extended his contract, Frisch put himself at the disposal of the Ministry of Labour,[10] though he expected to sit out the war working on experiments, doing a modicum of teaching and writing a book on nuclear physics. Some aspects of English life grated on him – the endless talk about the weather, the reek of boiled mutton in the restaurants and a public transport system that would have benefited from a little Austrian discipline.[11] A private and apolitical man, slow to form close friendships, he mastered English less through conversation than through visits to the theatre and evenings reading the novels of Aldous Huxley.[12] What mattered most to Frisch was music, not words – he once said that if a piano was being played softly in a noisy and crowded room, he would hear the instrument's sound above all the chatter. He was an outstanding pianist himself and gave several public concerts.[13]

By early 1940, Frisch had got to know Peierls well. The youngest professor of applied mathematics in Britain, Peierls was one of only a handful of European refugee scientists to hold a permanent post in the UK.[14] Even before Oliphant took up his post in Birmingham, he offered Peierls a job and they began work there on the same day. It was an inspired appointment: the young German was a physicist of proven versatility, a popular colleague and quick to strike up collaborations, just the sort of person to make Birmingham one of the leading centres of theoretical physics research in the country. Formal in his manner but with no trace of pomposity, Peierls was politically moderate with an exceptionally strong sense of duty and fair play. As Rutherford remarked, 'He would make a very good Englishman.'[15]

In August 1939, Peierls abruptly changed his priorities from building his career as a quantum physicist to helping to fight the Nazis. He wrote to government officials to underline his willingness to serve as a scientist, pointing out that he had left Germany when Hitler came to power in 1933 and had formally applied for British nationality in 1938.[16] He might have added that he had no trouble making himself understood in his adopted country: Peierls and his wife had different mother tongues and decided soon after they met to speak in the one language they both spoke passably, English.[17] In company, Peierls's coolness was more than counterbalanced by the warmth of his Russian-born wife Genia, a physicist who had given up research. A flamboyant, generous host, she was also a hands-on mother to their two small children and was always ready to welcome guests, to whom she dispensed liberal quantities of advice, whether they liked it or not. At least one evening a week, Frisch left the tranquillity of his bachelor digs and walked to the Peierls's noisy Georgian home, where the rich stew of their family life was always simmering.[18]

Birmingham did not appear to be as well prepared as London for war, but was certainly ready for air attacks, boasting that it had room in its shelters for nine hundred thousand people.[19] Food supplies were now rationed, citizens went about their business carrying a rubber gas mask, and everyone had to get used to the strictly enforced blackouts. Frisch was 'gradually developing the instincts of a cat', he told a friend.[20] In common with every other 'alien', he was obliged to carry identity papers everywhere, obey a curfew, and observe travel restrictions. He was also forbidden from owning a motor vehicle (Rudi Peierls easily got round that particular rule – he lent his car to an English friend and then borrowed it back).[21] Frisch was philosophical about these restrictions, aware that he was much more comfortable than the rest of his close family.[22] In mid-February 1940, he had to break the news to his parents that the British Home Office had refused them asylum.[23] Peierls was spared this disappointment: his parents had arrived in England the year before and were then safely on their way to the United States.[24]

The British authorities spurned all the offers made by Frisch and Peierls to help with the scientific war effort, though Peierls signed up as a volunteer firefighter. They knew that several of their colleagues were working on radar, including John Randall and Harry Boot, who were developing the cavity magnetron, a device that made it possible to generate short-wavelength radar beams in machines so compact that they could be installed in aircraft. Frisch and Peierls were for the most part onlookers on this project – they were officially forbidden to know anything about it or to enter the laboratories where the research was being conducted. So the two men had plenty of time to think about a subject considered by security officials to be irrelevant to the war effort: nuclear physics.

They had worked independently on nuclear chain reactions and knew that the overwhelming consensus among physicists was that building a nuclear bomb was not feasible, as it would require tons of uranium to produce sufficient ^{235}U to set up and sustain the reaction. But in a conversation that took place in the Nuffield Building in March 1940, probably early in the second week of the month, they realised that the conventional wisdom was almost certainly wrong. It was Frisch who posed the crucial question: if someone had a large quantity of *pure* ^{235}U, 'what would happen?'[25]

Szilárd had posed this question a year before, but he and his colleagues had left it hanging in the air. Frisch and Peierls were well prepared to solve it – Frisch contributed his uncanny ability to understand the workings of gadgets, while Peierls brought his wide knowledge of theoretical physics.[26] Scratching out their formulae on the backs of envelopes, the two men did the calculations quickly, sometimes guestimating quantities not yet measured, such as the likelihood that a fast neutron will interact with a ^{235}U nucleus. The conclusion left Frisch and Peierls 'quite staggered', as Peierls later recalled:[27] the critical size needed to set up a sustained chain reaction was not several tons but roughly eleven pounds, which would have about the volume of an orange. It should, they found, be possible to make a bomb by taking two hemispheres of ^{235}U, each with half the critical mass, and firing them directly at each other – when they formed the critical volume, a chain reaction would spread through it, making it about as hot as the core of the sun and giving rise to a pressure some ten billion times that of the Earth's atmosphere. The result would be a huge explosion, equivalent to the blast of about a thousand tons of dynamite.

Frisch and Peierls stared at each other in silence. Could it be that Hitler's scientists were already making the weapon? The risk was too great to ignore, even though it would cost a fortune to set up an industrial plant to produce enough ^{235}U to make the Bomb. Frisch and Peierls agreed: 'Even if this plant costs as much as a battleship, it would be worth having.' Unsure of how to bring all this to the government's attention, they consulted the streetwise Oliphant, who advised them to write a short document that he would then send to Henry Tizard, who would know the best levers to pull in Whitehall. Within only a few days, the two physicists had put flesh on the bones of the idea, and had also thought through some of the consequences of using the weapon they had conceived. They wrote up their findings in a two-part memo in flawless English that for the most part had a Lindemannian clarity. Despite a few venial errors, the document was a masterpiece of science writing.

In the main part of the memo, 'On the Construction of a "Super-Bomb"', Frisch and Peierls set out their idea using language intelligible to a high-school physics student and included only one, skippable equation. For officials intimidated by even a sprinkling of jargon, the memo's more colloquial second part, 'On the Properties of the Radioactive Super-Bomb', was more digestible. Here, the authors graphically described the destructiveness of one of these bombs and the radioactive fall-out that would ensure that 'even for days after the explosion any person entering the affected area will be killed'. The conclusion made sobering reading. Prefacing it with the appropriately modest statement that they 'do not feel competent to discuss the strategic value of such a weapon', Frisch and Peierls enumerated a list of points, one of them smuggling in a firm policy recommendation, based on the assumption that Germany already had the Bomb or soon would:

> The most effective reply would be a counter-threat with a similar bomb. Therefore it seems to us important to start production as soon and as rapidly as possible, even if it is not intended to use the bomb as a means of attack.

The whole matter was, Frisch and Peierls concluded, 'very urgent' and it was 'of extreme importance to keep [the] report secret'. They finalised it in Peierls's office, anxious that someone might overhear them – during one conversation, they were startled when a face suddenly appeared at the window, though the interloper turned out to be a technician innocently watering tomatoes.[28] Unwilling to entrust the document to a secretary, Peierls typed up the final draft, keeping only a single carbon copy under lock and key. By 19 March, the top copy was on Tizard's desk.[29] If the memo had gone to Lindemann, he would probably have made up his own mind after talking privately with a few trusted colleagues. But Tizard, typically, sought a collective decision by setting up a committee, which first met on the afternoon of 10 April at the Royal Society's headquarters, on the site today occupied by the Royal Academy of Arts. This was a day of grim news – the Nazis had begun their occupation of Denmark and Norway.[30] The British operation in Norway – 'half-prepared and half-baked' according to Lloyd George – was a humiliation for the Prime Minister, who was savaged soon afterwards in the Commons and in the press.[31] Although Churchill had, as First Lord of the Admiralty, championed the operation and been involved up to his ears in planning it, he escaped the worst of the blame.

Tizard arranged for the 'U-bomb subcommittee' to be chaired by his Imperial College colleague G. P. Thomson in any spare time he could muster.[32] The first meeting, on 10 April, was low-key and sparsely attended. Only Mark Oliphant, John Cockcroft and Philip Moon joined Thomson, who had to write his own notes, as he had not been provided with a secretary. After the formalities, he introduced a guest, Jacques Allier, a debonair Frenchman well versed in the chain-reaction research that had been pursued in France, then leading the world in the field. Allier, an intelligence officer, knew that the Germans wanted to get their hands on supplies of heavy water, a possible moderator of chain reactions, which was at that time produced only in one factory, in Norway. A month earlier, he had masterminded an audacious midnight raid, removing virtually the entire global supply of heavy water – 185 kilograms of it – from the factory. Later, he helped to ensure that the heavy water was eventually out of harm's way in Windsor

Castle, depositing it briefly in Wormwood Scrubs in the care of the librarian.[33]

The meeting helped to convince G. P. Thomson that it was worth taking seriously the idea of a bomb made from pure ^{235}U. A few days later, he wrote to Chadwick about 'Oliphant's suggestion' and commented that 'although at first it seems a bit wild, it is not so impossible when you come to look into it'.[34] But it was not Oliphant's suggestion, it was Frisch and Peierls's, and they were livid to have been excluded from the meeting. They had been deemed too much of a security risk to discuss the very idea they had conceived.[35] They had not received a written acknowledgement that their report had arrived in Whitehall and they were not even allowed to know the name of the chairman of the committee following up their work.[36] Peierls knew, however, that it was Thomson, and soon afterwards wrote him an exasperated note about the committee's progress. 'I feel I cannot permit myself the luxury of reserve,' Peierls began, writing with the slightly florid politeness of an English gentleman – which he now was, having been granted British nationality a few weeks before.[37] Thomson was sympathetic, and sought permission to involve the two Birmingham scientists.[38]

At around this time, Frisch became a lodger in the Peierls's home after their children had been evacuated.[39] The three adults got on well – at weekends, they often took the train and went out to the countryside around Birmingham to take long, restorative walks. One of these hikes was especially memorable for Peierls.[40] It took place in the late spring and there was a lot to talk about: Neville Chamberlain's tottering premiership, Mark Oliphant's packing his family off to Australia after concluding that Britain was two weeks from a Nazi invasion, the recent flurry of secrecy-busting newspaper articles on the American project to harness nuclear energy.[41] Worst of all, the

British government was running the scientific civil service with what their temperate friend John Cockcroft had described as 'incredible incompetence'.[42]

Frisch and the Peierlses had counted on staying overnight in a village hotel, but they were turned away by the proprietor, perhaps because of their foreign accents. Having a drink afterwards in a local pub, they heard a politician talking impressively on the radio, giving a speech so compelling that no one left the room until it ended. That unmistakable voice was, as the writer A. P. Herbert once remarked, 'like an organ filling [a] church', sending the congregation out 'refreshed and resolute to do or die'.[43] It was their new Prime Minister, Winston Churchill.

At last, here was a leader of authority and boldness, one who prided himself on his understanding of the value of new science to the military effort. Britain had become the first country in the world to be led by someone who had shown clear signs of being a nuclear visionary.

MAY AND JUNE 1940

Churchill has more pressing problems

'Overnight it was revealed that . . . Winston was the Angel Gabriel and not, after all, Beelzebub. There has been no more startling transformation since the Creation.'

JOSEPH MALLALIEU, politician and author, 1950[1]

When Churchill became Prime Minister in May 1940, the war was going so badly for Britain that he had to invest all his energies in the desperate task of defeating the Nazis. In the first months of his premiership, while a few scientists worried frantically that Hitler might be acquiring nuclear weapons, only a few memos on the subject reached the Cabinet office, and it is conceivable that Churchill read none of them. He was, however, as determined as ever to support the military with the latest science and technology, especially through devices that promised to give a quick and effective advantage.

In the scientific field, Churchill had only a few close advisers, with Lindemann by far the most influential. It was ultimately the narrowness of his advice on these matters that led Churchill to make such an uncharacteristically flat-footed response to the most powerful explosive his scientists devised.

Churchill could easily have failed to become Prime Minister. At eight in the morning on Friday 10 May, a few hours before he was appointed, he was at a meeting of the War Cabinet and the Chiefs of Staff. The grim news arrived of the Nazis' blitzkrieg attack on neutral Holland, Belgium and Luxembourg, which were overrun. France was next – the panzer divisions began to move across its border that morning. With war now blazing in Europe, Chamberlain had decided to resign, and the most popular choice to take over from him was Lord Halifax, favoured as leader by all three of the main political parties, by the King and most members of the House of Lords.[2] After Halifax had turned down the job, Churchill emerged as the only acceptable candidate.

A few hours later, after a tense Cabinet meeting at 8 a.m., Churchill became, in C. P. Snow's phrase, 'the last aristocrat to rule – not just preside over, rule – this country'.[3] It was not a popular appointment in the Conservative Party, as Churchill saw three days later when he walked into the Commons chamber for the first time as Prime Minister. Labour and Liberal MPs applauded him, while his Conservative colleagues remained almost silent. In the White House, Churchill's promotion was welcomed warmly, with one reservation – Roosevelt remarked privately that the new Prime Minister 'was the best man England had, even if he was drunk half of his time'.[4] The President almost certainly knew that he had been the subject of one of Churchill's essays in the collection *Great Contemporaries*, where he had been gently criticised for several of his New Deal policies. Churchill had, however, concluded that 'it is certain that Franklin Roosevelt will rank among the greatest [Presidents]'.[5]

Churchill formed his government with a caution that belied his impetuous reputation. His administration was a coalition, always a congenial arrangement to him, with Lord Halifax as Foreign Secretary and the Labour leader Clement Attlee as Lord Privy Seal. Neville Chamberlain was given the face-saving role of Lord President of the Council, officially responsible for science, but everyone knew that Churchill – assisted by Lindemann – was to all intents and purposes going to run this area of policy, in addition to the other responsibilities he had given himself: Leader of the Commons and the new post of Minister of Defence.

One of the few pieces of cheering news for Churchill during his first month as Prime Minister was the success of the scientists and mathematicians at Bletchley Park in breaking the Germans' Enigma codes. Later, the defence of the Allies' vital North Atlantic sea-lanes depended almost entirely on the Bletchley teams' ability to read the instructions to U-boat commanders in the area.[6] Churchill, who had for decades championed the value of military intelligence,[7] was overjoyed. He made the Cabinet's Joint Intelligence Committee responsible for coordinating all the relevant information pouring into Whitehall, with excellent results.

Most of the news he received that month was, however, terrible – on 15 May, only five days after he took office, he first fully appreciated the scale of the crisis he was facing. At 7.30 in the morning he was awoken by a phone call from the French Prime Minister Paul Reynaud, evidently under stress, declaring, 'We have been defeated. We are beaten; we have lost the battle.'[8] Scarcely able to believe that the French had been able to summon so little resistance, Churchill was dumbfounded. The news from France was no better in the coming weeks, as Hitler's armies blasted their way towards the coast, backed by heavily armoured tanks and an imperious Luftwaffe. The British Expeditionary Force, more than three hundred thousand-strong, was by late May bottled up around Dunkirk, and extremely vulnerable.

Churchill brought a new belligerency to the British campaign and a much-needed urgency to the running of the government, firing off a stream of pugnacious memos, many of them bearing the red label 'ACTION THIS DAY'.[9] Most importantly, he was able to turn his long-honed literary gifts into a powerful weapon of war – Churchill the writer was supporting Churchill the politician.[10] Through the relatively new technology of 'the wireless' – not available as a national medium during the previous conflict – he could speak directly into the homes of the British people and across the Empire. From the day he became Prime Minister until the end of the year he made seven broadcasts, most of them brief, all of them inspirational.[11] Many of his parliamentary colleagues had been sniffy about the amount of time he had spent away from Westminster – reading, writing and giving speeches – but these diversions now paid handsome dividends. Although his language was in many ways old-fashioned and sentimental, it struck a chord with his listeners in 1940. Betraying no sense of the doubt he sometimes felt, he brilliantly made his case over the heads of his dithering and sometimes defeatist colleagues, in words as imperishable as his self-belief. Gibbon and Macaulay would have been proud of him.

Yet many MPs were unconvinced that Churchill, with his long record of impulsiveness and poor judgement, was capable of providing the steady, determined leadership the country obviously needed. Out of his earshot, resentful colleagues and eye-rolling officials muttered that he might be the most popular politician in Britain, but he was only a stopgap Prime Minister.[12]

With the arrival of Churchill in Downing Street, Frederick Lindemann became the most influential scientist ever to work at the heart of the British government. Each of the two men had the role he had wanted for many years, and each had achieved his political ambition without being elected to his high office. Unlike the more generous-spirited Churchill, however, the Prof was not disposed to meet his old enemies and heal the breach. When his protégé R. V. Jones passed him another request from Tizard to work together amicably for the duration of the war, Lindemann declined icily: 'Now that I'm in a position of power a lot of my old friends have come sniffing around.'[13]

Among Churchill's earliest instructions to Lindemann were orders to 'scrutinise and push forward small inventions' and to establish a committee 'to investigate scientific and technical war devices'.[14] The idea of setting up such a group had little appeal to Lindemann, who replied that it would be better if he were given personal responsibility to call for experts as and when he saw fit. As usual, the Prof got his way. Churchill encouraged him to interrogate every minister and all their officials on every aspect of policy except military strategy, a licence that Lindemann used without the tact needed to make the role welcome. Whitehall insiders regarded him as Churchill's snooper.[15]

The Prof sat for hours in his armchair with a large blotting pad on his knees, amending draft minutes to Churchill, excising every unnecessary word and every dispensable qualification.[16] His hard-pressed staff in the Statistical Department supported him loyally, keeping him stocked with new material that he used to supply Churchill with a constant stream of facts, figures, graphs and concise commentary.[17] Outside his own bailiwick, entire government departments spent weeks preparing reports that Lindemann would condense for the Prime Minister into half a typewritten page. By the time the war ended, Lindemann had written some two thousand of these notes – an average of almost one for every day of the conflict – on a wide range of subjects, from shipping and troop movements to economics and food supplies.[18] In this correspondence, the possibility of a nuclear bomb was the subject of only a few dozen minutes, many of them critically important influences on British nuclear policy and, indirectly, on the American programme.

Lindemann was especially interested in a report in the 7 May edition of *The Times* suggesting that the Germans were working on nuclear weapons.[19] The piece, drawn to his attention by an alarmed colleague, was based on a front-page report two days before in the *New York Times*, which claimed that leaks had revealed that the Nazi government had ordered German scientists in the field 'to drop all other experiments and devote themselves to this work alone'.[20] He seems not to have taken any specific initiatives as a result of this article, but he kept an eye on the reports of the U-bomb subcommittee. Always a conscientious reader of such documents, he is likely to have perused the Ministry of Supply's report on the consequences of dropping a uranium bomb on a large British city. But he does not seem to have been much concerned, his intuition telling him that nuclear explosives were unlikely to be viable.[21]

The Prof devoted more of his time to new weapons that were, in his view, likely to make a significant difference to the war. This was why he encouraged Churchill's interest in the small government department MD1, set up in early 1939 at Princes Risborough, Buckinghamshire, to work largely outside the usual civil-service system on the development of unorthodox new weapons. Churchill behaved 'like a small boy on holiday' when he visited it, the technical staff saw, and was always ready to protect them from the interferences of the Whitehall bean-counters.[22] He could not resist the pleasure of seeing an ingenious new gadget demonstrated in front of him and was known to make bulk orders for such devices on the spot.[23] It seems that his temperament was less well suited to the long haul of developing the uranium bomb.

The second half of May 1940 was no less disastrous for Churchill than the first. Allied troops were under such pressure from the Nazis that the Prime Minister agreed to a mass evacuation of British and French soldiers from Dunkirk and adjacent beaches: between 26 May and 4 June, some 335,000 troops were withdrawn on a motley armada of some nine hundred small boats, most of them designed for quite different purposes.[24] The British government was riven with disagreements about the way ahead, with Halifax and others showing clear signs of defeatism. Churchill would have no truck with this: presented with a draft recommendation that the warring nations might meet in conference to negotiate a peace settlement, he crossed it out in red ink and dismissed the idea as 'rotten'.[25]

Churchill was determined to 'drag the United States in' to the conflict, as he commented to his son.[26] It was not going to be easy: Roosevelt had any number of warm words for the Allies but offered little in the way of concrete support. On the night of 19 May, Churchill wrote to the President pleading with him to help Britain by supplying desperately needed fighter aircraft. 'Here's a telegram for those bloody Yankees,' the Prime Minister told his Assistant Private Secretary Jock Colville.[27] Churchill was in no mood to agree with the prominent scientist A. V. Hill, who was arguing that Britain should donate technical information to the United States in the confident hope that it would encourage a spirit of cooperation, much as his friend Rutherford had done during the First World War.[28] Although Lindemann seemed sympathetic to Hill's case,[29] Churchill flatly refused the request, as he had done earlier when asked to share British radar secrets. Britain seemed to be some way ahead of the US in many aspects of military technology, so it made no sense to him simply to give away secrets 'unless we can get something very definite in return', as one of his assistants put it.[30] After further pressure, he climbed down a few weeks later in a single-sentence memo that included a grudging 'I concur', though he soon regretted it.[31]

The resistance of the French was crumbling before Churchill's eyes. After a meeting with demoralised French leaders on

12 June, he told his military Chief of Staff 'Pug' Ismay that it looked like Britain would be fighting the war alone. When Ismay commented, 'We'll win the Battle of Britain,' Churchill gave him a look and said, 'You and I will be dead in three months.'[32] The German army marched into Paris unopposed two days later, and Churchill was infuriated to hear that the remaining British forces in the country were in retreat. After the French surrendered on 22 June, it was obvious that Britain would be next in line for invasion. With fears of an attack escalating, Churchill's government intensified its policy of interning without trial any foreigner deemed to be a security threat, including many Jewish refugees, sending the internees to detention camps or deporting them to Canada and Australia. Privately he lamented the crude management of this policy, though not enough to put a stop to it.[33]

Churchill was now under more pressure than any British Prime Minister since William Pitt took on Napoleon. The stress was beginning to show. In late June, his wife Clemmie wrote him a letter – as affectionate as it was cautionary – warning him that his 'rough sarcastic and overbearing manner' might lead to his being 'generally disliked' by his colleagues and subordinates.[34] Yet he retained the respect and admiration of his staff, including his Private Secretary John Martin, who one night visited him in his room and witnessed an example of his sometimes disconcertingly informal behaviour: 'I found him dressed only in a vest, pacing up and down. I gave my report. He turned angrily away from me, picked up the pot from under the bed and made noisy use of it.'[35] After a bout of obnoxious aggression, the Prime Minister could defuse resentment with winning charm, as Martin saw at the end of one especially fraught evening, when Churchill put a hand on his shoulder and said: 'You know, I may seem to be very fierce, but I am fierce with only one man – Hitler.'

While displaying a rather cloying loyalty to his king, Churchill himself in many ways behaved as a monarch, expecting unswerving loyalty from every one of his huge complement of staff. They had to dance to his tune, at every hour of the day, often well into the night, in meetings that his colleagues called 'midnight follies'.[36] The responsibilities he shouldered, together with his punishing workload, also entitled him to live royally, eating and drinking as well as he did in peacetime. Churchill knew that power is a borrowed robe, but he was going to make full use of all his entitlements while they lasted. One of the privileges he thought he deserved was his right to have his courtiers, such as Lindemann, at the heart of his government, even if they occupied posts that would have been better suited to others. The Prof was never going to be comfortable until his arch-enemy Tizard had been shorn of all his influence, and it was only a matter of time before Churchill administered the *coup de grâce*, six weeks after he took office.

After Lindemann summoned Tizard from the Air Ministry to Downing Street for a discussion about their 'overlapping responsibilities', Tizard soon realised that he was being frozen out.[37] On 21 June, his humiliation became too much to bear. At a tense meeting in the Cabinet Room, Churchill and several colleagues – including Lindemann and the demoralised Tizard – were discussing R. V. Jones's well-reasoned belief that Nazi bombers were being directed towards British targets by radio beams the Germans had set up in the sky.[38] The idea was wrong, Tizard believed, but the young Jones prevailed, having dazzled the Prime Minister and others with his reasoning.[39] At the end of the meeting, Churchill declared (rightly, as it soon turned out) that they should assume the beams existed, angrily banging the table and denouncing Tizard's department: 'All I get from the Air Ministry is files, files, files!' Later, having found that Lindemann had replaced him as chairman of a meeting at the Air Ministry, Tizard resigned.[40] Britain had lost 'the greatest genius at applying science to [military] tactics this country has ever known', in the judgement of G. P. Thomson, one of the few leading academic scientists friendly with both Tizard and Lindemann.[41]

Churchill probably gave the consequences of this no thought at all. He was living through some of the most harrowing weeks of his life, watching the Nazis sweep British forces from the continent and snuff out all opposition. Although Britain's leading academic scientists supported Churchill as leader, many of them were angry about his studied indifference to their advice and his treatment of Tizard. The difficult relationship between the Prime Minister and his scientists would normally be a matter of no consequence, but this was an unusual time. Several of the physicists were coming to believe that an explosive of unprecedented power, and possibly huge strategic significance, might soon be available.

In one respect, Churchill and his colleagues running British science had reason to be relieved when work on nuclear weapons began to take off in 1940. In a private letter ten years later, Lindemann's colleague R. V. Jones revealed why:[42]

> . . . we were very short of physicists, and nearly all our best men were working in radar . . . At the same time the large body of refugee scientists in Britain were an embarrassment to us, because we felt that we could not trust them with matters of immediate defence, and we [could] not see a safe way of making use of their talents. When, however, there appeared a remote hope that a uranium bomb might be made, I think many people saw with relief a chance of clearing up two awkward situations at the same time by putting the refugee scientists on to working for the uranium bomb; it gave them something worthwhile to do, and it enabled the country as a whole to do something worthwhile [about the] uranium bomb.

At the beginning of what threatened to be the most destructive war ever seen, human beings were on the verge of creating their most devastating weapon. Scientists were no longer 'fumbling with the keys' of the nuclear chamber 'hitherto forbidden to mankind', as Churchill had written nine years before in 'Fifty Years Hence'. His own scientists now believed they had the key in their hands, and would soon be clamouring to bring this to his attention. Would Churchill the politician have the foresight of Churchill the writer?

JUNE TO SEPTEMBER 1940

Thomson and his MAUD committee debate policy on the Bomb

'It was agreed that Dr Frisch should be informed of the importance of avoiding any possible leakage of news in view of the interest shown by the Germans.'

G. P. THOMSON, minutes of first meeting of the MAUD committee, 10 April 1940[1]

In the past two years, nuclear physicists had been forced to change their priorities. They had previously been like any other curiosity-driven scientists, discussing new results openly and paying little attention to the possible consequences of their findings. But now they were obliged to practise their still-young discipline underground and work secretly on what was expected to be a new weapon of mass destruction.

The U-bomb subcommittee's proceedings had put its members – including most of the leading lights of the British nuclear community, all of them political ingénues – under severe pressure. It was their task to advise Churchill's government, at what was probably the beginning of a long war against a brutal fanatic, about whether to invest huge resources in a bomb that might not be feasible or that might already be in the enemy's hands.

The professor charged with collating the advice, G. P. Thomson, was in some ways an unlikely choice as the committee's chairman. He was not a nuclear scientist of international standing and, as his colleague John Cockcroft observed, was not in the class of the two finest experimental physicists of their generation, Patrick Blackett and James Chadwick.[2] Nor was Thomson a talented manager of people: after clearing up a personnel mess he left at the Cavendish a few months later,

Cockcroft griped: 'G.P. is very tactless.'[3] Thomson showed bounteous tact, however, in his dealings with Lindemann, who will have approved of his politics and known that he posed no threat. If Lindemann mistrusted Thomson, there was little chance that the committee's conclusions would be taken seriously by Churchill.

Thomson had many other virtues, too. He was open-minded, well respected by his peers, good at persuading them to talk through their disagreements, and quick to size up complicated arguments by summarising the salient points crisply and fairly. By turns flippantly humorous and penetratingly subtle, he sometimes surprised his colleagues with his pugnacity, though he rarely crossed the boundary into bad manners. He was more interested in things than in people, preferring to tinker with his model ships than to read poetry or listen to music, which he regarded for the most part as ordered vibrations of air molecules.[4] For him, the only music worth listening to was the operas of Gilbert and Sullivan.

Able to spare only a day or two a week from running the physics department at Imperial College, Thomson was always pressed for time on the committee. His chairmanship came with no executive authority, so he had to rely on goodwill, and his largely untested powers of persuasion in Whitehall. His challenge was to coordinate the nuclear work at several laboratories – most importantly Birmingham, Bristol, Cambridge, Liverpool and Oxford – and work constructively with nuclear physicists who had only meagre technical support and diaries crammed full of lectures and university meetings.[5] Most inconvenient of all for Thomson, several of the best physicists in the field were (or had been) 'enemy aliens' and so were objects of suspicion to the authorities, often making it difficult for them to travel. Thomson made light of the official warnings and discreetly shared top-secret material with Frisch and with

Peierls,[6] who was working full-tilt on his chain-reaction calculations, having borrowed the services of a secretary and secured permission from his university to jettison his teaching duties. Peierls stepped down from faculty meetings after hearing mutterings that it was not quite right that former 'enemy aliens' were attending them.[7]

During the life of Thomson's committee, discontent with the government's handling of science rumbled like distant thunder in the nuclear community. In the Athenaeum Club and Blackett's apartment in Westminster, scientists, civil servants and service chiefs moaned over their brandies about the government's mishandling of their scientists.[8] Lindemann was the grumblers' Aunt Sally. Their vehemence probably stoked by jealousy, they blamed him for Tizard's ejection and for joining with Churchill in the hare-brained pursuit of wheezes and gadgets rather than concentrating on strategic initiatives.

Foremost among the critics was A. V. Hill, still bearing the scars of his battles with Lindemann on the Tizard Committee. A sprightly fifty-four-year-old, he was a handsome figure with his wavy silver hair and matching moustache, both always trimmed neatly. He was one of the most powerful and popular figures in British science, a straight-dealer who was by nature an activist, by conviction a sceptic. Always ready to prick the egos of the pompous, he intensely disliked and mistrusted Lindemann, whom he regarded as a gifted explicator but a second-rate scientist and a malign influence on Churchill. A few months before, Hill had been elected as one of the Members of Parliament representing Cambridge University, describing himself as an Independent Conservative (he was the only Nobel Prize-winning scientist ever to be elected to the Commons, a record that still stands).[9]

Like his late acquaintance Rutherford, Hill wanted British

scientists to collaborate with their American colleagues and to share their military work unstintingly. In January 1940, he had begun planning a personal mission to America with this agenda – in the same spirit as Rutherford's in 1917 – and spent two months touring the eastern US and Canada, privately making his case.[10] During his stay, Hill had talked with nuclear physicists at Columbia University – apparently including Fermi but not Szilárd – about their chain-reaction experiments. Hill sent a memo, 'Uranium – "235"', to Tizard, declaring it a waste of time to do this research in Britain:[11]

> If anything likely to be of war value emerges they will certainly give us a hint of it in good time. [The American physicists I've met] feel that it is much better that they should be pressing on with this than that our people should be wasting their time on what is scientifically very interesting, but for present practical needs probably a wild goose chase.

This advice was out of date. Although Fermi and his colleagues were making swift progress in setting up nuclear chain reactions, they had no idea that Frisch and Peierls had hit on a way of making a nuclear weapon.[12]

One of the problems Thomson's committee had to tackle was whether to share its knowledge with the Americans or try to go it alone. This would certainly be a great challenge for Britain, which was hard pressed for resources and not a sensible place to test the weapon. Although Lindemann urged the committee to be scrupulously apolitical – that is, to stay off his territory – it was going to be difficult to make a clean separation of science and politics.

As the threat of invasion grew starker by the week, Thomson worried about the safety of his family and, with expected increases in rationing, his children's diet. He and his wife Kathleen were alarmed by a chance conversation at the Carlton

Club with the MP Rab Butler, who warned that their home in Surrey was likely to be on the inland path of the Nazi army.[13] After weeks of agonising, the Thomsons decided that the three children should go to America accompanied by Kathleen, who would return after they had settled into their new homes and schools. On a sunny summer morning in late June, they set off, Thomson driving them across the North Downs to Guildford railway station. An hour later, he was sitting at his desk in silence. After years of living with the noise and pandemonium of family life, he was probably reflecting on the melancholy thought that for the foreseeable future he would be living alone. The Thomsons were just one of tens of thousands of families in Britain to be broken up by the war.

Only a few weeks later, he heard that his elderly father – one of the greatest British physicists of the past century – had died. There was, however, plenty of work that summer to distract him from his grief and loneliness. As he and his colleagues became increasingly confident that a nuclear weapon could be built, so the secrecy of their work intensified. It was time to give their committee a name that would be unlikely to catch the interest of prying eyes. Thomson agreed to name it MAUD, after an incomprehensible word in a garbled telegram Lise Meitner had sent to reassure one of Cockcroft's colleagues that Niels Bohr and his wife were safe.[14] Meitner had asked for the news to be passed on to 'MAUD RAY KENT', which Cavendish physicists fretted might be a botched anagram disguising an urgent message for the Allies to separate ^{235}U at the earliest opportunity (MAKE UR DAY NT) as the Germans already had it.[15] The truth was rather more banal: the words referred simply to a former governess of the Bohrs' children, Maud Ray, who lived in Kent.

Thomson's job was made no easier by the authorities' refusal to allow him to recruit any scientist who was an 'enemy alien', or who had ever been classified as such. Frisch and Peierls were increasingly frustrated by the way they were being treated – they had been refining their idea but had heard nothing from Thomson or an Air Ministry official to acknowledge their work. At the end of July, Peierls's patience ran out and he accosted Thomson in London.[16] Soon afterwards, Frisch and Peierls sent him a ten-page summary of the 'uranium problem', showing again that they still knew more about the prospects of nuclear weapons than anyone else in Britain.[17] A few weeks later, Thomson, realising that it was ridiculous to exclude the two pioneers from his deliberations, arranged a compromise by setting up a 'Technical subcommittee', deemed to be so harmless that it was safe for even former 'aliens' to join it.

At the Cavendish Laboratory, the French refugees Hans Halban and Lew Kowarski were even unhappier than Frisch and Peierls about their exclusion from MAUD's inner circle.[18] The two Frenchmen were trying to set up a chain reaction using slow neutrons, with a view to generating nuclear power rather than producing a weapon – a priority that made their experiments less important, from the British point of view. Based on the work they had done in France, they had brought with them a sheaf of patent applications filed in Paris shortly before the Nazi invasion. They then filed corresponding claims in London. Thomson and his colleagues thought these patents on priority were a trifling, time-wasting distraction from the business of building the Bomb – he later described them as 'the most unmitigated nuisance'.[19] But to the Frenchmen's eyes, the claims were perfectly reasonable ways of preserving their commercial rights in the post-war future of nuclear energy.

The mood of Halban, Kowarski and the other foreign physicists working on the MAUD project darkened when the government stepped up its internment campaign. In late July 1940, every 'alien' in Britain – even those who had been British citizens for many years – received a curt letter with an instruction to report to a police station.[20] The Royal Society pressed the government to exempt some of the scientists from internment and, in the case of Otto Frisch, was successful.[21] While Thomson was discretion itself in arguing the case for foreign nationals to be allowed to work on the project, A. V. Hill was contemptuous of the government's high-handed policy, which he believed denied many refugees the opportunity to fight the Nazi regime they hated.[22] This protest was part of his barrage of attacks on the government's handling of science, culminating in late June in a long, damning memorandum, 'On the Making of Technical Decisions by HM Government'.[23]

Hill began by pointing at the main culprit: 'It is unfortunate that Professor Lindemann . . . is completely out of touch with his scientific colleagues . . . his judgement is too often unsound.' Most serious of all was 'the fact that he is unable to take criticism or to discuss matters frankly and easily with those who are intellectually and technically at least his equal'. Hill deplored the Prof's 'ill-advised adventures which slowed down the production of tried weapons', almost certainly an allusion to Lindemann's support for MD1, often referred to disparagingly by outsiders as 'Churchill's Toyshop'.[24] Hill attached two appendices, the first a list of Lindemann's wackiest ideas, each followed by a crisp refutation, the second a denunciation by Tizard, complete with examples of the Prof's past bad behaviour. Yet Hill accepted that the government's scientific incompetence was not exclusive to Lindemann: 'A system has grown up of taking sudden technical decisions of high importance without, or against, technical advice.' It was striking that Hill did not question Churchill's authority, 'now so important to the nation', and accepted that Lindemann's presence may be 'indispensable' to him. Yet he was adamant that unless the system used by the Prime Minister changed,

'The situation will become highly dangerous.'

At the time Hill wrote this philippic, a group of twenty-five scientists led by the anatomy professor Solly Zuckerman was working on the Penguin Special *Science in War*, urging 'the effective utilisation of scientific thought, scientific advice, scientific personnel' in the war. The book, conceived, written and published in only a month, sold modestly, but successfully drew attention in Whitehall to 'the impression of vast potential forces insufficiently coordinated or inadequately marshalled', as a *Nature* editorial noted approvingly.[25]

None of the complaining scientists heard a word from Churchill, whose attention was focused on fighting the Battle of Britain, which had begun on 10 July. Using their newly acquired airfields in northern France, the Germans began raids on the west of England and South Wales, the Luftwaffe's fighters and bombers engaging in brief but brutal encounters with the Royal Air Force's Spitfires and Hurricanes. This was the beginning of the first major battle to be fought entirely in the sky, with 258 civilians killed that month in Britain.[26] Much worse was to come.

In the final week of July, a dextrous move by the Prime Minister blunted the impact of Hill's protest.[27] Churchill invited Tizard, then packing up his files in the Air Ministry, to lead the mission to give the Americans all the British technical secrets that might be useful to its military, as Hill and others had long urged. Tizard paid several visits to 10 Downing Street, where he found the Prime Minister still in two minds about the mission's wisdom, and uneasy about the idea of giving the Americans secrets in the naive hope that they would reciprocate this generosity.[28] Exasperated, Tizard agreed to lead the mission provided that he was allowed to run it with a free hand, which Churchill refused twice before finally succumbing.[29]

The plan was for Tizard to be accompanied by representatives of the services and by two radar experts, 'Taffy' Bowen and John Cockcroft, who would have to take time off from his work on the MAUD committee, widely agreed to be more important. In the two weeks before they set off for the United States, Cockcroft and Bowen collated plans, blueprints, circuit diagrams and samples relating to all the most sensitive British secrets: plans for the jet engine, gyroscopic gun sights, submarine detection devices and – the prize exhibit – one of the twelve prototypes of the cavity magnetron invented in Birmingham. This device generated ten-centimetre microwaves with ten thousand times the power of older devices, a revolution in radar technology and certain to be snapped up by the Americans.[30] This and other material were stored in a black box, a japanned tin trunk Cockcroft had bought at the Army & Navy Stores in Westminster.[31]

There was a chance that the idea of a close nuclear collaboration with the United States could be seeded during the Tizard mission, though Sir Henry was sure to focus on promoting Britain's radar technology. Although he knew of Frisch and Peierls's findings, he believed they would not be useful in the war – for him, it was essential to concentrate on the technologies most likely to make a difference to the military outcome.[32] The best hope for the advocates of building the Bomb was Cockcroft, who had attended four of the first five MAUD meetings. Yet even he was sceptical that such a weapon could affect the outcome of the war, as he told a group of sailors on board the *Duchess of Richmond* on his way to join the mission. Asked to entertain the troops with an impromptu lecture, he decided to talk about radioactivity and the energy stored in atomic nuclei. In one part of his presentation he used an argument advanced earlier by Rutherford, that there was sufficient energy in a cup of water to raise a battleship high enough out of the sea to break its back.[33] This was 'a very safe subject to talk about', Cockcroft assured his audience, as 'There was no hope at all that such a thing would be achieved during the present war.'

Tizard and his colleagues first gathered together on 6 September at the Shoreham Hotel in Washington DC, as they prepared to give away some of their country's most valuable secrets, asking nothing in return except American goodwill. The morning had brought the news that the Luftwaffe was now bombing London in massed raids, killing hundreds of civilians, in retaliation for British bombers' recent targeting of Berlin. The *Washington Post* featured three articles on the 'Battle of Britain', reporting that Nazi bombers had just staged their longest raid on London, lasting over seven hours.[34] A German pilot had used a vapour trail to draw a swastika in the sky over Westminster, thumbing his nose at Parliament. After Tizard read the reports he feared the worst, and bet Cockcroft five dollars that Britain would be invaded within a month.[35]

The *Washington Post* and most other leading American newspapers reported on the speech Churchill gave after the German bombers had left. He told Hitler, 'We can take it,' and went out of his way to thank America for its recent support – a feeble deal whereby Roosevelt supplied a few dozen ancient destroyers, in return for which Britain made available sites for US naval and air bases. The President was now beginning his re-election campaign and the Prime Minister took care not to embarrass him. He referred to America's 'non-belligerency' rather than to its neutrality, but avoided publicly pressing for it to enter the war. The reports of his speech drew another surge of sympathy for the British and helped to make Tizard and his colleagues even more welcome. It seemed to the British physicist Ralph Fowler, now running the British Central Scientific Office in Washington DC, that the flow of information was going to be largely one way. He confided to A. V. Hill, then under the bombs in London, that the Americans talked big about their military technology but 'have damned little to offer'.[36]

While Tizard and his colleagues were on their mission in America, the MAUD committee was approaching the end of its investigations and preparing to set out its conclusions. One of the crucial roles Thomson played was to prepare the ground for the committee's report, to try to ensure that it was given a sympathetic hearing. In a well-coordinated programme of lobbying, Chadwick and Peierls spoke with Lindemann – sure to be the most influential reader of the report – to explain why the nuclear bomb had to be taken seriously.[37] Peierls had explained the case to the sceptical Prof and urged him to send copies of all the files on the project to North America, in case Britain was overrun. Lindemann's responses were limited to a series of inscrutable grunts.[38]

There was controversy among the MAUD committee members on the central question of whether, if the Bomb were to be built, Britain should go it alone or try to work with the United States. If collaboration was agreed to be the best way ahead, Tizard's mission might prepare the ground, always assuming that it was successful. The strategy had seemed rather implausible to Churchill, though he was far too busy to pay it much attention now that the Nazi blitzkrieg had begun.

AUGUST 1940 TO AUGUST 1941

In his finest hour, Churchill begs America for help

> '[Churchill] has pulled himself together. He is pulling us all together . . . It is like awakening from a nightmare to think of what might have happened to my country without him.'
>
> H. G. WELLS, November 1940[1]

It was now total war. The London sky was the theatre for the highest-technology aerial conflict the world had ever seen, the terrible climax of the Air Ministry's research and development programme that Churchill had been promoting for years. It was also the fruition of poetic prophecy, he pointed out on the rooftops of Admiralty House to his staff and two American guests he invited along 'to watch the fun', as Jock Colville put it.[2] He quoted some of his favourite lines from Tennyson's 'Locksley Hall', written over a century before:

> Heard the heavens fill with shouting, and there rain'd a ghastly dew
> From the nations' airy navies grappling in the central blue

But even Tennyson's imagination had not envisaged the sheer violence of the spectacle: the drone of the bombers, the hacking cough of anti-aircraft fire, the dancing pencils of the searchlights, the bone-shaking thud of the explosions, the stench of the burning buildings, the fires blazing in Turneresque patches of colour across the city.[3] As Churchill surveyed the skies with his binoculars, he may have given a thought to H. G. Wells's description of aerial battles in *Anticipations*, written three years before the first human flight: 'Everybody, everywhere, will be perpetually looking up.'[4]

The harder German bombers hit the city, the more Churchill's popularity grew. Within a month of his becoming Prime Minister a cartoon in the *Daily Express* portrayed him as a bulldog, his prominent jowls declaiming his defiance – the felicity of the image had quickly made it a cliché.[5] Reservations about him melted away as he repeatedly demonstrated that he was confidently in command and could keep his country's chin up – he had the ability of a star actor to hold the stage, whether in Cabinet, in the Commons or in a bombed-out community in East London.

He gave several of the finest speeches the Commons had ever heard, precisely when he needed to. Among the best was the one he delivered on 20 August 1940 in praise of Britain's pilots: 'Never in the field of human conflict was so much owed by so many to so few,' recycling a phrase he had first used forty-one years before.[6] Radar helped the overstretched squadrons of the Royal Air Force during daylight hours, when operators could locate the incoming aircraft, but in darkness the new technology was a blunt instrument. The Luftwaffe soon got wise to this and made most of their raids at night, much reducing the accuracy of their bombing.[7] Later, however, the radar technology matured, and by 1944 it would have made a second Blitz impossible.[8]

In the middle of September, Hitler stopped trying to win control of London's skies, and what became known as 'the Battle of Britain' ended in a draw.[9] The Luftwaffe nevertheless continued its bombing campaign, attacking some twenty cities besides London, including Birmingham, Bristol, Coventry, Glasgow, Liverpool, Plymouth and Southampton. There had been fears that the German bombs might be spiked with radioactive material, but these proved groundless. By the autumn, the attacks on Britain were killing an average of four thousand civilians a month, a terrible figure but nowhere near the predictions of the Home Office experts. In the brutal political calculus of war, Britain could sustain losses on this scale quite comfortably – the Luftwaffe had done little to hobble the country's ability to fight.[10]

Churchill's leadership won the admiration of many who previously doubted him. H. G. Wells declared himself entirely won over: 'I will confess I have never felt so disposed to stand by a man through thick and thin as I do now in regard to him.'[11] Like most Londoners, Wells did his best to ignore the bombing ('Why should I be disturbed by some wretched little barbarian in a machine?'[12]). He refused to move from his boarded-up home near Regent's Park, from where he occasionally fired off letters of advice to Churchill, and soon afterwards began his final bid for respect as a scientist, by researching 'The Illusion of Personality' for a doctorate, later awarded to him by the University of London.[13]

For almost a year after the Battle of Britain, Churchill appears to have given no thought to nuclear weapons – he was much too busy to spend time on what seemed to be a minor matter, satisfactorily delegated to and overseen by his trusted science adviser. During this period, however, he took several initiatives that later proved important when the plans to build nuclear weapons took shape, especially his project to bring the United States into the war. Much less significant, but still important, was his tense relationship with many of his academic scientists, who wanted as much as he did to secure American involvement, but who favoured – in the field of science – a different way of going about it.

Still under pressure from the mortally ill Neville Chamberlain and A. V. Hill to give more breadth and depth to scientific advice in Whitehall, Churchill agreed to set up a

Scientific Advisory Committee. It was to be chaired by Lord Hankey, a contemporary of Churchill and Britain's most respected senior civil servant, with shrewd grey eyes, a huge domed forehead and the manner of a country solicitor coasting to retirement. He was often called 'Whitehall's Man of a Million Secrets'.[14] It was, however, no secret in Westminster that he had little time for Churchill and disapproved of Lindemann's influence over him. Despite this, Churchill enabled Hankey to become one of the chief nurturers of Britain's plans for a nuclear weapon.

Hankey knew the Prime Minister was not going to tolerate a bunch of scientists intent on muscling their way to the top table. Churchill implied precisely this when he wrote a short minute to Chamberlain: 'As I understand it, we are to have an additional support from the outside, rather than an incursion into the interior.'[15] He almost certainly sensed a plot to subvert or dilute the Prof's influence, and it was canny of him to give his discontented academic scientists the forum of an officially recognised committee, albeit one with more prestige than power. The committee, comprising mainly Royal Society officials such as A. V. Hill and Rutherford's successor William Bragg, first met on 10 October and was welcomed by *The Times* as 'a kind of scientific powerhouse' from which great things could be expected.[16] Behind the scenes, however, it was proving no easier for these scientists to influence the Prime Minister.

A few days later, Churchill received a report on the work of the Tizard mission, which had scarcely touched on the possibility of a uranium bomb during its discussions with American officials.[17] In the document, Tizard urged the government to capitalise on the success of the visit by redoubling its efforts to collaborate with American scientists, who wanted to help.[18] Churchill was unimpressed – in his opinion, there was no need to help the Americans develop military technology. In the War Cabinet's discussion about disclosing secret information to the United States, the focus was not on the central recommendations of Tizard's report but on his aside that 'on the whole, the United States had much less to tell us than we had to tell them'.[19] After the Foreign Secretary remarked that it would be 'disastrous' if the American authorities believed Britain lacked confidence in them, Churchill concluded that the British Ambassador Lord Lothian should be quite frank with President Roosevelt: 'We should say that we do trust them, but they would appreciate that we were fighting for our lives,' so 'There was some information which we could not possibly divulge.'

These words were unlikely to go down well in the United States. The Tizard mission had been an unqualified success – American scientists had been 'extraordinarily appreciative' and had acknowledged that the British were 'at least two years ahead' in developing new military technology, as Lindemann's brother Charles had seen at the British Embassy in Washington, where he was working.[20] So when Roosevelt heard that Britain had to keep some technical secrets from America, it probably seemed that Churchill was blowing hot and cold about scientific collaboration. From Churchill's point of view, however, America's support for Britain's entire war effort was disappointing – half-hearted and parsimonious. The inactivity in the White House since Roosevelt's re-election in November had left Churchill 'rather chilled'.[21]

The tide appeared to turn at the very end of the year. On 30 December, Churchill heard reports of Roosevelt's latest fireside chat, broadcast over the radio networks, promising that the United States would become 'the arsenal of democracy', and making clear his support for Britain.[22] In the following year, the Prime Minister tried still harder to draw Roosevelt and his country into the war, putting out the flags for every American visitor who showed the slightest sign of having an influence on Capitol Hill. The first guest, Roosevelt's adviser and confidant Harry Hopkins, was especially welcome – after his visit was announced on 3 January, it drew a comment from 10 Downing Street that this was the 'next best thing to Mr Roosevelt himself coming'.[23]

Formerly a social worker, and later Roosevelt's leading New Deal enforcer, Hopkins was a slight, shambling figure whose frailty belied his intensity of purpose and the huge influence he had in the White House. During his month-long stay in London – twice as long as planned – he stayed at Claridge's Hotel in Mayfair. He was escorted from one lavish dinner to another, and was even presented to the King.[24] Hopkins became a popular figure, especially with Churchill, who regaled his guest with compliments and panegyrics about the President's greatness, and emphasised the huge stock he put in American support. Having braved one alcohol-soaked midnight gathering with Churchill in commanding form, Hopkins returned to his room at 2 a.m. and sat by his fireside exhausted, muttering 'Jesus Christ! What a man!'[25]

Convinced of Britain's need and that Churchill was the only person the President should bother with, Hopkins wrote to Roosevelt with recommendations of support for Britain that were so strong and concrete that Churchill could almost have written them himself. A few weeks later, Churchill waited anxiously to see if legislators on Capitol Hill would pass Roosevelt's Lend-Lease Bill, which sought to provide Britain and other allies with goods and supplies, ending the pretence of American neutrality.

A few days after Hopkins had checked out of Claridge's, another of Roosevelt's delegates checked in – Harvard's President James Conant,[26] recently appointed as Vannevar Bush's deputy on the US National Defense Research Committee. This time, the President's brief was narrower and less to Churchill's liking: to encourage 'the exchange of information on recent scientific developments of importance to national defense'. The gangly, tweed-suited Conant was a top-notch chemist, with the stooped shoulders of a man who had spent too many hours leaning over his test tubes. Although he had 'a smile as quick as a traffic cop can frown', as one journalist put it, Conant took life seriously and spoke out on difficult questions even when his views were bound to make him unpopular.[27] He was even known as a player on the national stage, mainly through an outspoken national radio broadcast in May 1940 urging Americans to support Britain immediately, to prevent the spread of Nazism to the United States.[28]

During his six-week visit, Conant met Churchill three times, though it seems they barely touched on the details of how British and American scientists might collaborate more closely. The Prime Minister had bigger fish to fry. Fixated on the passage of the Lend-Lease Bill at their first meeting, he grilled Conant when they lunched in the bomb-proof dining room in the basement of 10 Downing Street with Clemmie Churchill and Frederick Lindemann.[29] Ten days later, after the Lend-Lease Bill had passed, Churchill was back on his sprightliest and most agreeable form. During a Sunday-night dinner party at Chequers, he charmed Conant and several other American guests with entertaining reflections on the American Civil War ('The men who can win a war can never make a peace').[30] The Prime Minister shone at gatherings like these, the Pol Roger flowing freely, the conversation alternating easily between sensitive diplomatic matters and personal chit-chat. A grand, rations-busting dinner was usually followed by the playing of a few scratchy gramophone records and an undemanding movie, such as Churchill's favourite, the Olivier–Leigh vehicle *That Hamilton Woman* or a Donald Duck cartoon.[31]

Five weeks later, when Conant called on Downing Street to say farewell, he arrived during a crisis – the Germans had just invaded and overwhelmed Yugoslavia. After ushering Conant into the Cabinet room, Churchill said simply, 'Here we are, standing alone,' adding plaintively after a pause, 'What is going to happen?' When Conant, moved by all the destruction he had seen in London, returned to the US he became an even more outspoken supporter of the British cause.[32]

In the early summer of 1941, the war was going badly on every front. Churchill was still completely dominant, unwilling to delegate responsibility or to meet any visitor he deemed unimportant. He made one exception on 12 June – probably as a sop to his disgruntled scientists – when he joined a short ceremony in the Cabinet room, to be admitted as a Fellow of the Royal Society and thus become a member of Britain's scientific elite. A month before – backed by Hill, Tizard and J. B. S. Haldane – Churchill had been elected under the statute of the Society that enables it to appoint people it deems to have given special service to science or whose election would be of signal benefit to the Society.[33] This was good old-fashioned realpolitik, if perilously close to cynicism.

The Society did not film the Prime Minister signing its Charter Book, so we shall probably never have the pleasure of seeing the looks on the faces of the witnesses A. V. Hill and Henry Tizard. Hill was still fronting a campaign against what he saw as the government's inept use of its scientists;[34] Tizard had submitted his report on the mission to the US several months before but had not heard a word from Churchill since.[35]

The Prime Minister kept the Fellows waiting for forty minutes. Meanwhile, his Private Secretary Jock Colville chatted with them, probably touching on the news that the King's Birthday Honours List had elevated Lindemann to a peerage. The Prof took the opportunity to shed his surname, with its hint of German and Jewish origins, and took the name Lord Cherwell (pronounced *char-well*), after the river that meanders past Christ Church Meadow in Oxford. As Colville knew, this was an unpopular appointment. In private, many scientists defied polite convention and continued to refer to the Prof as 'Lindemann' (the name that will continue to be used here).[36] Tizard, who at the same time turned down Churchill's comparatively paltry offer of a junior appointment in the Order of the British Empire, could not resist commenting that 'the Cherwell is a small and rather muddy stream'.[37]

Ten days later, on the morning of 22 June, Churchill was awoken at Chequers with news he had been expecting for some weeks: Hitler had double-crossed Stalin by unleashing 148 divisions of the German army – some 3.2 million men – against the Soviet Union. In the first weeks of this vast eastern conflict, the Nazis appeared to be capable of storming all the way to Moscow.

In a radio broadcast that night Churchill announced that although he had spent twenty-five years opposing Communism, any state that fought Hitler would have Britain's aid, so 'We shall give whatever help we can to Russia and the Russian people.' Most of the top brass in the British military expected the Wehrmacht to crush Stalin's army within months, but the good news was that Britain and its Empire were no longer fighting alone. Though Hitler seemed to be turning his attention away from Britain, Churchill was taking no chances: he put British defences 'on concert pitch for invasion from September 1st'.[38]

Reports of the mood in Washington were encouraging for Churchill, who heard from the new British Ambassador in Washington, the former appeaser Lord Halifax, that Roosevelt privately no longer doubted that the US would have to enter the war.[39] During another visit to London in July, Harry Hopkins – arriving laden with supplies of ham, cheese and cigars – was even more upbeat, and soon afterwards Churchill received the news he had longed to hear: the President had consented to meet him.[40] It was agreed that they would convene in Placentia Bay, Newfoundland, so the Prime Minister would have to cross the U-boat-infested Atlantic.

On 1 August, the Prime Minister set off, unable to contain his excitement.[41] Brushing aside Roosevelt's suggestion that only the two of them needed to be involved, Churchill took along a retinue that would have been the envy of a medieval monarch – dozens of officials, servants, journalists, several ministers and, of course, Lindemann, who was about to have his first experience of international diplomacy at the highest level. Over the past few weeks, the Prof had been keeping a close eye on the MAUD committee, whose final report was about to begin its journey to Churchill's desk.

It had been almost seventeen years since Churchill had first alluded to the possibility of nuclear bombs in 'Shall We All Commit Suicide?' At that time, in 1925, he had struggled to maintain his usual optimism as he urged the leaders of the human race, able for the first time to exterminate itself, 'to pause and ponder upon their new responsibilities'.[42] Now, he was about to become the first head of government to decide whether to go ahead with building the most destructive explosives the world had ever seen. The report on which he would be obliged to base his judgement was being written mainly by James Chadwick, who had helped to realise the vision Churchill set out in 'Fifty Years Hence' by discovering 'the match to set the nuclear bonfire alight'.

JULY AND AUGUST 1941

Chadwick believes Britain should build its own Bomb

'. . . we entered this project with more scepticism than belief . . . As we proceeded we became more and more convinced that . . . conditions can be chosen which would make it a very powerful weapon of war.'

MAUD REPORT, finalised by James Chadwick, July 1941

A few years before his death in 1974, Chadwick told an interviewer that the spring of 1941 was still clear in his mind: 'I realised then that a nuclear bomb was not only possible – it was inevitable.'[1] He knew, too, that scientists everywhere would sooner or later find this out, that nuclear bombs would be built and that 'some country would put them into action'. In the interview, Chadwick said that the strain of having to keep this a secret led him to start taking sleeping pills. He added: 'I've never stopped.'

Like almost all the other MAUD scientists, Chadwick was living in a city under attack by the Luftwaffe. In early May, the German bombers had targeted Liverpool seven nights in a row, killing almost three thousand people and doing terrible damage, especially to the docks. Although the worst of the attacks seemed to be (and were) over, the Chadwick daughters were evacuated to Canada a few weeks later, leaving their parents to rattle around their blacked-out home, waiting every night for more bombs and parachute mines to fall. As soon as the air-raid sirens went off, the Chadwicks went down into their cellar, damp but well stocked, a private refuge from the bombs. Chadwick's house was never hit, but his department was less fortunate – its windows were blown out so often that he had them fitted with cardboard shutters.[2] He knew, however, that the harm done by the hundreds of explosives that had been dropped on the city was trifling compared with the carnage that would follow the use of a single nuclear weapon.

Most of the MAUD scientists agreed with Chadwick that nuclear weapons were destined to be more than speculation. Peierls had emerged as the project's leading mathematical physicist, coordinating and checking the results of his fellow theoreticians and working closely with experimenters. Grappling with dozens of thorny calculations, he was forced to cast around for help. High-quality British theoreticians were in short supply, but he found an excellent candidate among the few foreign-born scientists who had not moved on to America – the young applied mathematician Klaus Fuchs, a refugee from Nazi Germany, now making a name for himself at the University of Edinburgh.[3] MI5 began to worry when its investigators unearthed evidence that Fuchs had been an active Communist when a student in Germany, though there was no evidence that this sallow, taciturn man was any longer a security risk. Fuchs would be useful to the project only if he knew what it was planning to achieve, so – as Peierls pointed out to the dithering authorities in MI5 – he should either be rejected or cleared. Within a few days they agreed to give Fuchs security clearance, and by the end of May he was working at his desk in the Nuffield Building, and living with the Peierlses, who quickly made him one of the family.[4]

Although persuasive, Peierls's calculations and data from dozens of fission experiments could not prove that a uranium bomb would work. Without actually building one, no one could be certain that sufficient neutrons would be emitted quickly enough from a fissioning uranium nucleus, and in the necessary energy range, for there to be an explosion. Uranium atoms containing ^{235}U nuclei were so rare that no researcher had produced enough of them to be weighed by even the most sensitive equipment.

Lindemann's Oxford colleague Francis Simon believed it was best to isolate ^{235}U using the technique of gaseous diffusion, which involved forcing uranium hexafluoride gas through fine membranes, separating ^{235}U from the slightly heavier ^{238}U. To produce enough fissile uranium to make a weapon would require a large and complex industrial plant, which – according to Simon's early calculations – would occupy about forty acres and set the Treasury back some five million pounds. This cost was roughly a tenth of the UK's weekly expenditure on the entire war effort, so it would be a huge financial challenge to embark on the project, though not out of the question.[5]

To arrive at these estimates, Simon worked closely with Metropolitan-Vickers, Britain's largest firm of electrical engineers, and with ICI, which had set up a secret war committee to coordinate the company's work for the government.[6] Lord Melchett, a strong presence on that committee and a regular at MAUD meetings, had been in close touch with Lindemann since the war began. The two of them often talked over dinner at the Savoy about everything from Britain's agricultural policy to the latest gizmos from 'Churchill's Toyshop'.[7] G. P. Thomson made Melchett and his fellow industrialists feel as much at home on the committee as the professional nuclear physicists – no mean feat, as some of the physicists were ill at ease working with industrial leaders. 'This country has been sadly let down by its industrialists – who were, and in some ways still are, its government,' Oliphant groused.[8]

During the committee's discussions, Melchett and his friends at ICI saw a juicy business opportunity – if they funded Halban and Kowarski's experiments, the investment was likely to yield good returns after the war, when nuclear power would probably become commercially viable. Melchett soon offered to take over that part of the MAUD project lock, stock and barrel. Later, he went further and agreed to maintain the investment even if it proved necessary to continue the programme in North America – in the long run this was likely to be money well spent.[9] The question of whether to work closely with the Americans on nuclear weapons research was becoming more controversial by the week. Oliphant and Cockcroft could not wait to begin a collaboration, and Thomson was encouraging, once commenting with toe-curling condescension that it would both save time 'and give the Americans and Canadians experience'.[10] Chadwick was, as usual, cautious: 'The time has not yet arrived to take a decision on the question of moving to America,' he had commented at a MAUD meeting in April.[11]

Several of the MAUD scientists attended some of the dinners and receptions in honour of James Conant, whose eight-week courtship by his British hosts sometimes lapsed into an unseemly obeisance. In late March, he accepted an invitation to the Blacketts' apartment in Westminster, where the two men talked over dinner with Cockcroft and Philip Joubert de la Ferté, a senior commander in the Royal Air Force.[12] The Englishmen asked Conant about his meetings with their Prime Minister, but it is unlikely that they will all have shared their disappointment over what they believed to be the government's misuse of academic scientists in the war.

The only British scientist to sound out Conant about the possibility of building nuclear weapons was Lindemann. He surprised Conant by taking him out to dinner at a gentlemen's club and talking openly about nuclear chain reactions, a subject the American believed to be too highly classified to discuss, even in private. When Conant took the familiar line that such work might prove useful but only in the distant future, Lindemann abandoned secrecy and briefed him on Frisch and

Peierls's work. This was the first time Conant had ever heard it seriously suggested by a scientist that nuclear weapons might actually be built, though he returned home knowing nothing of the MAUD committee's conclusions.[13]

In the late spring of 1941, Thomson started writing his report and arranged for the committee to discuss it on 2 July. Chadwick worried that their findings were being brought together in too much of a hurry, and the first draft confirmed his fears. A few days before the meeting he sent the committee secretary a powerfully worded letter listing seven criticisms of its content and hinting that he regarded it as inadequate, though he was too much of a gentleman to say so explicitly.[14] The meeting, held in the Royal Society's headquarters, would turn out to be the MAUD committee's last. Apart from Blackett, all the leading members were there: the British scientists and their foreign-born colleagues, who were no longer regarded as serious security risks. Lindemann sat alongside his Oxford colleague James Tuck and Lord Melchett, who brought along two ICI scientists. Also present was the American physicist Charles Lauritsen, representing America's National Defense Research Committee office in London, set up by Conant during his recent visit.[15]

The MAUD committee agreed that its chairman's draft needed polishing and that Chadwick was the best person for the job. By some distance the best nuclear physicist among them, he was also a writer of muscular but precise prose and could be relied on to summarise their consensus. Even better, he got on well with the Prime Minister's *éminence grise* and respected him. Lindemann returned the compliment, admiring Chadwick's quick brain, his straightforward manner and, in all probability, his Conservative politics. Speaking near the end of the meeting, the Prof proposed that the final report should stick to physics, keep away from politics and economics, and focus on the viability of nuclear weapons rather than on the less pressing matter of nuclear power – advice Chadwick was content to follow.

With his customary health-threatening zeal, Chadwick applied himself to the task of collating all the research and turning it into a readable, authoritative document. Occasionally assisted by his colleague Jo Rotblat and by Otto Frisch in Birmingham, Chadwick wrote the report mainly in his office in the physics department at Liverpool University, often consulting his fellow scientists in university laboratories, ICI and Metro-Vickers. It was not easy for him to work at home – like every other building in the city, his house was blacked out after sundown, forcing him to read or write by candlelight. Sometimes, he would take a few hours off and listen to recordings of his favourite music – including Verdi operas and Mahler lieder – on the gramophone he had built at home in his workshop.

Following the guidance of the committee, Chadwick decided to prepare two reports – 'On the Use of Uranium for a Bomb' and the much shorter 'On the Use of Uranium as a Source of Power'.[16] Each would begin with a section briefing lay readers on the substance of the case he and his colleagues wanted to make, followed by technical appendices that spelled out the costs and other details. In mid-July, he was struggling to meet the deadline and was working on the manuscript twenty hours a day.[17] When he finally finished what became known as 'the MAUD report' (invariably in the singular), he was exhausted and felt 'very down' – all he wanted was a quiet weekend and then a restorative family vacation, involving nothing more onerous than 'a little gentle fishing'.[18]

The report concluded that a uranium bomb was possible and 'likely to lead to decisive results in the war', and it urged the government to pursue the project as a matter of 'the highest priority', building on the nascent collaboration with American scientists.[19] The effort of building the Bomb was not likely to be wasted, it argued, because no nation 'would care to risk being caught without a weapon of such decisive capabilities', except in the unlikely eventuality of worldwide disarmament. Chadwick wrote that the Bomb could be built in only two and a half years, a figure he attributed to the advice of Metropolitan-Vickers and ICI, which spoke with more authority than anyone else round the table about industrial planning.

On 29 July, the report began its meandering way towards the government's Scientific Advisory Committee, which had the job of reviewing it. Chadwick had put together a document that had the clarity and directness of the Frisch–Peierls memorandum, but was even more readable. It was quite an achievement to forge such a powerful case from the opinions of so many contributors. Even Chadwick, however, could not resolve the most vexed disagreements among his MAUD colleagues, and he had no alternative but to dodge the crucial question of whether Britain should attempt to build the weapon on its own soil. Predictably, debates about this unanswered question were raging in Whitehall.

Tizard said that it would be 'absurd' to try to make nuclear bombs in the UK during wartime and argued that Britain should collaborate with the United States.[20] Blackett agreed. In a dissenting note to the Air Ministry, he wrote that it was unlikely that such a weapon would be 'of use in this war' and that the authors of the report were wrong to believe that the first bomb could be ready by the end of 1943. His colleagues had not allowed enough time for the delays that were certain to slow down such a large and novel project, Blackett believed.[21] The full-scale plant should not be constructed in Britain, he recommended – a final decision on whether to build it in North America should be taken only after British scientists had visited colleagues in the US to discuss the project's viability.

During the week Blackett's letter was circulating among officials, a cat was thrown among the pigeons by Charles Darwin, a grandson of the great naturalist and a theoretical physicist then serving as Director of the British Central Scientific Office in Washington DC. In a handwritten note to Hankey, each of its five pages marked SECRET, he pointed out that it was high time for Britain and America to decide whether they were serious about developing nuclear bombs or, perhaps, conclude that these weapons were too destructive to contemplate.[22] He had been talking with Bush and Conant, who floated the idea that the two countries should go beyond coordinating their research on the Bomb and regard it as 'a joint project of the two governments'. Darwin recommended that Britain send a delegation of physicists – including Chadwick, Thomson and Simon – to try to agree on a joint policy with their American colleagues.

Hankey refused to be hurried by Darwin or anyone else.[23] He began to plan a careful review process, aiming to cross-examine all the scientists involved in order to probe every detail of the MAUD report and check the validity of its recommendations. The Hankey committee would have to do without the advice of Thomson, who had dropped everything and hurried to New York in mid-August, after hearing that his wife had been taken gravely ill. Before leaving, he asked Chadwick to take over the running of the project, commenting that it looked likely that a nuclear-bomb factory would be set up in Britain, a prospect Thomson regarded as so impracticable that it was 'deplorable'.[24] The handover to Chadwick would turn out to be important, as he was one of the few MAUD scientists who agreed with Lindemann and Melchett that the Bomb should be built on home soil. Now that he was no longer out of the picture, it was much more likely that Lindemann and Chadwick's view would prevail.

The committee's recommendations would count for little with Churchill compared with the opinions of Lindemann, who was unlikely to wait for the committee to plod through weeks of hearings before he gave his view to the Prime Minister. Hankey probably guessed this; if so, he was right.

AUGUST TO OCTOBER 1941

Lindemann backs a British Bomb

> 'Physicists and engineers in the US would be competent to develop this project, [but] it would seem undesirable to depend on them. In general, the Americans [are] slow starters.'
>
> FREDERICK LINDEMANN, 17 September 1941[1]

On 27 August 1941, the day Hankey officially received the MAUD report, Lindemann decided to cut the red tape and give his views directly to the Prime Minister.[2] In his first minute to Churchill on the advisability of building a nuclear bomb, the Prof began by referring to their conversations about a 'super-explosive' but, perhaps surprisingly, did not remind him that they had worked together on several articles pointing out that such weapons were probably on the way. The important news now, Lindemann wrote, was that 'it seems almost certain that this can be done'. He suggested that the government should fund research for another six months, when a final decision would be possible. In words that will have resonated with Churchill's concerns about sharing technical secrets with the Americans, Lindemann declared himself strongly in favour of building the Bomb 'in England or at worst in Canada'. 'However much I may trust my neighbour and depend on him,' Lindemann wrote, 'I am very much averse to putting myself completely at his mercy.' Britain should 'not press the Americans to undertake this work', he wrote, but simply 'continue exchanging information'.

Lindemann was sceptical that the Bomb could be produced within two years, as the experts were claiming. They were giving odds of ten to one on success, he reported, while he would not bet more than two to one against, or evens. But he believed strongly that the project should go ahead: 'It would be unforgivable if we let the Germans develop a process ahead of us by means of which they could defeat us in war or reverse the verdict after they had been defeated.'

When Churchill read Lindemann's note, he was in low spirits.[3] The talks with Roosevelt earlier that month went well, with the President reaffirming his support for Britain, though he made it clear that he had little time for the Prime Minister's enthusiasm for preserving the Empire in perpetuity. Churchill had returned to London with nothing concrete apart from the Atlantic Charter, a stirring post-war vision that later formed the basis of the United Nations. Whitehall mandarins muttered that the agreement was a glorified press release.[4] Worse, when the President arrived home he was as slippery as ever about his commitment to Britain and its allies: when asked in a press conference if America was any closer to entering the war, he said no, but refused to be quoted.[5]

On 28 August, probably the day Churchill received Lindemann's hand-delivered note on nuclear weapons, he cabled Harry Hopkins to say that reports of the President's comments had sent 'a wave of depression' through the Cabinet, which feared that the Russians would be 'knocked out' in a few months, leaving Britain and its Empire alone again. Churchill ended pitiably, asking Hopkins 'if you could give me any sort of hope'.[6] Lindemann's note appears to have cheered up the Prime Minister, as news of a fearsome new device produced by his scientists usually did. When he read that it was extremely likely that the Bomb could be built, he wrote in the margin, 'Good'.[7]

Predictably, Churchill went along with his chief science adviser's advice, and a few days later he became the first national leader to approve the development of a nuclear weapon. He wrote a brief minute to his Chiefs of Staff, beginning with words that he later said 'turned on the full power of the State to what was then a remote and speculative project':[8]

> Although personally I am quite content with the existing explosives, I feel we must not stand in the path of improvement, and I therefore think that action should be taken in the sense proposed by [Lindemann], and that the Cabinet Minister responsible should be Sir John Anderson.

A week later, at a meeting chaired by Churchill, the Chiefs of Staff enthusiastically supported the project, declaring that the government should spare no time, labour, materials or money in pushing it forward.[9] Furthermore, the Chiefs agreed 'that the development should proceed in this country and not abroad', while accepting that the final tests could be 'carried out, if necessary on some lonely, uninhabited island'.[10]

Lindemann had no interest in overseeing the nuclear project – he always preferred advising to delivering, exercising power and influence without the wearisome demands of executive office. It seems from comments made by Lindemann's friends that in his heart he could not quite believe that the Bomb would work – and he never risked association with an operational failure.[11] Lacking managerial talent, it suited him to work alongside the stolid Anderson, who had such qualities in abundance.

Less than two years before, Anderson was 'quite sure' that the uranium bomb was not feasible, but over the past few months he had changed his mind.[12] A trained scientist now approaching sixty, Sir John was already in charge of the rest of the government's science research and was well qualified to oversee the development of nuclear weapons. As a student at Leipzig University he had written a thesis on the chemistry of uranium, before becoming a career civil servant in the Hankey mould – reliable, hard-working and without the slightest trace of glamour. Some admired his gravitas, others thought him crusty and portentous, but everyone agreed that he was an administrator without peer and a formidable negotiator. ICI executives had learned this to their cost when he ran rings round them during an attempt to extract a subsidy from the government – the officials were so impressed that they appointed him a 'lay director' in 1938.[13]

No one admired Anderson's administrative skills more than Churchill, who delegated to him all the spadework of domestic government. Anderson absorbed huge quantities of work without complaint, with the concentration of a surgeon, from precisely 10.15 a.m. to 6.15 p.m. every working day, with an hour and a half for lunch, never taking work home.[14] With no following among the public or in any of the political parties, he was no threat to any of his colleagues.

Soon after his appointment as overseer of the British Bomb project, he wrote the first of his scientifically literate, thoughtful letters about it, immediately attaching 'enormous importance' to its post-war implications.[15] He proposed setting up a small Advisory Council, including Lindemann, Hankey and Appleton, but with no specialist nuclear physicist.[16] Although Anderson could be charged with failing to avail himself of the best scientific advice, no one could accuse him of following the majority. He agreed with Blackett that the MAUD report had seriously underestimated the time needed to develop the weapon and was quite clear that he disagreed with the Prime Minister and Lindemann: the Bomb should be produced 'not here but in America'.

Until the end of his tenure as Prime Minister during the war, Churchill treated the British nuclear project as something close to a private fiefdom. Within Whitehall, only he, Lindemann and Anderson knew exactly what was going on, leaving the Cabinet in near-total ignorance of Britain's role in making the most revolutionary weapon of the war. This way of doing business, very much of Churchill's own making, put him under considerable pressure as a strategist in a new field of weapons research. Success depended heavily on the quality of his advisers, who would influence Churchill in decisions that were to prove critically important for the country's nuclear policy for decades to come.

Lindemann's industry and commitment to the war effort were never in doubt. Even the Luftwaffe could not disrupt his monastic routine and his devotion to work – he disdained communal air-raid shelters, preferring the comfort of his apartment in Westminster, even when bombs were falling around his street. After an especially violent raid one night, his valet dashed to his side and found him in bed reading P. G. Wodehouse. On hearing the knock at the door, Lindemann looked up and enquired, 'Is there anything wrong, Harvey?'[17]

The Prof spent most working days in his room, a few paces from the Prime Minister's, in the Great George Street headquarters of the War Cabinet. At weekends, he returned to Oxford, his chauffeur delivering him in his new Packard Tourer limousine to the front door of the Clarendon Laboratory, where Francis Simon gave him an insider's perspective of the MAUD project. Lindemann stayed in his newly decorated rooms in Christ Church, complete with tasteful furnishings and a Bechstein piano.[18] Only here would he allow himself a few hours' relaxation, lying in his bath – the water at precisely 104 degrees Fahrenheit and no deeper than the government's permitted maximum of six inches – reading Dornford Yates thrillers, the latest *Quarterly Journal of Mathematics* and back copies of *Men Only*.

His one hobby was intellectual – thinking about prime numbers, an interest he had been pursuing in his spare moments since the early 1930s, when he published the first of four articles on the subject, some of them later cited by experts.[19] In mid-September, sitting at his desk in Christ Church, he drafted the second of his contributions, in the form of a short letter to the editor of *Nature*, a few days before he gave his views on the MAUD project to a meeting of the Scientific Advisory Committee held in Hankey's office. Five other scientists were present, including the President of the Royal Society Henry Dale and also A. V. Hill, who was still campaigning to lessen Lindemann's influence on the running of the war effort.[20]

It was extremely unusual for Lindemann to emerge from his Whitehall cave and allow his scientific colleagues to question him. He spoke only briefly and gave away little of his hand, making a few technical comments about the fission of ^{235}U, all of them common knowledge to experts. There was not much doubt that the Germans were interested in the subject, he said, pointing out that they had secured some supplies of heavy water.[21] Most revealing was the case he outlined – timidly, by his standards – for developing the Bomb in Britain, rather than handing it over to the slow Americans. To the question of how a factory in the UK could be protected from aerial bombing, he said that 'it should be possible . . . either by placing it underground or by the use of concealment or camouflage'.

The committee's report was ready a week later, on 24 September 1941.[22] It was a classic Hankey document – clear, well informed and careful to accommodate every shade of opinion, though it firmly rejected ICI's offer to take over the research into nuclear power. Hankey recommended uncontentiously that the venture should be transferred from the Air Ministry to the natural home for the government's science projects, the Department of Scientific and Industrial Research. Most important, the report proposed that the gaseous diffusion plant needed to produce the ^{235}U for the Bomb should be built in Canada, with the Americans treated as consultants.[23] The British and American project would therefore be separate but linked.

Later, after Churchill's meeting with Roosevelt, Hankey had heard excited talk from the Air Ministry that nuclear weapons might enable 'America and ourselves to control and police the world'.[24] He was sceptical. Thirty years of working with America had led him to doubt that it would ever join Britain in such a role.[25]

On Sunday 12 October, a message from Roosevelt arrived on Churchill's desk. This was not unusual: since their meeting in Placentia Bay, the President had written to him eight times, on each occasion about American support for the Allies. Churchill attended to these notes carefully – as he commented several years later, 'No lover ever studied every whim of his mistress as I did those of President Roosevelt.'[26] The communication that arrived that day deserved special attention – it was the first note to be exchanged between the leaders on the subject of nuclear weapons:[27]

> My Dear Winston, It appears desirable that we should soon correspond or converse concerning the subject which is under study by your MAUD Committee, and by Dr Bush's organization in this country, in order that any extended efforts may be coordinated or even jointly conducted . . .

The President had deemed the letter so important that he arranged for it to be carried to the Prime Minister by hand, by Frederick Hovde, head of the US's National Research and Defense Committee's office in London. Roosevelt concluded by inviting Churchill to contact Hovde directly 'to identify the subject more explicitly' and to answer questions about American research in this field.

When Churchill read this letter, he may well not have understood that it was about nuclear research, as he probably did not know what the 'MAUD Committee' did – it was one of dozens in Whitehall and not a prominent one. Soon, his aides will have apprised him of the note's content: Roosevelt was offering him a close collaboration – quite possibly a partnership on equal terms – in developing the Bomb. The benefits to Britain were potentially considerable: it would be able to capitalise on the lead it had established over the Americans, take advantage of their huge scientific and financial resources and begin to develop the weapon far from the Luftwaffe's reach. This was an exceptional diplomatic opportunity.

As he had demonstrated in his attitude to the Tizard mission, Churchill was chary of giving technical secrets to the United States, at least until it was rather more generous in its support of Britain's war effort. So it was far from clear whether he would set aside his unwillingness to work closely with the country whose partnership he valued most dearly, and grab the offer with gusto, as many of his nuclear scientists would have urged if they had known about it. One thing was certain: such an offer was not going to be on the table for long.

When Churchill, Lindemann and Anderson were considering how to respond to the President's note, they may have wondered what prompted him to make the offer, apparently out of the blue. They did not know that it was largely the result of the activities in the US of one of the MAUD physicists most eager to involve the Americans in building the nuclear bomb.

AUGUST 1941 TO JANUARY 1942

Oliphant bustles in America

> 'Mark [Oliphant] is getting very notorious for outspoken and quite unjustified statements in everything and sundry . . . "Oliphantic" has been coined to describe his statements.'
>
> JOHN COCKCROFT, February 1941[1]

Mark Oliphant was not among the prime movers of the MAUD committee but he was the most vociferous champion of its report. Speaking with a bluntness reminiscent of his hero Rutherford, Oliphant lobbied energetically in the United States, and then at home in Britain, urging officials to act quickly and effectively on the report's conclusions. His forceful candour, sometimes as embarrassing to his colleagues as it was to government officials, served to remind them how different the interface between politics and fundamental science would have been if Rutherford were still alive.

In his native Australia, Oliphant had first seen Rutherford lecture in 1925 and had been 'electrified' by the great physicist's 'words and personality'.[2] Within two years, they were working together at the Cavendish and getting on famously – later, they even went on vacation together with their families. Oliphant was a resourceful and hard-working experimenter, becoming the first to demonstrate that atomic nuclei could be fused, and that the process would release energy. Although he was not the most talented physicist in Cambridge, he was a valued member of the team and one of its liveliest characters, hot-headed and impulsive. 'He's a brash young man, but he'll learn,' Rutherford had chuckled.[3]

Despite his belligerence, Oliphant was always entertaining company. With his tight curls of greying hair, gold-rimmed circular spectacles and neat three-piece suit, he often resembled a lovable schoolmaster. But when some idiocy moved him to outrage, he became an angry proselytiser, buttonholing and telephoning every colleague who would hear him out. Although G. P. Thomson knew that Oliphant had a loose tongue, he nonetheless asked him to make 'discreet enquiries' during a long-scheduled visit to the US in the summer of 1941 – shortly before Thomson stood down as MAUD's chairman – to find out why American scientists were virtually ignoring the committee's reports.[4] As Thomson did not explicitly say that anything on the MAUD agenda was too secret to share, Oliphant believed he had been given a free hand to talk about the project to anyone he deemed worthy of his trust.[5]

Oliphant had planned to spend most of his time discussing radar, but when he saw the state of American research on nuclear weapons he made it his top priority to promote the conclusion of the MAUD committee. His single-handed campaign to shake things up in America began on a sweltering Washington day in early August. When he and Charles Darwin paid a visit to Lyman Briggs, chair of the Uranium Committee set up by Vannevar Bush to look into the possibilities of producing nuclear chain reactions, they found him to be distressingly dull and incoherent.[6] The bumbling Briggs was obsessed with secrecy – whenever the minutes of MAUD meetings had arrived from London, he had locked the papers in his safe, without copying them to his colleagues, who remained ignorant of the British initiative. Forty years later, Oliphant could still remember how 'amazed and distressed' he had been by Briggs's incompetence.[7]

Oliphant spent most of August and September badgering American scientists and officials, trying to persuade them of the urgency of beating the Nazis to the manufacture of nuclear weapons.[8] The best way to achieve the goal of being first to make the Bomb, he believed, was for the British and Americans to pool their resources and get cracking. At first, his pitch did not go down well. Over a tense lunch in Washington, Conant was tight-lipped, unwilling to discuss state secrets. In New York, Bush was no more forthcoming and Fermi was still unconvinced that a bomb was possible, though Szilárd – struggling to win influence – was desperate to breathe life into the moribund American project. Oliphant was most successful at Berkeley, where he met his old friend Ernest Lawrence, whose invention of the cyclotron two years before had made him America's youngest Nobel Prize-winner. He was also a member of Briggs's ineffectual committee.

It was in Lawrence's company that Oliphant first mentioned the Bomb project to Robert Oppenheimer, who knew nothing about it. Seeing Lawrence's discomfiture at Oliphant's indiscretion, Oppenheimer hinted that it might be best to end the conversation, but Oliphant ploughed on. 'That's terrible,' he told Oppenheimer. 'We need you.'[9]

At the end of Oliphant's stay in California, he gave Lawrence a pithy summary of the MAUD report. Lawrence was convinced of Oliphant's case and quickly began to use his influence to lobby for the cause. By 3 October, Bush's deputy James Conant had officially received the report from Thomson. Six days later in the White House, Bush used the report as the basis of his first meeting about the project with the President. After Bush sketched the current arrangement with Britain as 'complete interchange on technical matters', Roosevelt approved its continuation but saw the need to clarify how an Anglo-American project would be best managed.[10] On his most decisive form, he agreed to set up a secret 'Top Policy

Group' to guide the initiative, and ordered Bush to draft a letter 'to open discussion of the [nuclear bomb] at the top'.[11] Bush quickly drafted a short note to Churchill proposing that Britain and the US work together to develop the MAUD findings. After making a few trivial changes to Bush's wording, Roosevelt sent the message post-haste to Churchill, suggesting that the nuclear project 'may be coordinated or even jointly conducted'.[12]

The note testified to Oliphant's success in the United States. Leó Szilárd, who had been labouring in vain for years to get the project off the ground, later wrote that 'if Congress knew the true history of the atomic energy project, I have no doubt that it would create a special medal to be given to meddling foreigners for distinguished services, and Dr Oliphant would be the first to receive one'.[13]

In late October 1941, while nuclear weapons research was taking off in the United States, in Britain it was running into the sand. The MAUD scientists had been waiting for three months for the government to respond to their report, whose recommendations had obviously not struck a chord in Whitehall. On the last Monday of the month, Oliphant – now back in Britain – and other leading MAUD scientists received a letter from Edward Appleton at the Department of Scientific and Industrial Research. The news was not what they wanted to hear. The government had decided to hand over the running of their committee to ICI officials, Appleton told them in a note so tactless that many of its recipients considered it rude. The project would in future be shepherded by the company's research director Wallace Akers – a member of the team that had organised ICI's abortive attempt to take over the nuclear-energy project – assisted by Michael Perrin, one of the company's leading administrators. The ICI boss Lord Melchett had lobbied Churchill's officials so effectively that they had given him more than he had bargained for. The government had earlier refused to sanction ICI's proposal to run MAUD, and then placed it in the hands of the company official who had written the bid.[14]

Most of the scientists on the project were appalled not only by the decision but also by Appleton's discourtesy. To add salt to the wound, he had informed only a handful of the MAUD physicists, leaving the others to hear about it on the grapevine.[15] G. P. Thomson was not told officially of the fate of his committee's work until December.

Oliphant resigned immediately, in a biting letter to Appleton.[16] Every line of it was drenched in hurt and anger: hurt that he had not been appointed to the new, top-tier Policy Committee; anger that the project was now being run by 'commercial representatives completely ignorant of the essential nuclear physics upon which this whole thing is based'. The Americans were also delegating to know-nothings, he said, which is why their project was being 'badly mismanaged'. Later that day, when writing to Chadwick, he was even more outspoken: the appointment of the new leadership was 'disgraceful' and Akers would 'obviously . . . look after the commercial interests of ICI'.[17] It was hard to believe that Akers was unbiased in his treatment of ICI as he was still doing part-time work for them.[18]

Oliphant contemplated giving up his work for the government, he told Chadwick, so that he could organise a revolt among his fellow MAUD physicists and run a 'rival show'.[19] They could then do their research without referring to 'a lot of interfering busybodies who know nothing whatever about the problems involved'. Chadwick, who had seen almost a decade of Oliphanticism in the Cavendish, did his best to calm him down. During an exchange of letters, Chadwick agreed with his temperamental colleague that the handling of the transition to the new regime had been dictatorial and impolite:[20]

> I am most dissatisfied with the way [the new arrangements] have been carried out, without the slightest reference to the MAUD Committee. This treatment I consider both autocratic and discourteous.

But when it came to the rub, Chadwick took the establishment line, telling an incredulous Oliphant that 'the new arrangement is an excellent one'. Chadwick added that while the new leadership would continue to cooperate with the United States, the aim was to preserve an independent British initiative – 'We are some way ahead and we shall remain ahead.' To some extent pacified, Oliphant believed that Chadwick was underestimating the scale of the American project.

The small-mindedness of the new management especially irked Oliphant. When two senior American scientists arrived to talk over the nuclear project with British colleagues, Appleton told them that they should discuss only scientific and technical matters, and, according to Oliphant, arranged to have them accompanied by '"Gestapo" representatives' to ensure that the regulation was followed.[21] Yet Oliphant's opposition to ICI's involvement soon mellowed. In early January, he had lunch with Akers, who had quickly come to grips with the sprawling project and its political sensitivities.[22] In a letter to Chadwick written in January 1942, Oliphant conceded that Akers was 'an excellent person to be in charge of the work', but still worried that his loyalty to ICI would compromise his leadership of the nuclear project, now known by the nonsensical but harmless-sounding name of 'Tube Alloys'. During their conversation, Akers had confided in Oliphant that 'nothing could happen in this matter before 1944', though Oliphant thought that was optimistic by some two years.[23] The timetable was too long for him and he knew he had no way of influencing it, so it was time, he sighed, to take a step back from his self-appointed role as an onlooker and critic: 'In the future, I shall be neither critical nor useful.' Two months later, he sailed for Australia to join his family and to help with his country's war effort, though he would not be away from Britain for long.[24]

Oliphant had no idea that the Americans were by then planning a nuclear project that would deliver a weapon long before 1946. Nor had he any idea that, as a result of his proselytising in the United States, Roosevelt had in October 1941 proposed to Churchill a jointly conducted nuclear project along the lines several British scientists were hoping for. Had Oliphant been informed of Churchill's response, it is safe to predict that he would not have taken it well.

NOVEMBER 1941 TO JULY 1942

Churchill talks about the Bomb with FDR

> '[Winston told me that FDR] was a "charming country gentleman", but [the President's] business methods were almost non-existent, so Winston had to play the role of courtier and seize opportunities as and when they arose. I am amazed at [the] patience with which he does this.'
>
> ANTHONY EDEN, 1943[1]

When Churchill regarded a note from Roosevelt as pressing, he usually replied to it within a few days. Yet the President had to wait almost two months for a reply to his offer of an equal-harness collaboration to build nuclear weapons. By then, events had moved on, and Churchill had missed the great opportunity given to him by his nuclear scientists.

Churchill responded to Roosevelt's offer with a perfunctory cable, noting that he had – following the President's suggestion – delegated Anderson and Lindemann to explore the matter with the American scientists' representative in London, Frederick Hovde.[2] The hope was, Churchill wrote, that 'it will be possible for them shortly to hand Mr Hovde a detailed statement for transmission to America'. There is no enthusiasm in this cable and no sense that either Churchill or any of his advisers had grasped the significance of the proposal the President had made.

The meeting with Hovde that Churchill described had taken place on 21 November, seven weeks after the arrival of the President's letter. It was also attended by Anderson's Private Secretary, whose excruciating account of the discussion makes clear that Lindemann and Anderson believed they were holding all the aces.[3] They had read, and taken at face value, a briefing by British officials who had overestimated the ability of their country's experts to solve the scientific, technological and industrial problems posed by the Bomb.

After hearing that Bush and Conant were 'anxious' for a fuller collaboration and that the President wanted it to be pursued 'with all possible speed', Anderson countered with provisos: although he and his colleagues were also 'anxious' to collaborate, he said, the Americans needed to improve their security so that it was on a par with Britain's. His Majesty's Government would respond to the request only after it had seen and reviewed a statement of the present American organisation of the project. In the meantime, Anderson said he would advise the Prime Minister to write to the President giving him 'a general assurance of our desire to collaborate'.

To Roosevelt and his advisers, Churchill's half-hearted response to the offer of a nuclear partnership was in keeping with his reluctance of the past two years to exchange technical information. Any possibility that Roosevelt might give Churchill a second chance ended on 7 December, with Japan's attack on Pearl Harbor – the US joined the war and pursued the Bomb with an awesome energy, and with no interest in giving Britain any more of a role than was necessary to help the Americans achieve the goal of acquiring the Bomb. Churchill had missed the last bus and was soon running after it with an unseemly desperation, having turned down the driver's offer of a ride.

Churchill heard the news of Pearl Harbor during a quiet Sunday-evening dinner at Chequers, when he switched on his radio.[4] Minutes later, Churchill spoke on the phone with Roosevelt, who confirmed that he would be going to Congress the following day to declare war: 'We are all in the same boat now.' Hitler and Mussolini declared war on the United States on 11 December – sealing their fate – and Churchill embarked on his journey across the Atlantic the following day. His personal popularity would guarantee a warm reception, though feelings among Americans about the British were mixed, as he knew from dispatches written by Ambassador Halifax.[5]

Roosevelt granted Churchill the honour of meeting him on his arrival at Washington's new National Airport. The President could scarcely have made his guest more welcome, talking strategy and tactics day and night, dispensing pre-dinner cocktails, and eating with him and their advisers dozens of times, the liquor flowing much more copiously than the Roosevelts usually allowed.[6] On Christmas Eve, the two men stood together when the White House Christmas tree lights were ceremonially switched on and, two days later, a joint session of Congress cheered the Prime Minister as he gave a bravura speech of amity and resolve, sometimes blinking back tears. The excitement was too much – shortly afterwards he had a mild heart attack, the first sign of his coronary condition, but he quickly recovered. In a joyous telegram, he later told Clement Attlee of his great admiration for the President and how well they were getting on: 'We live here as a big family in the greatest intimacy and informality.'[7] Perhaps he was thinking of the incident when the President entered his bedroom and found him dictating memos while pacing up and down in the nude. 'You see, Mr President,' Churchill said, 'I have nothing to conceal from you.'[8]

The President and the Prime Minister had much in common – their patrician upbringing, their lust for life, their love of power, their aversion to abstraction and doctrine, their blazing self-confidence.[9] Yet there were important differences between them. Roosevelt was the more skilful politician, his gaze fixed optimistically on the future, while Churchill looked backwards to the British Empire when it was in its prime, fearful of what might lie ahead. Roosevelt was cunning and manipulative, whereas Churchill usually got his way through charm and determination. Both men sparkled in company, though for different reasons: Roosevelt was a relaxed and intimate conversationalist, whereas Churchill's oratorical flights sometimes lapsed into bloviation. It was a friendship of sorts, if a somewhat one-sided one – Churchill got rather less out of it than he put in.[10]

After a visit to Canada, Churchill took a five-day break in the Florida sun, at the same time running Britain's war effort from afar and monitoring the threatening news from South-East Asia. Japanese troops were advancing almost unopposed towards the fortress of Singapore and its hinterland, whose survival he regarded as 'vital'.[11] Shortly before he returned to 'the stir' of Washington,[12] he received a briefing note from his secretary John Martin about Tube Alloys 'in case the President mentions the matter'.[13] Martin reminded him that he had not met with Frederick Hovde, who had hand-delivered Roosevelt's offer, but had delegated Anderson and Lindemann to visit him. They had assured Hovde of the British wish to collaborate,[14] promising a statement in writing, though they had not delivered it before the Prime Minister left London.

It seems that Churchill and Roosevelt did not discuss the Bomb project to any significant extent – it probably seemed a far-fetched prospect compared with all the immediately pressing matters on their plate, including the Lend-Lease Bill, the Battle of the Atlantic and the advance of the Japanese. When Churchill's visit ended on 14 January, the President crafted his parting words perfectly to touch the Prime Minister's heart: 'Trust me to the bitter end.'[15] When he returned to London, Churchill told the King of the triumph – Britain and the United States, after many months of courting, 'were now married'.[16] This was a landmark in relations between Britain and America, the birth of the much-vaunted Special Relationship that Churchill held so dear – they would never again be so close.

Roosevelt and his advisers had got the message about Churchill's attitude to the Bomb – that, for him, it was not especially important. Five days after the leaders parted, Roosevelt approved the top-secret proposal to build the weapon, suggesting to Bush that he keep the document in his own safe. The President did not inform most of his senior White House colleagues, nor does he appear to have mentioned it to Churchill. Yet Roosevelt clearly saw the strategic value of the Bomb. Six weeks later, he told Bush that he wanted the programme 'pushed not only in regard to its development, but also with due regard to time. This is very much of the essence.'[17] Meanwhile, Churchill was allowing himself to be guided by the solid but unimaginative Anderson and by the clever but supercilious Lindemann, who both underestimated American ability and resolve. One of the scientifically-minded officials left out in the cold was Lord Hankey, who nonetheless kept Churchill well informed with pertinent advice. In early December, Hankey wrote a paper to brief him on the secret programme to prepare bacteriological weapons, and suggesting that Britain might want to poison German cattle with feed cake laced with anthrax.[18] He stressed that it would not be wise to trust the Nazis' promise to abide by the 1925 Geneva Protocol, and that Britain should secretly begin to produce the cattle poison 'for purposes of retaliation'. Churchill endorsed the project, on the understanding that it was kept top secret.

For Churchill, 1942 was to be the unhappiest year of his wartime premiership. Reports of the Japanese army's advances and other bad news led him in late January to demand a vote of confidence, which he won by 464 votes to 1, the challenge leaving him chastened though unbowed.[19] Much worse was to come in mid-February with the humiliating surrender of about eighty thousand British, Indian and Australian soldiers in the colony of Singapore. This was, he later recalled, 'the worst disaster and the largest capitulation in British history'.[20] For the first time in his premiership, he saw that it might be time for someone else to take over as leader – the ascetic Stafford Cripps, former ambassador to the Soviet Union and a friend of physicist Patrick Blackett, began to emerge as a credible replacement.[21]

Churchill was now deeply concerned about the quality of his army, which appeared to be less well led and less effective than Germany's, and less resolute than the Soviet Union's vast armed forces. Sustained attacks in the press and the Commons put him under pressure to change his team and review the roles of his experts, including Lindemann. Churchill had been unrepentant in Parliament a few weeks before: after an MP publicly questioned his chief science adviser's integrity, Churchill murmured to his Private Secretary, 'Love me, love my dog, and if you don't love my dog you damn well can't love me.'[22] Hankey, long unhappy about what he saw as Churchill's dictatorial leadership, confronted him about the Prof's influence and was rewarded in March 1942 with a letter of dismissal written with exceptional cruelty.[23]

Churchill next visited Roosevelt three months later, a few weeks after the Americans' victory at the Battle of Midway ended any chance that the Japanese could win naval supremacy in the Pacific. The German armies had been halted in front of Moscow the previous winter, and the Eastern Front was now draining Nazi resources. The Allies' fortunes were beginning to turn for the better.

The President and the Prime Minister discussed in the White House whether to plan a landing in France later in the year, or whether the Allies should focus on fighting the Germans in North Africa. After the meeting, the two leaders spent thirty-six hours together at the President's family home of Hyde Park in New York State, a fine property with furnishings that smelt of old money, set in grounds with glorious views across the Hudson Valley. Hyde Park was the President's Chartwell, a place of repose, a country home well away from the hurly-burly of the capital.

Here, they discussed nuclear weapons for the first time. They both knew that the Allies' work in this field was not running smoothly, and that American scientists were now surging ahead, though Churchill appears not to have appreciated the extent of their lead. The two men began their discussion after lunch on Saturday 20 June, with Harry Hopkins at the President's side, but Churchill was alone, presumably because he felt he had no need of an adviser.[24] The meeting took place on the ground floor of the main house, in the President's study, a small dark room dominated by his desk, where he sat untroubled by the intense heat. Churchill was wilting.[25] They seem to have come to an amicable agreement – Roosevelt told Vannevar Bush three weeks later that he and Churchill were in 'complete accord' – but it was not clear what they had agreed.[26] Several years later, Churchill wrote in his war memoir of how he had urged at this meeting that Britain and America 'work together on equal terms' and had reached 'a basis of agreement' with the President. However, this account included no details, as nothing had been written down.[27] Most likely, Churchill had been told what he wanted to hear by Roosevelt and had then been a victim of the President's 'almost invariable unwillingness to dictate any memoranda of his conversations with foreign statesmen', as Under-Secretary Sumner Welles later noted.[28]

In the Oval Office the next morning, Churchill was hit by one of his most humiliating blows of the war: the President passed him a pink piece of paper bearing the unexpected news that the town of Tobruk in Libya had 'surrendered, with twenty-five thousand [of our] men taken prisoner'.[29] Only a few days before, Churchill had impressed on the garrison's commander the vital importance of holding the port, commenting that 'Defeat would be fatal.'[30] Roosevelt's response was just what the wounded Churchill wanted to hear: 'What can we do to help?'[31]

Churchill returned to face a Commons vote 'of no confidence in the central direction of the war', put down by a cross-party group of MPs in early July. He won it handsomely, but rebellion was in the air. In the days before the debate, Hankey was doing his best to inflict the greatest possible damage on the Prime Minister, while protesting unconvincingly that he had welcomed his own dismissal. At a meeting in the Dorchester Hotel, he went through the familiar litany of charges against Churchill's leadership: 'too self-confident and so too prepared to gamble; no team work in government; tried to put too much glamour into the war, which is really a tough business proposition'. After concluding 'Churchill Must Go', he warned. 'If you keep him, we shall not win the war, but shall lose it.'[32]

Tube Alloys was among the least of Churchill's worries for most of that grim July, until he received a 'very urgent' memo from Sir John Anderson informing him that the agreement he thought he had made with Roosevelt appeared to be all but worthless.[33] Anderson had been told the bad news by the head of Tube Alloys, Wallace Akers, who was now struggling to keep the project alive.

JANUARY 1942 TO JANUARY 1943

Akers attempts a merger

> 'We have got to make up our minds whether this project is to go on as a minor and spare-time occupation [of the government's science ministry] or if we are to try to make some sort of show in comparison with the Americans.'
>
> WALLACE AKERS, 21 December 1942[1]

When Wallace Akers took over the MAUD project and named it Tube Alloys – words he coined to deflect unwelcome curiosity – he had good reason to believe he would soon be running one of Britain's biggest wartime science projects.[2] In the coming year, however, his vision would crumble into a handful of dust.

After winning the confidence of Oliphant over the time it took to eat lunch, it was only going to be a matter of time before Akers had the respect of all the other scientists on the project. So it proved. It was hard to take exception to this thoughtful, donnish man. He was a good listener, slow to pass judgement, and was an engaging conversationalist, quick to share his enthusiasms, perhaps for a Velázquez portrait he had recently seen at the National Gallery or an especially fine new recording of the Emperor Concerto.[3] His long, open face radiated approachability and he was quick to take the initiative in cultivating friendships among his new acquaintances, often treating them to dinner in the restaurant overlooking the Serpentine at the Royal Thames Yacht Club, where he lived in bachelor rooms. After a glass or two of Château Lafite Rothschild on the restaurant terrace, with swans fluttering nearby on the riverbank, even the harshest of professional disagreements would abate, if only temporarily.

It was easy to see why Anderson and Lindemann had chosen him to run the Tube Alloys project – after he graduated from Christ Church three decades earlier, he had carved out a successful career as an industrial chemist at ICI. In one tussle between the company and the government, Akers's integrity and organisational talent had impressed Anderson, who regarded him as a kindred spirit. Akers had the knowledge, energy and experience to realise the potential of the MAUD report and oversee the huge industrial project that would be needed to build the Bomb in Britain. The plan had been based on wishful thinking, as Blackett had foreseen, so Akers had to make the best of a misconceived policy and try to merge Tube Alloys with the American project on the best possible terms. By all accounts he did a creditable job of this, energising his small team in the project's headquarters in a multi-storey, seventeenth-century house on Old Queen Street, a dark lane in Westminster, about ten minutes' walk from both the House of Commons and Downing Street. From his office, he had a fine view across St James's Park, the vista spoiled only by the sandbags and rolls of barbed wire in the foreground.

The year 1942 began well for Akers. Invited to the United States to liaise with leaders of the American project, he arrived in New York in January with Francis Simon and was welcomed warmly. The French researcher Hans von Halban joined them soon afterwards, followed by Rudi Peierls, who combined business with pleasure by visiting his children in Toronto. The British team members spent several weeks touring the main centres of research and policy-making – including Berkeley, Chicago, Washington and Virginia – and were impressed by the resources now being invested in the project and the friendliness of their American colleagues. Nothing, it seemed, was too secret to discuss. Within a few days of his arrival, Akers knew that it was time for the British to reassess their project's relationship with the American effort: 'One thing is clear,' he reported to London, 'an enormous number of people are now on this work so their resources for working out schemes quickly are vastly greater than ours.'[4]

The British project had a series of research and development problems, besides suffering from a lack of decisive leadership. Officials could not decide whether to get the Tube Alloys programme moving by setting their theorists to work and by commissioning a chemical plant to produce ^{235}U. Meanwhile, the Americans were planning to leave nothing to chance by investing in several ways of separating out ^{235}U – they had substantial plans to build no fewer than four plants, two of which would use methods that the British had scarcely contemplated.[5]

One surprise for Akers and his colleagues was that the Americans were concentrating on the newly discovered element plutonium as a suitable material with which to make a bomb. Its most accessible form, ^{239}Pu, was fissile like ^{235}U, and in some ways was better, as less of it would be needed to set up a nuclear chain reaction.[6] Although Chadwick had understood this some months before, the Americans were making all the running: by March 1941, a team led by the chemist Glenn Seaborg at the University of Berkeley had made ^{239}Pu and had confirmed that it was fissile. Although the British still had a modest lead on the theory of nuclear weapons, the Americans were now way ahead on the experimental side, propelled by an unyielding determination to beat Hitler to the Bomb.[7] The intrepid spirit that had once energised the MAUD project had migrated to the other side of the Atlantic.

Although Akers was a polished and hard-working ambassador, his presence could not make up for the absence of two leading figures in the Tube Alloys group, Sir John Anderson and James Chadwick, who did not set foot in the US that year. Anderson's authority as a member of the War Cabinet, and Chadwick's as a world-class nuclear physicist, could have made a significant difference to the case for a greater involvement in the American venture. By the time Akers returned to Britain in March, he was convinced that there was no alternative but to fuse Tube Alloys with the American machine and transfer Halban's nuclear-power project to North America.

In June, Anderson's policy committee gave Akers's proposal a frosty reception, Lindemann maintaining that Simon's separation plant should indeed be built on the British mainland. The committee fudged the decision, concluding that it could not reach a verdict until an agreement had been reached with the Americans about the handling of nuclear patents. Churchill's two nuclear advisers were still in denial about American industrial and organisational capacity. A few days after they dismissed the palpably strong case for a merger, more ammunition arrived on Akers's desk. His assistant Michael Perrin had just arrived in the US, and wrote to say that he had already come to believe 'very strongly' that there was 'probably less than a month' left to draw up plans for coordinating the American and British projects.[8] The progress the Americans were now making was startling.

By midsummer, Akers had won over Anderson and Lindemann, telling them that to refuse a merger would risk giving the Germans the chance to make the Bomb first.[9] Lindemann still declined to allow Halban's reactor project to be transferred to Canada, though he conceded that his opposition was principally 'sentimental' – not a quality he usually admitted to – an opening that Akers quickly exploited. He took the Prof aside and persuaded him to back down, clearing the way for Anderson reluctantly to recommend a merger with the American project, in a 'very urgent' memo to the Prime Minister on 30 July 1942. Sir John could not have expressed the British predicament more bluntly:[10]

> We must . . . face the fact that the pioneer work done in this country is a dwindling asset and that, unless we capitalise it quickly, we shall be outstripped. We now have a real contribution to make to a 'merger'. Soon we shall have little or none.

Churchill agreed immediately, without comment.[11] Unfortunately, he gave his approval too late to enable his chief negotiator to have much leverage with the Americans. As Akers later told Perrin, the lifeblood of collaboration was haemorrhaging, with American officials making no secret of their suspicion that 'the wily British' would exploit 'the secrets and "know-how" of the innocent American inventors'.[12] From September, the Americans began their crash programme to build nuclear weapons, handing over its management to the military. Akers was now looking on helplessly as the Americans' ocean liner of a project surged forward – his only option was to do his best to tether to it the British dinghy.

This was forcefully brought home to Akers on the first day in December, when he visited Enrico Fermi's nuclear-reactor experiment in Chicago, then hunkering down for another Great Lakes winter.[13] Nuclear physics had never before been advanced in such an unlikely place, an unheated makeshift laboratory under the west stands of the University's Stagg Field stadium, on the floor of a disused squash court. Having shown his credentials to the armed security guards, Akers made his way to the site of the experiment, a dark and ill-lit space, the air thick with graphite dust – 'We breathed it, slipped on it and it oozed out of our pores,' one of the physicists later remembered. Fermi's team was building the reactor, brick by brick, a wooden framework supporting a roughly spherical structure assembled from some three quarters of a million pounds of ultra-pure graphite, about eighty thousand pounds of uranium oxide and roughly twelve thousand pounds of uranium. The only moving parts were a few control rods.

Akers had a good view of the experiment. Standing on what had once been a spectators' balcony, he introduced himself to some of the physicists who were measuring the reactor's output with their chart recorders and meters. Fermi was directing the operation with the authority of a five-star general and a sureness of judgement that had earned him the nickname 'the Pope'. He was not one for small talk during experiments, so Akers probably got most of his information from others. He saw the scientists put some of the final graphite block into place and saw that the assembly was already breeding neutrons – the physicists appeared to be only hours away from setting up the self-sustaining, controlled nuclear reaction that Szilárd had envisaged nine years before. If Fermi and his colleagues were successful, plutonium would be produced in such reactors as a by-product and the Americans could use it to make a weapon.

Akers was not present to witness the project's climax. The day after his visit, shortly before four in the afternoon, the pen on the chart recorder traced an exponential increase in the flow of neutrons that ended only when Fermi ordered one of the control rods to be released into the core, immediately slowing the flux of neutrons. There would be no meltdown. Fermi pulled out his slide rule and did a quick calculation that left him smiling broadly, before he announced, 'The reaction is self-sustaining.' After a gentle round of applause, Eugene Wigner shook his hand and presented him with a bottle of Chianti, soon shared in paper cups with the rest of the team, who sipped in silence, their eyes fixed on their pontiff. They knew that he, officially an enemy alien, had done as much as anyone to allow the Americans to build the Bomb in what they could only hope was enough time for the Allies to beat Hitler to the punch. After the team had dispersed, Szilárd remained alone with Fermi, shook his hand and told him: 'This will go down as a black day in the history of mankind.'

A little over three weeks later, Roosevelt secretly agreed to fund the Manhattan Project, named after the location of its first headquarters. He had already appointed the soldier and engineer Leslie Groves to direct it, after its first leader Colonel James Marshall had shown himself lacking in the required urgency.[14] Groves was a graduate of West Point and MIT and a butt-kicking project director, who already had several huge ventures under his belt, including the building of the Pentagon in Washington, the world's largest office complex. A brusque forty-six-year-old, still with the uncompromising ambition of someone half his age, he brushed aside or crushed anything or anyone that stood between him and the achievement of the task in hand. In many ways, he was the perfect person for the job – smart, efficient and resolute. At his first meeting with scientists at the project's Chicago laboratory – including Szilárd and three Nobel laureates – he put them firmly in their place and made clear his indifference to their fancy qualifications, telling them that his studies had given him 'the equivalent of about two PhDs'.[15] His audience sat in silence.

Within six weeks of formally taking charge of the project, he appointed Robert Oppenheimer as its director of science.[16] This was a stroke of brilliance. It was not obvious that 'Oppy' was the right person for the job: something of an intellectual popinjay, he was a theoretical physicist of international standing but had no experience of management, industry or the military. Although he was a physicist of a high calibre, Oppenheimer was well aware of the sustained brilliance it took to join the elite of fundamental physics and, at the age of thirty-eight, he knew he was never going to reach such heights. So, when Groves offered him the chance to make his mark in another way, he grabbed it, leaving behind promising work on the theory of sub-atomic particles and on the astronomical objects later called 'black holes'.

Groves was impressively decisive, though not all his decisions appeared to be grounded in logic and rationality. Among his prejudices was a marked Anglophobia that he had acquired twenty years before during a stay in London. This did not take long to surface. Soon after his appointment, Bush and Conant agreed with his view that 'the British effort would probably be limited to the work of a very small number of scientists without any significant support from either the British government or industry'.[17] Convinced that the cunning British were trying to freeload on a venture funded entirely by American money, he trampled all over Akers and his finely honed arguments, playing on concerns about lax security on the MAUD project and the time-wasting disputes with the French over their nuclear patents.[18] Groves was angered, too, by the Churchill government's appointment of two ICI officials to run the British project. He suspected that Akers and his deputy Michael Perrin wanted to exploit new American technology to create a profitable nuclear-power industry for their company – 'very nearly a monopoly', Conant pointed out – after the war.[19] Akers's use of the ICI office in New York and reports in the press of anti-trust cases involving the company did nothing to quell the General's suspicions of the British corporation.[20] After he in effect halted the exchange of information in early 1943, the British authorities retaliated by forbidding their scientists from attending meetings in the US, leading Groves to enforce even more draconian restrictions.[21] By the spring of 1943, the pious ideal of unrestricted scientific 'interchange' was in tatters.

Having no idea of why the flow of information from their American colleagues had dried up, the Tube Alloys scientists in Britain and Canada were puzzled and frustrated.[22] They knew that Roosevelt had offered to set up a 'jointly conducted' project twelve months previously,[23] and they had worked with their American colleagues like brothers only a few months earlier – so why had the collaboration suddenly stopped? Akers could offer no explanation. The British government was still funding its nuclear project to the tune of a quarter of a million pounds a year, and money was still pouring into the laboratories. But the project desperately lacked focus and leadership.[24]

As morale in Britain soared after the military successes in North Africa, especially after victory at El Alamein had been secured in November, the *esprit de corps* of the Tube Alloys staff crashed through the floor.[25] In Liverpool, James Chadwick was 'extremely anxious' about the project's future; in Birmingham, Rudi Peierls, who had long urged his British colleagues to be more open and less condescending to the Americans, knew the collaboration was on its last legs; G. P. Thomson, a regular at Washington meetings on the Anglo-American project, realised he was being 'shut out'. The dream of a balanced UK–US nuclear partnership was now all but dead.

The scientists on the Tube Alloys project were not the only ones struggling to understand what was going on in Whitehall. A. V. Hill was still condemning the government's failure to use its weapons advisers effectively and had hammered away at this theme in July, after the Tobruk debacle. In a letter to *The Times*, he suggested that the government should take a leaf out of Roosevelt's book and set up a committee of the kind Vannevar Bush was running in Washington.[26] Why, Hill wondered, was Churchill so weak at taking decisions about deploying science and technology in the war? Although Hill often saw him perform in the Commons, Churchill's face-to-face dealings with technical experts was a mystery. To some extent it was illuminated by an indiscreet letter from an official at the Ministry of Supply, Campbell Clarke, who painted a picture of the Prime Minister that was less than flattering:[27]

[Churchill] has a histrionic instinct for the centre of the stage, whenever and wherever there is an audience – and an audience may consist of not more than two other persons . . . while he would be quite amenable to ideas and counter-arguments put to him privately by any one man, the presence of a third-party distracts him . . . and he at once takes the centre of the stage. Any difference of view then becomes a lack of respect for his office . . .

According to Clarke, when experts gave Churchill their opinions on a subject outside his expertise, his vanity came to the fore:

If [the expert] agrees, [or] fails to express disagreement, [the Prime Minister] is pleased. If [the expert] expresses disagreement – and still worse, if he proceeds to give reasons for his contrary opinions – [the Prime Minister] feels personally rebuffed and humiliated in the presence of others. He has not been treated with the respect due to his leading position. It is injured vanity that hampers his judgement.

All the technical advisers knew they were liable to be second-guessed by Lindemann. He and others had persuaded Churchill to agree to the carpet-bombing of German cities in order to wear down the enemy's morale and do serious damage to their industrial capability. Lindemann's *bête noire* Patrick Blackett was an implacable opponent of this policy, arguing that it was both immoral and futile. The RAF should concentrate on attacking German submarines and their harbours in the crucial battleground of the Atlantic, Blackett believed. He was in the vanguard of this conflict, working closely with naval and air-force colleagues as a virtuoso 'operational researcher', helping to improve the design of weapons and increase their effectiveness, using a combination of simple mathematics, scientific reasoning, practical nous and creative thinking.[28] These skills, honed at his bench in Rutherford's laboratory, were now helping to give Britain and its allies the edge in the Battle of the Atlantic.

Blackett, like Hill, was at heart a problem-solver, one who preferred in wartime to concentrate on a small number of promising technological ideas rather than fritter away resources on long shots. As Hill remarked, the Prime Minister was behaving like a 'little [boy] of sixty-seven who will play with toys',[29] spending 'tens of millions [on] chasing wild geese'.[30] Blackett suspected that the nuclear bomb was one of these toys, extremely expensive to build but unlikely to help Britain win the war. Akers and many of his colleagues – especially Chadwick, Frisch and Peierls – had good reason to disagree: according to intelligence reports from Germany, it was still possible that the Nazi state might yet be able to acquire the weapon. If that happened, Moscow would surely be obliterated and Hitler's bargaining power would soar.

Partly to soothe his senior scientists, the Prime Minister appointed a few full-time scientific advisers to the Ministry of Production. If Hill and his colleagues thought they were making progress, they were soon disabused – a few weeks later, Churchill gave Lindemann a seat in the Cabinet with the role of Paymaster General. In Whitehall, the promotion was marked with this verse, written in the style of William Gilbert's pattering libretto for *The Pirates of Penzance*:[31]

> My secretariat scrutinises memoranda topical,
> Elucidating fallacies in detail microscopical;
> I plumb the depths of strategy, I analyse ballistics,
> Reform the whole of industry, or fabricate statistics;
> My acumen's infallible, my logic irrefutable,
> My slightest proposition axiomatic, indisputable;
> And so in matters vegetable, animal and mineral,
> I am the very model of a good Paymaster General.

Akers knew that it was essential to keep Lindemann sweet. It was the only way to preserve the confidence of the Prime Minister, who most of the time was no more than a bystander on the Tube Alloys project.

On New Year's Day 1943, Akers walked to his Washington office – without wearing an overcoat, as was his habit – to put the finishing touches to a five-page letter to General Groves, setting out the case for merging Tube Alloys with the Manhattan Project.[32] As Akers knew, he was playing a weak hand against a cardsharp who held all the trumps, and the end of the game was near. The subtext of the Americans' diplomatic manoeuvres, Akers believed, was their 'desire to build up a monopoly in this field', but he could not help liking Groves, who 'is honest, if very misguided'. Some of Groves's plans were as odd as they were unwise – he even proposed to lock up his nuclear theorists in isolation, perhaps allowing one final discussion with Chadwick and Peierls before the jail gate shut.

Almost two weeks later – on the freezing, windy morning of 13 January – Akers heard the Americans' decision. James Conant had already written to the chief executive of the Anglo-Canadian nuclear-reactor project – without copying his letter to Akers – laying down the law on the terms of their collaboration, much to the Canadians' disadvantage.[33] The auguries for the Anglo-American nuclear project were poor.[34] Later that day, the *coup de grâce* was delivered: with Bush in the room as a witness, Conant read Akers a single-page note setting out the principles of American cooperation on the project, and then handed it to him.[35] The news was even worse than he had expected.

The memo covered seven aspects of the project, including the production of heavy water, its use in chain reactions and methods of separating ^{235}U from ^{238}U. The first two lines of the memo, about the so-called electromagnetic method of separation, set the tone, declaring peremptorily: 'No further information to be given to the British or Canadians.' It continued in this vein, spelling out with brutal clarity the arrangements the Americans were now imposing. The message could not have been clearer: the British researchers were being shut out, except in a few areas where their knowledge might be useful to the Americans. In case anyone doubted that they had the authority to impose the new arrangements, Bush and Conant pointed out that they were following 'orders from the top'.[36]

Akers had spent months labouring in vain. The Americans' nuclear project had been jump-started by the MAUD report and they had benefited from the open exchange of nuclear expertise, but now they were unilaterally abandoning the policy of cooperation that Churchill and Roosevelt had agreed. Akers had the grim task of making the best of things with his colleagues in Whitehall, of explaining the Americans' volte-face to his disgruntled Tube Alloys colleagues,[37] and of trying to do business with the scientist who had done most to betray his trust, Vannevar Bush.

OCTOBER 1942 TO JULY 1943

Bush aims for an American monopoly

> '[Churchill] had an ego that has never been matched anywhere . . . he did not hesitate to consider himself an expert on the application of science to weapons.'
>
> VANNEVAR BUSH, 1972[1]

Vannevar Bush and his deputy James Conant had been stringing the British along for months. The two men, almost always of one mind about American nuclear policy, had reassured Akers that they wanted to work closely with the British, while working sedulously behind the scenes to reverse the policy of collaboration that Roosevelt had apparently supported. In a series of long, courteous and factually accurate letters to an increasingly anxious Sir John Anderson, Bush skilfully played for time, as the American project overtook the British one. After an especially adroit letter to Anderson that contained no untruths but neatly avoided addressing any of his concerns, Conant congratulated Bush on a 'masterly evasive reply'.[2] It was not long before Bush and Conant had persuaded their President to abandon the collaboration policy, enabling them to pass the buck and assure poor Akers that the new arrangements came 'from the top'.[3]

In the always-measured words of John Cockcroft, Bush was now 'the king of US scientists'.[4] Since the President had given him the go-ahead to set up the National Defense Research Committee, he had built a formidably well-resourced and effective system, linking some two hundred thousand scientists, from the humblest lab technician to Albert Einstein, to the needs of the American military. Away from the public eye, he directed the organisation from his wood-panelled office in the Carnegie Institution on P Street in north-west Washington. Some of the military chiefs were uneasy about working alongside civilians and did not trust them to keep secrets, but 'Van' waved these reservations away, dealing majestically with objectors. He took them to dinner, patiently spelled out his case to them between puffs of his pipe and sips from a glass of milk, and dropped heavy hints about his close links with the President. Most dissenters quickly shuffled into line; those who did not risked being skunk-sprayed with profanities.

During the week, Bush worked long hours to push his agenda, but he also knew how to relax. Some weekends, he left the local hotel where he lived and headed out to his New Hampshire farm, where he took it easy for a day or two, feeding his turkeys, playing the flute and sailing along the coast.[5] He retained an agrarian earthiness even while bestriding Capitol Hill, looking every inch the Washington insider, wiry in frame, wry in demeanour. A eulogy in *Collier's* magazine, written a few weeks after Pearl Harbor, captured his personality well and invited readers to 'meet the man who may win or lose the war'.[6] Bush probably grimaced when he read that the author had quizzed his office staff about progress on one of their most secret topics: using ^{235}U to build a nuclear bomb that 'would make Berlin one vast crater'. The staff handled the question well, telling the author that they 'cringe when they hear such stories' – their fundamental interest was in '^{235}U as a source of power', they insisted, the opposite of the truth.

Within a year or so of mocking early press reports that linked the discovery of nuclear fission with the threat of imminent Wellsian cataclysm, Bush had begun to take seriously the possibility of nuclear weapons. He quickly became convinced that the United States should take the initiative to build such a bomb on its own, and was soon disenchanted with the British negotiators, their wearisome pushing for equal participation, their feeble resources, and their blithe indifference to the interests of American taxpayers. Nor could Bush understand how the British system, with its committees entangled like a heap of spaghetti, ever got anything done.[7] Frisch and Peierls's seminal memorandum, Chadwick's masterly MAUD report, Anderson's gentlemanly pleadings for fair play – none of them cut any ice with Bush. All that counted for him was the long-term future and interests of America.

All the leading players agreed with the Bush–Conant line. Among their leading supporters was the Secretary of War Henry Stimson, who had been an advocate of Britain's interests since the beginning of the war but firmly believed that it was America's destiny to lead the world.[8] The problem was that by far the most important person in the loop, Roosevelt, was also the most unreliable, inclined to change policy in a trice and able to hold several opinions at the same time. In late October, he remarked that he and Churchill had discussed the Bomb project only 'in a very general' way – quite a change from the 'complete accord' he had described three months before.[9] Roosevelt soon became sympathetic to Bush and Conant's view that America no longer needed British help on the nuclear project, and that it was time to exchange only purely scientific knowledge and to share only a very limited amount of engineering and manufacturing information. That way, America could keep for itself a weapon that, in Bush's phrase, may be 'capable of maintaining the peace of the world'.[10]

At a meeting in late October 1942, Stimson told the President that the US was doing 'ninety per cent of the work' on the Bomb and suggested that they should not share 'anything more than we could help'. Roosevelt concurred, but proposed they confer with Churchill. Bush wanted to take a tougher line – to restrict the exchange with British scientists only to information they could use during the war. Furthermore, he believed there should be 'no interchange on research or development . . . on bomb design'. In mid-December, all this was approved without demur by the influential Military Policy Committee, whose members included Stimson and Vice-President Henry Wallace. In all significant details, they agreed to shut Britain out of the development of the Bomb, even though they knew that putting an end to the collaboration would delay the project and might conceivably mean that Hitler got hold of the weapon first. For Bush, Groves and their associates, the challenge was to persuade their unpredictable President to rescind the cooperation agreement he had made with Churchill and then to stick to his decision. Roosevelt's colleagues had no difficulty, however, in persuading him to authorise the multi-billion-dollar investment needed to deliver to his military a nuclear weapon as soon as humanly possible.[11]

Any misgivings Roosevelt may have had about curtailing cooperation with the British vanished after Stimson showed him a secret Anglo-Russian agreement on the exchange of new and future weapons, implying that Britain might share American information with Stalin. This was odd – anyone who had even a passing acquaintance with Churchill knew that he would have dropped the agreement immediately if he thought it stood the least chance of jeopardising Anglo-American relations. In his cables to Churchill soon afterwards, the President made no mention of the policy reversal, probably because he was confident that he could talk his way out of his volte-face when they met in the sunshine and warmth of Casablanca a few months later. The difficulties with the nuclear project scarcely figured in their discussions, Roosevelt palming off Churchill with an assurance that Harry Hopkins would sort out any problems after the meeting. Hopkins then remained silent for months, provoking a series of cables from the increasingly anxious

Prime Minister: 'That we should each work separately would be a sombre decision,' Churchill wrote in April 1943.[12] Hopkins was stonewalling, as Bush had recommended.

Now that Tube Alloys had lost most of its momentum, Britain's bargaining power was rapidly dwindling. The strategy of virtually excluding Britain from the American project was paying off well for Bush, yet he seemed to be out of favour in the White House – he scarcely heard from Roosevelt. It seems that the President was allowing his chief scientist not only to freeze out the British but also to take the flak for it. If and when matters came to a head, Roosevelt could then claim he was not privy to the details and could renegotiate from the position of great strength guaranteed by his officials' temporising. At the same time, Bush was twisting in the wind, worrying whether the President had grasped the strategic significance of nuclear weapons.[13] It still rankled with Bush and Conant that they were discussing Britain's nuclear policy with employees of ICI rather than with academic scientists with no axe to grind. In March 1943, Conant told Bush that the controversy over the UK–US collaboration on the production of nuclear weapons 'would never have arisen if the negotiations had been in the hands of British scientists comparable to yourself and if [they] had had the same voice in determining policy in Great Britain as you have had here in the United States'.[14]

Bush feared that Roosevelt would make another of his policy somersaults during his next meeting with Churchill. From 12 May, when the talks began, 'Van' was standing by in his office at the Carnegie Institution, ready to be summoned to the White House, but he heard nothing for almost two weeks. Then, on the last day of the talks, Harry Hopkins called him in the early afternoon: would he come over and see if a meeting of minds could be reached between him and Lindemann?

The three men got together an hour or so later in Hopkins's office, in his second-floor White House suite overlooking the Washington Monument and the Jefferson Memorial.[15] Lindemann was not the softly spoken Prime Ministerial counsellor familiar to Bush, but a rasping accuser, demanding to know why on earth the Americans had unilaterally ended their collaboration. With Hopkins looking on, puzzled, Bush put the Americans' case.

Lindemann firmly rejected it. The reason why the British wanted to be involved in building the Bomb was, he said, that it wanted the weapon *after* the war.[16] Commercial considerations about nuclear power were unimportant. If the technical information on how to build nuclear weapons was not forthcoming from the Americans then – in his view – Britain might have to divert some of its war effort in order to get it, though the decision was ultimately Churchill's. As the meeting wound up, Hopkins said that he now understood for the first time the point of the disagreement. For the time being, Bush thought it best to 'sit tight', he told Hopkins.[17]

No one in the American administration bothered to tell Bush that, on the day he defended the Americans' case against Lindemann's attack, the President decided to accede to Churchill's pleas to resume full exchange of nuclear information with the British. Roosevelt did not mention it even when he met Bush over lunch a month later, on 24 June. Bush informed the President that Britain's number-one priority was – according to Lindemann – to have the Bomb after the war. Roosevelt was 'astounded', murmuring that Lindemann was 'rather a queer-minded chap'.[18]

Bush left the White House believing that the President 'had no intention of proceeding further on the matter of relations with the British'. Yet, if the experience of the past few months was anything to go by, these were unlikely to be the wayward

Roosevelt's last words on Anglo-American nuclear exchange. Bush could only follow what he believed to be the President's most recent line. This is what he did two weeks later, when he began a long-planned visit to London to talk with British experts about military technology that the two countries could share. Inevitably, Bush would have to face the music during this visit and explain America's policy directly to the Prime Minister.

Late in the afternoon of 15 July 1943, Bush walked into Downing Street for a meeting of the War Cabinet's anti-U-boat subcommittee, expecting to spend a few hours talking strategy with Blackett, Lindemann and some two dozen others. No sooner had Bush entered the building, however, than he was ushered privately into the Cabinet room for an audience with the subcommittee's chairman, Winston Churchill.

Bush expected to exchange a few pleasantries, but moments later he was like a chicken roasting on a spit.[19] The Prime Minister, just up from his afternoon nap, was sitting at the huge Cabinet table, livid. He spent almost a quarter of an hour bawling out his guest, spluttering vituperation about the failure of the interchange agreement while fitfully trying to light his cigar, intermittently tossing burnt matches over his left shoulder towards the fireplace. Bush said nothing. But when Churchill brought up the matter of Conant's January memo spelling out the drastic limitations on the exchange of nuclear information that the Americans were now imposing, Bush spoke up, doubting the document's existence. When Churchill brought out a copy, Bush claimed he had never seen it, evidently forgetting that he had, a few months before, watched his deputy Conant reading it aloud to Wallace Akers.

Bush went on the offensive. It was plain, he said, that Britain's main interest in the project was to lay the foundations of a post-war nuclear-power programme. Churchill butted in, saying that he couldn't give a damn about that. Later, after another of the Prime Minister's tirades, Bush reminded him that the US military was now handling the Bomb project so he should take up the matter with Henry Stimson, who happened to be in London. 'I certainly do not propose to discuss this subject in his absence,' Bush huffed. He left the room shaken but determined to win the next round.

During an expedition to Dover two days later, Stimson was taken aside by Churchill. 'He was most anxious that I should help him by intervening in the [matter of the Bomb],' Stimson wrote in his diary.[20] They arranged to meet in London shortly afterwards. Stimson knew little of the subtleties of nuclear negotiations and understood less, so went to the meeting accompanied by his colleague Harvey Bundy – a cautious lawyer familiar with the Manhattan Project – and by Bush, the real expert. Less than an hour before, over lunch at Claridge's, Bundy had warned Stimson that if Churchill had his way, the President might inadvertently be persuaded to overreach the constitutional powers he had in wartime by subsidising the post-war British nuclear industry. As they walked into Downing Street, Stimson told Bush: 'Van, I want you to handle this.'

Churchill had also left nothing to chance. He was accompanied by his two nuclear confidants, Lindemann and Anderson, who said little during the meeting, as did Stimson and Bundy.[21] According to Bush's later account, the conversation was another tussle between him and Churchill, who again stressed that he was not interested in post-war nuclear energy. In that case, Bush countered, why did the British select as their representative on the project someone who had been an ICI engineer? Churchill looked to Lindemann and Anderson, but they said nothing.

Churchill eventually ran out of steam and realised he had no choice but to compromise, Bush observed. 'I will make you a proposition,' the Prime Minister said, before setting out a four-point plan. First, the Bomb would be built as a joint venture with free interchange of information; second, neither country would use the weapon against the other; third, neither would pass information to other countries without the other's consent – a move that would remove any chance of Roosevelt sharing the science with Stalin, with whom the President was more sympathetic than Churchill thought wise. Finally, recognising that the Americans would have to bankroll the project, Churchill agreed that the President would have a veto on any post-war venture Britain might develop to use nuclear energy as a source of electrical power.

As the meeting drew to a close, Stimson agreed to put the four points to the President. Bush and his colleagues had achieved much of what they wanted, most importantly a commitment to scientific interchange so vague that it would be easy to block. Bush left Downing Street with the impression that Churchill was not best pleased with the outcome of the meeting and would not be in a hurry to do him any favours.

A few days later, when Bush was still in London, his relief curdled into anxiety. He read a garbled cable from the President, written before the second meeting in Downing Street, instructing him 'to review' full interchange of information with the British. In Washington the following week, Bush learned that the President's cable had been wrongly transcribed – Roosevelt had actually dictated an order 'to renew' the agreement. But it was too late: the result of the Bush–Churchill meeting stood, and exchange of information was not renewed. If H. G. Wells had come up with this storyline in one of his political satires, it would have been condemned as laughably improbable.

It was now time to seal an agreement on collaboration. Bush, Anderson and a few colleagues were tasked with drafting a text for their leaders to consider at their forthcoming meeting in Quebec. Although Bush was optimistic, he still could not be sure what his President really believed the American policy on nuclear collaboration should be, and how Roosevelt would react when the bulldog began snapping at his heels again. 'Van' did not have long to wait to find out, however – Churchill and Roosevelt were scheduled to meet only a few weeks later.

Churchill's nuclear deal with FDR

> 'But what a cultivated animal FDR is . . . and a cute, cunning old bird – if ever there was one. But I still know who gets *my* vote . . .'
>
> Churchill's daughter MARY, 3 September 1943[1]

Churchill had first fully appreciated the significance of the British nuclear-weapons project at the beginning of April 1943, almost two years after his scientists on the MAUD committee had assured him that the Bomb was viable. He had realised that Britain was in serious danger of being shut out of the Manhattan Project, and had at last tackled the problem head-on, commissioning reports, calling meetings and talking with his advisers. Britain's negotiating position on this question was by that time dire.

Lindemann had drawn Churchill's attention to the seriousness of the problem in an angry note written after the Prof heard, almost certainly in January 1943, that Britain had been all but shut out of the American nuclear project.[2] There was a risk, he wrote, that 'we might lose [the War] if Germany completes the work first'. Assuming that the Allies prevailed, then there was still a danger that Britain would emerge from the conflict grossly handicapped, without access to the Bomb, which promised to be the world's most powerful weapon:

> The principles and possibilities are known to scientists throughout the world. In five years undoubtedly the leading powers will possess these weapons unless forcibly prevented. Can England afford to neglect so potent an arm while Russia develops it?

That mention of Russia in this context must have unsettled Churchill. Lindemann's note prompted the series of memos the Prime Minister sent to Hopkins, whose repeated evasions galvanised Churchill into a concerted attempt to pin down the Americans on the matter of nuclear collaboration.[3] In early April, he commissioned a report from Lindemann, who summarised the project and its underlying physics in five pages of opaline prose, noting that 'the specialists are prepared to lay 100 to 1 on success', though he believed their odds were too optimistic by a factor of ten.[4] Churchill, saying that he now understood 'the broad outline of the story', ordered his intelligence staff to investigate whether there was 'the slightest chance' that the Germans were building the plant needed to make a nuclear bomb. Eight days later, he asked Anderson to investigate the cost of developing a nuclear-weapons project 'at full speed ourselves'.[5] The results made sobering reading: it made no sense to lavish so many precious resources on building the weapon on British soil – it would be far more efficient to work closely with the Americans and, if they refused to share the weapon, use the experience of working on the project to build a bomb in the UK after the war.[6]

This was why Churchill had been so determined to persuade Vannevar Bush and Henry Stimson of the need to resume the Anglo-American collaboration, when they met him in the early summer of 1943. Although those talks had been difficult, Churchill believed they had been successful from the British point of view. But Anderson warned him that there was still a risk that the American generals might cause trouble. Anderson asked Churchill soon after their meetings: 'Is there not a danger that General Groves, at any rate, will simply tell Stimson and Bush that, like all Americans who come to our misty island, they have been taken in by our hypocritical cunning and carried away by our brilliant Prime Minister?'[7] Within hours of receiving this note, Churchill read a friendly but vague cable from Roosevelt, who wrote that he had 'arranged satisfactorily for Tube Alloys' and suggested that he send over his 'top man' to sort out the details.[8]

It had been many months since the Prime Minister had received such an encouraging note from the President about the nuclear project. Anderson, wanting to strike while the iron was hot, packed his bags and prepared to visit Washington to draft an agreement with American officials, so that Churchill could close the deal at the leaders' next conference in Quebec.[9] Lindemann and Anderson were still, in effect, Churchill's only nuclear counsellors, neither especially adept at dealing with American officials. Anderson excelled at negotiating compromises, so he could be trusted to come up with wordings acceptable to both parties, though the experience and diplomatic skills of Tizard – exceptionally popular with the Americans – would surely be missed. Sir Henry would play no part in such sensitive matters, however, as Churchill ensured by sending him on a three-month mission to Australia.[10]

When Anderson arrived in Washington on 5 August, a telegram was awaiting him – 'Best of luck – Winston.'[11] Sir John, dressed like an Edwardian butler, cut an anachronistic figure on Capitol Hill, though his steady manner made him agreeable company.[12] He spent five days tweaking his draft agreement with Bush, Conant and other officials, with Akers on hand to advise on the practicalities from the British scientists' point of view. Conant was convinced that the American project was going swimmingly – the Bomb was 'in the bag', he thought – while his boss Vannevar Bush was more cautious. Although the Americans were 'spending money like water', he doubted whether the weapon would be ready in time to be used in the war.[13]

The negotiations went more smoothly than Anderson had expected. Neither Groves nor his army colleagues made any trouble, and the American negotiators requested only slight changes to his draft agreement, though Conant baulked at the arrangements Anderson suggested for the interchange of technical information. Rather than risk a contretemps, it was agreed that a new Combined Policy Committee – chaired by Stimson and consisting of British, Canadian and American officials – would oversee the policy's implementation. Anderson trusted the Americans to turn over a new leaf and collaborate in a friendly spirit, knowing that Churchill would be happy to back him up. The arrangement was, however, a classic bureaucratic fudge.

Churchill and Roosevelt's agenda in Quebec was sure to be dominated by debates about the opening of the second front in northern France. There would be time to consider the nuclear project briefly and the draft text agreed in Washington promised to make it easy for the leaders to reach a final agreement. If everything went to plan, Churchill would win the prize he had sought for months: to open up an uninhibited exchange of information on the Bomb project with the Americans.

When Churchill set off for his next meeting with the President, shortly after midnight on 5 August, he was in a jaunty mood. Marching up and down the platform of Addison Road railway station in West London, he sang a William Gilbert ballad he had known since he was a boy:[14]

> I go away this blessed day
> To sail across the sea, Matilda!

The next day, he arrived at the Clyde in Scotland and boarded the *Queen Mary*, together with some two hundred staff, his wife and their daughter Mary. The visit promised to be both a productive business trip and an agreeable family vacation.

Churchill had good reason to feel chirpy, as the war was

now going well. In May, North Africa had been cleared of enemy armies, the surrender of Italy was expected soon – following the deposition of Mussolini by the Fascist Grand Council in July – and the Russians were inflicting vast casualties on Hitler's armies, now in retreat. Churchill had met Stalin for the first time in Moscow in August 1942 and was agreeably surprised. Stalin was not an imposing figure – only a few inches above five feet tall, with pock-marked skin, bad teeth and a withered left arm – but he had an impressive directness. After a six-hour man-to-man talk with him over a merry, alcohol-fuelled banquet – a roasted suckling pig crowned their table – Churchill left shortly after three in the morning with an enduring belief that he could do business with the Soviet dictator. Two months later, the Prime Minister dismissed the story doing the rounds in Moscow that he had described Stalin as 'that monstrosity' as nothing more than 'a silly lie'.[15]

The Russian army had turned the tide at Stalingrad and by early February 1943 the entire German 6th Army had been defeated or captured. The Russians were now on the brink of fighting the Battle of Kursk, which was to be the biggest tank battle the world had ever seen. Although the Germans would also lose that battle, the Soviets' losses were now terrible, far greater than those of all the other Allies combined, and Stalin resented what he regarded as the tardiness of the British and Americans in opening another front in Europe, to help reduce pressure on the Russian forces. Roosevelt wanted to placate him and wrote to suggest that just the two of them talk privately.[16] After an American official let slip that this overture had been made, a disturbed Churchill raised the matter with the President, who shamelessly denied making any such approach. Stalin eventually declined to meet, but Roosevelt's mendacity drew attention once again to the great difficulty of doing business with him in a straightforward way. If Churchill was hurt and wounded, he did not show it: he liked nothing more than to be in Roosevelt's sparkling company, and had arranged to spend a week in Hyde Park before the conference began.

The conference was hosted in Quebec by the Canadian Prime Minister William Mackenzie King at the seventeenth-century Citadelle, which had fine views over the local fortifications and the St Lawrence River, hundreds of feet below, making its majestic way towards the Atlantic. Within a few minutes of arriving, Churchill was working in his mobile map room, talking with officials in Whitehall and giving directions to his generals.[17] Sir John Anderson brought him the draft agreement on nuclear collaboration, but it seems that they did not discuss it for long – Churchill and his daughter soon checked out and made their way to the Roosevelts' home in New York State, a day's train ride away. Warmly welcomed by the relaxed and cheerful President, Churchill swam in the outdoor pool, donned a ten-gallon Stetson and picnicked in the late-summer heat with the President and guests on hamburgers, corn on the cob, fish chowder, water melon and hot dogs.[18] Daisy Suckley, one of Roosevelt's cousins, was struck by Churchill's obeisance – he 'adores the President, loves him, looks up to him, defers to him, leans on him'.[19]

The leaders' talks in Quebec began on 17 August and lasted a week. Churchill was on his most dynamic form, performing far into the small hours and leaving Roosevelt feeling 'nearly dead'.[20] Among the early fruits of their talks was an agreement to invite Stalin to meet them both in Alaska[21] – Churchill had put a stop to Roosevelt's desire to meet the Soviet leader alone. The fear of being sidelined by the emerging superpowers was always present in Churchill's calculations. Although the Alaska conference did not come off, there was enough goodwill between the Big Three for them to meet for the first time a few months later, in Teheran.

The high point of the Quebec Conference, especially for the Americans, was that Churchill agreed in principle that an Allied landing in France would take place in the following year under US command, though he still left himself wriggle-room. The Americans had long resented what they – and the Soviets – had regarded as Churchill's obsession with wearing down the Germans by bombing their cities and attacking what he called the 'soft underbelly' of Europe, rather than attempting a frontal assault.[22] Churchill feared that such an operation would be a bloodbath and a strategic disaster.

At a private meeting, Roosevelt finally gave Churchill what he wanted – confirmation that the Americans would resume collaboration on the Bomb. During 19 August, on the second day of their meetings, they signed a document, typewritten on four pages of Citadelle notepaper, which closely followed the Anderson–Bush draft, though Churchill added a few rhetorical flourishes to give it a more elevated tone.[23] The agreement began with a statement that the speedy completion of the project was 'vital to our common safety', referring to the threat that Hitler might have the Bomb first. Yet the leaders knew from intelligence reports that this was unlikely.[24] The point of the American Bomb project was beginning to change from beating the Germans to the weapon, to winning global dominance after the war.

One section of the agreement made clear – almost embarrassingly so – that Britain was the junior partner and that, to be granted even that role, the Prime Minister was prepared to make what could well be a substantial concession:

> The Prime Minister expressly disclaims any interest in these industrial and commercial aspects beyond what may be considered by the President of the United States to be fair and just and in harmony with the economic welfare of the world.

Churchill regarded British participation in the American-run project as the only way of ensuring that his scientists learned enough to build the weapon after the war. To achieve that goal, he gave the President an unprecedented veto on the development of nuclear power in Britain, a potentially profitable and strategically important industry.

The two leaders had also given themselves a veto on the other's use of the Bomb against any other country, relinquishing an element of their sovereignties. Had American lawmakers known of this decision, they would almost certainly have declined to support it, and it is possible that Churchill's own government would have found such a veto hard to follow, too. In the long term, after one or both of the leaders had left office, it was hard to imagine that their successors would want to be bound by it. Lindemann appreciated this, to the irritation of Churchill, who always insisted that the agreement was the best possible one that Britain could have obtained in the circumstances.[25] Had he been quicker off the mark and more shrewdly advised two years earlier, when Roosevelt first offered to develop the Bomb jointly, Britain would have been well placed to negotiate a considerably better deal.

Churchill was delighted with the Quebec Agreement: it had for the first time committed Roosevelt – in writing – to a policy of collaboration, and immediately achieved its aim of involving British physicists in the Manhattan Project. Within a few hours of the document's signature, Rudi Peierls, Francis Simon and Mark Oliphant arrived in the US to recommence collaboration, Chadwick following soon afterwards.[26] These were the first of some two dozen British scientists to play a part in building the Bomb.

One serious weakness of the document was the vagueness of its wording – Anderson predictably had delivered a compromise but had not demanded the specificity needed to make the agreement watertight. As a result, the question of what kind of information was to be shared was unclear and there were several elastic loopholes in the text, ready to be exploited by Groves and his colleagues if they so wished.

The agreement did not explicitly mention the Soviets for a reason that was too obvious to put in writing: the two leaders wanted the Bomb kept secret from them. Churchill was finding it hard to deal with Stalin, who was becoming increasingly importunate. On the final day of the Quebec Conference, after an angry telegram arrived from the Soviet leader about the Allies' response to Italy's surrender, Churchill commented darkly that 'Stalin is an unnatural man', and worried that 'grave troubles' were ahead.[27] Yet Churchill knew that Britain and the United States had to deal with what was sure to be 'the greatest land power in the world after this war' and wanted to be on 'good terms' with Russia after the conflict.[28] After an official visit to Canada, on the sleeper train to Boston, Churchill dictated a speech he was to give the next day, Monday 6 September, at Harvard University about how to strengthen the bonds between the English-speaking peoples. He gave the speech, broadcast on both sides of the Atlantic, at a ceremony at which he was to be presented with an honorary degree by James Conant, who had worked hard for many months to marginalise Britain's involvement in the American project to build the Bomb.

After Churchill was presented with a red morocco tablet bearing the degree, he advanced to the lectern looking like Holbein's Henry VIII, dressed in an Oxford gown.[29] With loud applause echoing around the hall, Churchill put on his glasses, walked to the lectern and began his speech with a confidence that befitted the boldness of this theme – the need for a permanent alliance between Britain, its Empire and America. He stressed the countries' common values and the 'gift of a common tongue . . . a priceless inheritance' before floating the idea that one day the countries might share 'common citizenship'. In the climax of the speech, he looked forward to the war's endgame, perhaps also thinking of the sacrifices of sovereignty he believed that he and Roosevelt had made in signing the secret Quebec Agreement:[30]

> I am here to tell you that, whatever form our system of world security may take, however the nations are grouped and arranged, whatever derogations are made from national sovereignty for the sake of the large synthesis, nothing will work soundly or for long without the united effort of the British and American peoples.

The audience rose as one to applaud and cheer, leaving him as the only person in the theatre sitting down.

After what had been his longest visit to North America during the war, Churchill arrived back in London on 19 September. Lindemann was far from happy with the Quebec Agreement. As he wrote to Churchill a few weeks later, he thought it entailed 'not an altogether satisfactory version of cooperation', adding tactfully, 'It seems likely to be the best we can get for the time being.'[31] In the following months, as the agreement's flaws became plain, Lindemann was to take an even less rosy view of it. The matter became a bone of contention between him and the Prime Minister, who regarded it as the granitic foundation stone of Anglo-American nuclear politics.

Although both Churchill and Roosevelt now understood that nuclear weapons would be an important asset after the war, they do not appear to have thought through the implications of developing them. Nor did they or their advisers seem to have reflected on the consequences of excluding Stalin, who was chafing at the disproportionate price his country was paying to win the war and already looking forward to a territorial pay-off when it was over. Churchill's apparently weak grasp of the effects the new weapons were likely to have on geopolitics contrasts sharply with the far-sightedness he had demonstrated in 'Fifty Years Hence' more than a decade before. In that essay, he had worried that democratic governments would shy away from taking a principled view of their new military capabilities and, instead, muddle along from one compromise to the next. Now, in one of the awkward concessions inevitable during the hurly-burly of war, he had made a narrow and woolly agreement with his closest ally to develop one of the new technologies he had once dreaded, and to keep it secret from their governments. Neither Churchill nor Roosevelt nor any of their close associates appears to have thought deeply about what would happen when other countries acquired the Bomb after the war.

This complacency would soon be challenged. Seven weeks after Churchill and Roosevelt signed the Quebec Agreement, a refugee arrived in London and encouraged fresh and provocative political thinking about the implications of nuclear weapons. The interloper, who knew nothing about plans to build the Bomb, was not even a politician, but a physicist.

SEPTEMBER 1943 TO MAY 1944

Bohr takes a political initiative

'Bohr was like a great big cuddly teddy bear, and was always inside a cloud of smoke.'

Chadwick's daughter JOANNA, November 2012[1]

The prospect of Niels Bohr's arrival in London excited the normally imperturbable Sir John Anderson so much that it turned the ink in his pen from black to purple. Bohr was worthy, Anderson told his colleagues, 'to rank in every respect with a Newton or a Rutherford'.[2] If not at their level of achievement, Bohr was famous as both a scientist and a sage, so he seemed certain to be a great asset to Tube Alloys. He would probably give the British government some much-needed influence on the Manhattan Project, too. Before long, he was pressing for access to the Prime Minister, determined to alert him to the threat that nuclear weapons might pose unless Anglo-American thinking about them changed. It was not clear, however, that Churchill would be prepared to listen to him.

The story of Bohr's escape from Denmark would have sounded far-fetched in a John Buchan thriller.[3] Sitting out the war at his Institute in Copenhagen, under the relatively benign rule of the local Nazi satraps, he had been able to work almost normally. In mid-September 1943, the mood in the city darkened: anti-German feelings were hardening and rumours began to spread that the Nazis were planning to deport local Jews – as Bohr's mother had been Jewish, he would be affected. On 29 September, after hearing a tip-off that Jews in the city were about to be arrested, he and his wife – carrying a few belongings – sneaked out of their home early in the evening, hid in a cottage near the coast and, with a few other escapees, travelled by boat under the moonlight to Limhamn in neutral Sweden, somehow avoiding the roving German patrol vessels. There, Bohr and his wife crawled across the beach on their hands and knees to safety. His four sons followed a few days later. By then, although he was talking about the plight of Danish Jews with the Swedish King and Foreign Minister, he believed he would be more useful in Britain, which he thought of as his second home. When Lindemann renewed an earlier invitation to fly him to London,[4] Bohr left his family in the care of the Swedish authorities and boarded an unarmed Mosquito bomber, where he was strapped into the unpressurised bomb bay, as the plane had no seats for passengers. For the first time in his life, Bohr's huge head presented a problem and made the journey unpleasant – the helmet he had been given was fitted with earphones, but it was much too small for him. As a result, he did not hear the captain's instructions to wear an oxygen mask and so spent most of the flight unconscious with altitude sickness.

After landing in Scotland on 6 October, he was whisked onto a flight south and was met at Croydon Airport by his old friends James and Aileen Chadwick, before being supplied with a set of official papers.[5] Bohr arrived knowing nothing about Tube Alloys and the Manhattan Project. He had guessed two years before, however, that a nuclear weapon might be about to become a reality after he received a secret invitation from Chadwick to work in Britain (Danish resistance fighters produced a typescript of the message from a microdot in the end of a door key slipped to them by the British security services). Chadwick escorted his guest to London and checked him into the Savoy Hotel, where they were joined for dinner that evening by Sir John Anderson. It was during these conversations that Bohr heard about the Allies' work on the Bomb.

He was astonished to hear that the project he had once confidently dismissed as being too enormous to contemplate was now in full swing.

The Whitehall mandarins quickly appointed the man they called 'the Great Dane' as a special consultant to Tube Alloys.[6] It was obvious that he was no ordinary scientist – true, he had all the unworldliness expected of an accomplished intellectual, but he was also surprisingly down to earth. Now approaching his fifty-eighth birthday, his thin hair was greying but he still ran up flights of stairs two steps at a time, displaying the last remnants of the athleticism he had shown as a soccer player in his youth, when he was an effective goalkeeper.[7]

His personality was a unique mix of companionability, wisdom and an endearing incoherence. Though a generous spirit, he was a poor listener. After a long and tortuous exchange of views – mainly with himself – his face would light up with a sunburst of a smile as he returned to terra firma and cadged another match for his pipe, which somehow always seemed to need rekindling. Bohr was a man of words but he had two serious problems with them – as he put it, he was determined not 'to speak more clearly than I think', and 'If I cannot exaggerate, I cannot talk.'[8]

He had no idea how to manage his affairs, as Wallace Akers's secretary quickly realised. Once, before he left for an important meeting, she packed him off with six typed copies of travel instructions, telling him: 'If you put one of these into each of your pockets, you are sure to find one when you need it.'[9] Help arrived a week later when his undergraduate son Aage – later a Nobel Prize-winning nuclear physicist himself – joined Niels and served as his minder, secretary and amanuensis. The two men soon felt at home, despite the sandbags, the rolls of barbed wire on the streets and the air-raid warnings, whose tiresome wailing they had not had to endure in occupied Denmark.[10] Anderson arranged for them to be issued with ration books and gave them an office in the Tube Alloys headquarters in Old Queen Street and an apartment nearby in St James's Court.

Bohr talked about the weapon at length with British physicists, but what interested him most were its political implications, especially in the long term. During hours of discussions with Lindemann and Anderson, his thoughts on the nuclear project coalesced into a way of looking at the Bomb that took his colleagues by surprise.

Bohr believed that if the Bomb could be made, it was foolish to pretend that the underlying science and technology could be kept secret: sooner or later – more likely, sooner – the government of any industrially advanced country could instruct its scientists to find out how it was done and then direct them to manufacture the weapon. A terrible arms race was therefore inevitable, with leading countries trying to outdo each other by building ever more powerful weapons, and the world would have to endure one nuclear conflict after another. It would be better, Bohr believed, for the Allies – including the Soviet Union – to share the idea and thus begin a new era of harmony and trust. This vision was long on idealism and short on detail, with no plan for beginning these negotiations, nothing about dealing with terrorists who might get hold of the weapon, and no indication of how acquiring it would affect conventional warfare. Yet Anderson – and even, it seems, Lindemann – saw the beginnings here of a new way of thinking about the latest military technology, and they decided it was worth trying to talk to the Prime Minister.

Bohr arrived in London eager to contribute to the war effort, having spent so long out of touch with most of his friends in the community of British scientists. The government in London wanted to keep his activities secret, but the *New York Times* blew the whistle on him three days after his arrival. It then compounded the embarrassment with a garbled report that he had brought to London plans for a new invention 'of the greatest importance for the Allied war effort' involving 'atomic explosions'.[11]

Morale was low in the Tube Alloys head office. After Churchill and Roosevelt had signed the Quebec Agreement, the normally effervescent Wallace Akers had been 'feeling a bit depressed'[12] and alarmed that progress was so slow. The resumption of cooperation was, he knew, 'very much at the discretion of the Americans', and signs were not encouraging.[13] He probably felt no better when, under pressure from Churchill, he was obliged to stand down from his role in the US after Bush and Groves objected that he was irredeemably tainted by his connections to ICI.[14] Akers's place was taken reluctantly by the Americans' favourite British nuclear scientist, James Chadwick.

There had been a surge of optimism in the office after the Quebec Agreement was signed, though it was fading now that the British scientists were finding it hard to work closely with their American counterparts. Groves talked with Chadwick, Oliphant and their colleagues with disarming frankness, but it was plain that the General wanted to involve only the few British scientists who could be useful to him. He had no intention of sharing information on the Manhattan Project with the British as openly as Churchill had expected.[15] The loopholes in the agreement were starting to gape, and Groves had no compunction about slipping through every one of them.[16]

Bohr brought encouraging news to his new Tube Alloys colleagues. Two years before, in August 1941, he had been visited in Copenhagen by his friend Werner Heisenberg, then working behind Nazi lines, who brought a sketch of a nuclear device that Bohr believed the Germans were working on. Bohr was unsure what the device was but now believed, according to the later recollections of Michael Perrin, deputy director of Tube Alloys, 'that the Germans had concluded that the Bomb project was impracticable'.[17] This chimed with the findings spelled out in intelligence reports, though it was still conceivable that Hitler's scientists were working on the Bomb and had somehow managed to keep it secret.

When Bohr toured the project's research centres in England, he felt the full extent of the disillusionment he had heard about in Whitehall. British scientists bent his ear for hours about how the government had wasted the lead that the MAUD committee had established, leaving them with a role in the construction of the weapon so feeble that it bordered on the humiliating. In the United States, Blackett was using his influence among British scientists there to try to persuade them not to join the American project, and was now arguing that trying to use a nuclear chain reaction to make a bomb was doomed to failure.[18] Of the scientists who looked on the bright side, Chadwick was the most upbeat. As usual, he agreed with the line taken by Churchill – it was simply not feasible for Britain to build a British version of the weapon during the war, so he believed that he and his colleagues were fortunate to be involved in the Manhattan Project, as the knowledge they would gain would enable them to build the weapon when the war was over. Neither Chadwick nor any of his fellow scientists had been told that the original purpose of the project – to beat Hitler to the Bomb – was now almost defunct. They would soon find themselves working on nuclear weapons for an entirely different, and undeclared, political purpose.[19] It was a classic example of what is now known as 'mission creep'.

It was not sensible to keep Bohr confined to the Tube Alloys head office – he wanted to be at the heart of the project and to make his voice heard where it would count. Groves wanted him in the US without delay and wholly on the American pay roll, though without full access to all the project's secrets.[20] Bohr 'would not allow himself to be drawn in any way into the American orbit', he promised Anderson, adding that he wanted to help ensure that the association between the British and Americans on the Bomb project was 'a real partnership involving full and reciprocal sharing of scientific and technical knowledge'. This was just what Anderson wanted to hear, and he quickly arranged for Bohr to travel to the US to meet the leaders of the Manhattan Project and to visit its nerve centre, at a secret location known as 'Site Y'.[21] It would prove to be an enlightened strategy.

On 29 November, Bohr sailed for America, under a smokescreen helpfully created by the government, which told the press that he was to be involved in 'post-war planning of international scientific cooperation'. At the dockside in New York a week later, British security officials handed Bohr and his son over to a posse of FBI agents, who issued them with official papers and pseudonyms, Nicholas and Jim Baker, before driving them off in a cab to a nearby hotel. The agents thought they had done an immaculate job of preserving the secrecy of the visit, until one of them noticed on the side of an item of luggage, stencilled in large black letters: NIELS BOHR.[22] The two Danes were allocated a relay of armed detectives, who shadowed them during their travels across America, and even slept outside their bedroom door.[23]

In Washington DC, Bohr visited Groves in his office on the fifth floor of the new War Department Building. In all meetings like this, the General behaved like the quintessential no-nonsense CEO, underscoring his determination to deliver his agreed mission: 'to produce a practical military weapon in the form of a bomb in which energy is released by a fast neutron chain reaction in one or more of the materials known to show nuclear fission'.[24] A delivery date for the weapon had not yet been agreed, but the General will have stressed the project's great urgency, the absolute priority of secrecy and that he – and only he – was in charge of it. He had no truck with the traditions of openness practised by those 'crackpot' scientists at Site Y and did not even seem to accept that he reported to a civilian Commander-in-Chief. He told Chadwick a few months later that 'the President had no powers and no authority to give away military secrets'.[25]

It speaks well of Groves that he was not put off by Bohr's philosophical mumblings, but took them in his stride. The General trusted his scientists' overwhelming recommendation – no doubt forcefully articulated by Oppenheimer – that Bohr would be a huge asset to the Manhattan Project and would be welcome on the staff at Site Y. Apparently not wanting to be shackled, Bohr declined Groves's offer and continued with his freewheeling consultancy with Tube Alloys.[26] The Americans were under the impression that he would ideally like to be based at the Institute for Advanced Study in Princeton, a sanctuary of academic enquiry where he could work with Einstein and other scholars, contribute to the Bomb project and be within convenient travelling distance of Washington.[27] Thrilled at the prospect of recruiting another illustrious member, the Institute's director Frank Aydelotte arranged for the Rockefeller Foundation to fund a temporary post for their visitor, and had a generous offer on the table for him when he arrived in Princeton four days before Christmas. Disappointingly for Aydelotte, the deal turned sour a day later.

Bohr was looking forward to meeting his old friend Einstein. Separated by Hitler for over a decade, they settled down for a talk over a cup of tea in the common room of the Institute, overlooking the lawn stretching out towards the distant woods. Minutes before, Aydelotte had offered to employ Bohr, perhaps hoping that he and Einstein might collaborate for the first time. But it was not to be. In the crowded common room, Einstein greeted Bohr and quickly began to bombard him with criticisms of the Manhattan Project, as if it were no more secret than the front pages of the day's newspapers. Einstein regurgitated the views of his friend Szilárd – the 'American army was making a frightful mess of the uranium work' and it was a thoroughly good thing that Bohr had come to the United States as he would surely 'be able to put this right'.[28] According to the account of this encounter given by Akers a few weeks later, this was a 'devastating experience' for the well-intentioned Dane, who knew that if he became associated with such indiscreet talk about the Bomb, the authorities would not allow him anywhere near the Manhattan Project. Bohr now knew that joining the Institute was 'quite impossible'.

Two days after Christmas, Bohr and his son began their journey to the Manhattan Project's headquarters at the mysterious Site Y.[29] They took the train to Chicago, where they met Groves and his science adviser Richard Tolman, who accompanied them on the sleeper. En route, the General preached for hours to the Great Dane about the crucial importance of not disclosing even a morsel of classified information and of saying nothing about the supposed German Bomb programme. The journey was a trial for the poor General, who knew he had to treat his guest with the softest of kid gloves. As the vast, empty landscapes of the Midwest spooled past, Bohr distractedly looked out of the train window, mumbling as if he had a mouthful of marbles, while Groves leaned over to get close enough to his guest to understand what on earth he was saying.

Late at night on New Year's Eve, their train pulled into a quiet little town in the New Mexico desert. After stepping out of their railway carriage into the biting cold, they were picked up in an army car and driven to a nondescript office in Santa Fe run by Dorothy McKibbin, 'the atomic lady', as insiders dubbed her. After they had been issued with security passes, the party was taken off on the final stage of its journey, a thirty-five-mile drive, the car making its way along a single-access road zigzagging up a desert mountainside they could scarcely see, the air thinning and cooling by the minute. Niels Bohr was about to see for himself a project he had dismissed on his previous visit to Princeton four years before as 'impracticable'.[30]

At Site Y, Oppenheimer and his colleagues greeted Bohr like a monarch returning from exile. Here was a scientific hero who had been determined to stay in his own country for the entire war until the Nazis' actions forced him to leave. Now, he was back with his colleagues to work with them, listen to their doubts about the project and buoy their spirits. Within five minutes, he was talking with Oppenheimer, spilling the beans about the Anglo-American project he had heard about in London and Washington.

By the end of the next day, Bohr had seen that this was a laboratory unlike any other in the history of science, surrounded by plunging canyons, in a wilderness planted only with sagebrush and piñon pines, and strewn with jutting formations of sandstone. It was like the setting of a John Ford Western, the kind of place that brings home the immensity of the entire North American continent. The site, which had 'opened' nine months before, still had a makeshift feel, a combination of a temporary army base, a jerry-built workplace for experimenters and theoreticians, and a mountain resort.[31] About 3,500 people were working on 'The Hill', behind two barbed-wire fences, where Groves was assembling the most intense concentration of top-class physicists the world had ever seen. Normally it was easier to herd kangaroos than to persuade such people to follow executive orders en masse. But here Oppenheimer had them all lined up and ready to work for as long as it took to solve the problem of building what they euphemistically referred to as 'the gadget'.

The technical facilities on The Hill were beyond munificent. Scientists expected virtually any item they ordered to be delivered within days from universities, industrial firms and the military, which jumped at the arrival of every order from the site. One group of researchers, later claiming that they were too busy to have their hair cut, tested the system by ordering a barber's chair, which arrived with minimal delay.[32] The price paid for access to this extraordinary largesse included unprecedented secrecy, a loss of some personal freedom and the imposition of having virtually all mail vetted by the military censors, though this too was done in secret. Despite the irritations, the scientific climate on the site was more than congenial for its physicists, and for many of them their time there was a highlight of their professional lives. This was made possible mainly by the tactful and effective leadership of Oppenheimer, who seemed to be omnipresent on the site – a tall, slim, flat-footed chain-smoker in a sweat-stained pork-pie hat.

As Oppenheimer explained to Bohr, the laboratory – near the town of Los Alamos – was the nerve centre of the Manhattan Project but only a small part of a network of sites. About 1,250 miles east of the site was a huge complex at Oak Ridge in Tennessee, a town that had been turned over almost entirely to the production of fissile materials, mainly plutonium and ^{235}U. Its population would ultimately rise from three thousand to seventy-five thousand, almost all of them working at – or associated with – one of the town's facilities, which eventually consumed about a seventh of the electricity generated in the whole of the United States. Among the four facilities then under construction in Oak Ridge was a gaseous diffusion plant of the type favoured by the British, a vast U-shaped facility with arms about half a mile long. A few weeks later, Bohr and his son paid a short visit to this site and saw, as Aage later wrote, 'almost unbelievable dimensions . . . like a glimpse into a new age'. They saw here that much of the heavy lifting on the project was being done by the chemists and engineers who laboured to produce every speck of fissile material so that the physicists at Site Y could carry out experiments on them.

Almost as large and impressive as Oak Ridge was another facility, a thousand miles from Los Alamos, on the Columbia River at Hanford, in the north-western state of Washington. This site was devoted entirely to the production of chemicals containing weapons-grade plutonium and was enveloped in even greater secrecy, out of bounds to the Bohrs and to every other member of the British mission except Chadwick.[33]

Bohr – now known by his pseudonym, Nicholas Baker – told Oppenheimer that he was more shaken by the vastness of the Manhattan Project than by anything since Rutherford's discovery of the atomic nucleus, thirty-two years before.[34] Nuclear science had previously been the province of physicists and chemists working in a few laboratories with the modest budgets needed to fund desktop experiments. Now, the drive to build nuclear weapons was well on its way to becoming the fourth-largest industry in America. Edward Teller, one of the physicists who had long thought that the Bomb project was viable, was looking forward to telling Bohr 'I told you so,' but before he could open his mouth the Dane butted in: 'You see, I told you it couldn't be done without turning the whole country into a factory. You have done just that.'[35]

Bohr took an interest in the technical work the scientists were doing and he contributed to it, but his main role was to be a kind of father confessor to his colleagues, their 'Uncle

Nick', as they called him. The switch from curiosity-driven science to mission-driven weapons development had been difficult for many of them, including Bohr's old friend Otto Frisch, one of the first to understand nuclear fission and co-author of the memorandum that had kick-started research on the Bomb three years before. Frisch and his former Birmingham colleague Ernest Titterton were the first members of the British team to arrive, seventeen days before Bohr. They were soon joined by others, including James Chadwick, Rudi Peierls and Mark Oliphant, whose reputation as a gabby critic was probably responsible for his being allowed on the site for all of one day, with only limited security clearance ('no access').[36] Together with their American colleagues, the British scientists were grappling with the challenge of projecting two lumps of fissile material towards each other at high speed to form a critical mass – as Frisch and Peierls had envisaged – after the device had been released by an aircraft. The entire project had turned out to be much more complex and difficult than almost anyone had foreseen. For some of the scientists, the excitement of the challenge eclipsed the horror of the potential consequences of their labours – the ultimate purpose of 'the gadget' was to atomise thousands of human beings in a matter of seconds.

As Oppenheimer later recalled, 'Bohr at Los Alamos was marvellous . . . he made the enterprise, which often looked so macabre, seem hopeful.'[37] The first serious question Bohr put to Oppenheimer about the Bomb was 'Is it big enough?' Was it so destructive an explosive that no sane leader would ever be able to use it, for fear that the enemy would retaliate in kind? Oppenheimer may well have told Bohr that even if the fission weapons they were building were not 'big enough', then another type of nuclear bomb that his colleague Edward Teller was already contemplating would indeed fit the bill. Teller was thinking about thermonuclear weapons, which use the heat generated by an ordinary fission bomb to release energy by fusing low-mass nuclei. Such weapons, later to be exemplified by the hydrogen bomb, would be far more powerful than any explosive Oppenheimer and his team were then hoping to build.

Bohr's optimism was founded in his philosophical approach to the Bomb. It was based on his favourite intellectual idea and tool, the principle of complementarity, which he first set out in 1927 as a way of looking at quantum physics, though he subsequently applied it outside science. He appears never to have published a definition of the general principle, though he made its content reasonably clear – roughly speaking, it says that every intellectual challenge should be viewed in terms of at least one pair of complementary perspectives, neither of which has a monopoly on truth. So, for example, when considering how institutions should develop, it could be argued that the only thing that counts is the preservation of tradition or, alternatively, that fostering innovation is all that matters. Bohr believed that in this and every other case neither extreme is ever wholly correct; rather, both are needed to explore the truth. For him, truth could be glimpsed only stereoscopically, never in perfect focus.

Bohr used this type of reasoning as he thought about the potential impact of nuclear weapons – they could be seen as a terrible threat to humanity, but also as a boon, which is why he wanted the Bomb to be as big as possible. He also wanted the Allies to be as open as practically feasible about this new military development: Britain and America should inform Russia of the Manhattan Project in order to help achieve a higher level of trust, he believed, as this would help to avoid a post-war conflict as well as a nuclear-arms race.[38]

During Bohr's stay, he made a huge impact on Oppenheimer's thinking about the weapons he was developing and the effects they might have on global politics after the war. After Oppenheimer and his colleagues had talked with Bohr, it seemed increasingly implausible that Hitler would be able to construct the Bomb before the Allies. That, however, certainly did not mean that work on the Bomb at Site Y would stop – as Oppenheimer's assistant David Hawkins later recalled, 'We were committed to building the bomb regardless of German progress.'[39] Having recruited the scientists to construct it, only Roosevelt and Churchill would decide how it would be used, as Bohr knew. Yet he had seen no sign that either leader had given deep thought to the significance of the weapon they would soon have at their disposal. This is why he took it upon himself to try to bring some of the same leadership to the new nuclear politics as he had given to the community of physicists twenty years before, during the birth of quantum mechanics, though he was ill equipped for the task.

He had been preparing the ground for his interventions even before he sailed for the United States. Through Anderson, he had secured an appointment with the British Ambassador in Washington, Lord Halifax, and he also arranged to meet his old acquaintance Felix Frankfurter, a Supreme Court Justice and a friend of Roosevelt.[40] At his second meeting with Bohr, at the Court Building on 15 February 1944, Frankfurter disclosed that he knew about the Bomb project, opening the way for Bohr to describe the diplomatic opportunities that he believed the Bomb offered, although he tiptoed around the military secrets. 'Let us hope that this will be a memorable day,' Frankfurter said when they parted, hinting that he would raise the matter discreetly with the President.[41]

In London, Anderson took the plunge and put Bohr's ideas on nuclear strategy to Churchill, not mentioning their provenance.[42] The Prime Minister's response was predictably terse, firm and dismissive: when Anderson suggested 'collaboration' with the Russians, Churchill circled the word and wrote in the margin 'on no account'.

Although fast becoming a close friend of the Great Dane, Anderson chose not to tell him about the rebuff, perhaps because it would almost certainly have taken the wind out of his sails. So Bohr continued to advance his case with impressive vigour, oblivious of the formidable obstacle now ahead of him. During the early Washington spring, he marshalled his thinking in a long report for Anderson with the title 'Confidential comments on the project exploiting the latest discoveries of atomic physics for industry and warfare'. Although the document begged for the services of a good editor, his circumlocutions contained a powerfully original argument for thinking about the Manhattan Project as a global opportunity rather than a threat. In the final section, Bohr stressed the long-term importance of his cause:

> Such an initiative, aiming at forestalling a fateful competition about the formidable weapon, need in no way impede the importance of the project for the immediate military objectives, but should serve to uproot any cause for distrust between the powers on whose harmonious collaboration the fate of coming generations will depend.

About two weeks after Bohr sent his statement to London, he received encouraging news from Frankfurter. He had met with Roosevelt, who told him that the nuclear bomb 'worried him to death', assuring Frankfurter that he was receptive to Bohr's ideas and that 'he would welcome any suggestion to this purpose from the Prime Minister'. Having spoken with Frankfurter, Bohr understood that he was to act as an emissary of Roosevelt on the need to look again at nuclear policy, so the Great Dane pressed for an urgent meeting with Churchill, with support from his colleagues in Whitehall. Soon, Bohr was packing his bags, preparing to leave the cherry blossom and creature comforts of Washington DC for another stay in dreary, war-torn London.

Bohr returned to find the door to 10 Downing Street firmly closed against him. Anderson had approached Churchill again in late April, having heard that Roosevelt was 'giving serious thought' to the possibilities of international arms control and 'would not be averse' to discussing it with him.[43] Anderson had even drafted a message to the President, concluding suggestively that the matter 'seems to me to require deep thought'. Once again, Sir John was rebuffed – 'I do not think that any such telegram is necessary,' Churchill replied by return. Anderson said nothing about this to Bohr, who was at a loss to understand why he was not being allowed to see the Prime Minister. The Dane was now more confident of his case: he was now sure that the Soviets were also working on the Bomb, so it was pointless to try to keep it secret from them. The evidence came from his imaginative interpretation of comments in a warm letter he received from Peter Kapitza, formerly one of Rutherford's favourite 'boys', now working in his Soviet homeland.[44] Bohr showed the letter to the British security services and, with their agreement, sent a non-committal reply that would soon get him into deep trouble.[45]

Behind the scenes, Anderson lobbied hard for Bohr to be given a chance to make his case in Downing Street, eliciting a testimonial from Churchill's friend Jan Smuts, the Prime Minister of South Africa, who likened Bohr to 'Shakespeare or Napoleon – someone who is changing the history of the world'.[46] Normally, this was just the kind of man Churchill liked to get to know.

Anderson's persistence paid off in early May, when the Prime Minister finally agreed to the meeting. Apparently, everyone involved except Bohr believed it was a disaster waiting to happen, Anderson and Lindemann doing their best to lower the risk of failure. R. V. Jones sat down with Bohr to agree a précis of his argument, which Bohr undertook to learn by heart.[47] The President of the Royal Society, Sir Henry Dale, was conscripted to write to Churchill to commend the importance of what his guest was likely to say: 'It is my serious belief that it may be in your power, even in the next six months, to take decisions which will determine the future course of human history.'[48] Dale produced two pages of verbose pleading that made the serious error of drawing attention to Bohr's attempts to win the ear of Roosevelt in Washington – this kind of ultra-high-level political dabbling was certain to put Churchill's nose out of joint. Dale's letter was 'not a happy effort', Anderson sighed, but Lindemann agreed to pass it to the Prime Minister, with a crisp reminder of the meeting's purpose.[49]

Churchill's secretary finally scheduled the meeting for Tuesday 16 May. The auguries could hardly have been worse. A few days before, Sir Henry Dale was bracing himself for the Prime Minister's encounter with Bohr's 'inarticulate whisper' and 'mild, philosophical vagueness'.[50] Impervious to the others' pessimism, Bohr showed no signs of wavering. An admirer of Churchill's vision and courage, he had been heartened by the head of the Secret Intelligence Service, Stewart Menzies, who told him how fortunate Britain was to have a Prime Minister with the imagination to grasp the long-term implications of possessing nuclear weapons.[51]

APRIL TO SEPTEMBER 1944

The Bulldog meets the Great Dane

> 'It is said that the Prime Minister has a great regard for scientists and employs [them]. I am very glad to hear it because there could be no greater antithesis than between the brilliant mind of the Prime Minister and the scientific mind.'
>
> LORD STRABOLGI, House of Lords debate, 29 July 1942[1]

'I'm through,' Churchill sighed to Lord Beaverbrook in the spring of 1944.[2] Almost half a decade of hyperactivity and relentless pressure had taken their toll on the Prime Minister, now looking every one of his sixty-nine years. A serious illness in late 1943 had left him tired, yawning in meetings, bereft of his usual drive, feuding acrimoniously with his ministers and Chiefs of Staff, some of them teetering on the edge of resignation.[3] His decline was painful to behold in some fumbling Commons performances and lacklustre radio broadcasts.[4] In late April, at the end of unpleasant meetings, one of his most senior civil servants thought the Prime Minister was breaking down and doubted if he could continue.[5]

Though victory over Hitler was now all but assured, Churchill was worried about the war's endgame. 'If Russian barbarism overlaid the culture and independence of the ancient states of Europe', he had written in October 1942, 'it would be a measureless disaster.'[6] Yet Roosevelt, less fearful of Soviet intentions, was striving to overcome their insularity and suspiciousness, and to draw them into a stable post-war community. The first meeting of the three leaders, at the Teheran Conference late in 1943, had been a turning point for the Alliance – Churchill was shocked that the President treated him on much the same terms as Stalin.[7] The Soviet leader had played his new status for all it was worth, as the American General Marshall observed: '[Stalin] was turning his hose on Churchill all the time . . . [Roosevelt] used to take a little delight in embarrassing Churchill.' One consolation for the Prime Minister, however, was that the President had never shown any interest in sharing with their Soviet ally the secret of the nuclear bomb.

At the Teheran Conference, Churchill had finally agreed, in sufferance, to the opening of a second front in northern France. He had no choice, as Roosevelt and Stalin had run out of patience with his duplicitous attempts to string them along with reasons to postpone the operation.[8] At the conference table, it was painfully clear that Britain was no longer an equal partner with the Soviets, with their huge army and Stalin's willingness to suffer casualties, and the Americans, with their colossal firepower. It was probably the anguish of seeing the power of Britain and its Empire diminishing by the month that led him soon afterwards to contract pneumonia and to have two minor heart attacks. His doctor had even feared for his life.[9]

Two weeks before the scheduled date of the landings, 6 June, Churchill was under huge pressure, with a workload as heavy as it had ever been during the war, and jumpy at even the thought that the operation might be a bloodbath on the scale of Gallipoli.[10] Most worrisome for Churchill was Roosevelt's apparent indifference to the possibility that the Soviets would dominate much of Europe after the war – 'I don't care two hoots about Poland,' the President had commented. 'Wake me up when we talk about Germany.'[11]

It was at this tense juncture that Churchill was due to meet Bohr, the first time the Prime Minister had met one of the Tube

Alloys nuclear scientists. The meeting was scheduled for the day after the final conference in London on the planning for D-Day.

Churchill apparently had no recollection of his first acquaintance with Bohr's science at Chartwell eighteen years before, when his excitement over the planetary model of the atom distracted him from his forthcoming Budget. Apart from Lindemann, Churchill resented the presence in government circles of scientists who had elbowed their way into Whitehall and claimed to offer politicians wisdom more profound than common sense could supply. Bohr stood a chance of getting his message across only if the Prime Minister had the patience to listen to his ramblings for long enough to sift the gems from the clay. The timing of the meeting, due to take place in Downing Street, was not propitious: it was to begin at three in the afternoon, when the Prime Minister normally liked to begin his daily nap.

The encounter got off to a bad start and then went rapidly downhill.[12] Bohr and Lindemann arrived early and were sitting together when Churchill walked in, having just read Sir Henry Dale's windy appeal 'on behalf of the scientific community' for him to meet the guest now in front of him. The Prime Minister appears to have sensed an ambush by the scientific elite and was probably indignant to read that the Dane had been meddling in Anglo-American politics. Ignoring Bohr, Churchill laid into Lindemann, accusing him of arranging the meeting only to 'reproach me for the Quebec Agreement'.

The Prime Minister's resentment over opposition to the agreement burst like a boil and infected the entire meeting, which degenerated into a long, private altercation between Churchill and the Prof. Bohr would have struggled to make himself understood even if the other two men were silent, so he was fighting a losing battle. Having found nothing of value in the tortured words that Bohr managed to utter, Churchill lost patience. 'I cannot see what you are worrying about,' he said – all they were talking about was a new and bigger bomb that 'made no difference to the principles of war'. The prospect of post-war nuclear proliferation that so frightened Bohr obviously meant nothing to the Prime Minister, who assured his visitor that there were no long-term problems with the weapons that could not be 'amicably settled between myself and my friend President Roosevelt'. For Churchill, policy on nuclear weapons was a private matter; nuclear scientists should mind their own business.

As he left, Bohr offered to set out his views in a letter, drawing a sharp reply from Churchill: 'It will be an honour for me to receive a letter from you, but not about politics!'[13] A week later, the Prime Minister received via Lindemann a long-winded, fawning letter from his unwelcome visitor. In it, Bohr said nothing new but underlined his regret if he had caused any offence.[14] Churchill appears not to have replied. Unimpressed with Bohr's arguments, he had no time for him personally – four months later, he wrote to Lindemann:[15] 'I did not like the man when you showed him to me, with his hair all over his head . . .'

Lindemann's disappointment with the Quebec Agreement was still festering. Nine days after the unfortunate meeting with Bohr, Churchill sent the Prof a note in a sealed envelope, defending the decision to sign the agreement, with arguments based on a defensive pragmatism that sometimes drifted close to sentimentality:[16]

> I am absolutely sure we cannot get any better terms by ourselves than are set forth in my secret Agreement with the President. It may be that in [later] years this may be judged to have been too confiding on our part. Only those who know the circumstances and moods

prevailing beneath the Presidential level will be able to understand why I have made this Agreement. There is nothing more to do now but to carry on with it, and give the utmost possible aid. Our associations with the United States must be permanent and I have no fear that they will maltreat us or cheat us . . . The great thing is to get on with the job and keep it absolutely as secret as we can.

Churchill resented what he saw as the tendency of some scientists to regard a secret as information that could be disclosed to only one person at a time. Later, when he dismissed Tizard's appeal to be formally allowed to know about the Manhattan Project, Churchill told his Chief of Staff: 'For every one of those scientists who is informed, there is a little group around him who also hear the news.'[17]

When Churchill wrote his defence of the Quebec Agreement, British and American officials were discussing the supplies of uranium ore they would need after the war in order to produce nuclear weapons. The Americans were pressing hard for British cooperation, as the amounts of these minerals in the US were small, and the British had ready access to huge amounts of them in territories of its Commonwealth and Empire. Vannevar Bush, James Conant and General Groves knew the strategic importance to America of making a deal and worked for months to secure it, talking over the details with Sir John Anderson, who kept Churchill well briefed.[18] Here was an opportunity for the strong negotiator to play hardball.

But these talks were a sideshow compared with the colossal enterprise of planning the D-Day landings. On the night of 5 June, Churchill remarked to his wife before they went to their beds, 'Do you realise that by the time you wake up in the morning, twenty thousand men may have been killed?'[19] The operation went far better than he dared hope, and as well as Roosevelt expected: under the command of the American

General Eisenhower, a hundred thousand British, American and Canadian troops landed on the fire-swept shores of Normandy. Three thousand were killed on the day, but this was not another Gallipoli. By 10 June, four hundred thousand men were ashore, smashing Hitler's Atlantic fortress.

After a short visit to the Normandy bridgehead, Churchill returned to Downing Street just in time to see Hitler's reply to the landings – the first German V1 flying bombs roared across the sky and fell on London (V stood for *Vergeltungswaffen* – 'weapons of revenge').[20] Intelligence reports had given the Allied governments months of warnings of these attacks and Churchill had expected them to begin in early 1944.[21] The bombardment came as a shock to civilians, however, thumping their already low morale and wounding their trust in the government. Three months later, another weapon sped into the skies over London, the supersonic V2 rocket, again terrifying people below. Yet, from a strategic point of view, these imprecise weapons were only a distraction – each one was extremely expensive, costing far more to manufacture than an aeroplane, but killed on average only a single person.[22]

Churchill was, for once, disappointed with the advice that Lindemann had given him about these German weapons.[23] The Prof had been right to insist that the experts were exaggerating when they predicted that V2s would each carry a ten-ton warhead. But he had been wrong to advise that reports of the Nazis' development of long-range rockets were a hoax to distract attention from their manufacture of flying bombs.[24] Almost a year before, when War Office scientists tried to warn ministers of what seemed to be afoot, Lindemann had dismissed their interpretation of what appeared to be (and were) launch sites for the rockets on the Belgian and French coasts.[25]

The attacks rattled and infuriated Churchill. Encouraged by 'most secret' briefings from Lindemann, he considered retaliating with biological (anthrax) weapons, which – as scientists at Britain's Porton Down research centre pointed out – were 'infinitely easier to make' than nuclear weapons.[26] Britain ordered anthrax bombs from the United States, and a small batch of them had arrived in May 1944. Two months later, Churchill asked his Chiefs of Staff to think 'very seriously' about using poison gas, telling them that he wanted 'the matter studied in cold blood by sensible people and not by . . . psalm-singing uniformed defeatists'.[27] He made his own views clear:

It is absurd to consider morality on this topic . . . in the last war the bombing of open cities was regarded as forbidden. Now everybody does it as a matter of course. It is simply a question of fashion changing as she does between long and short skirts of women.

The Chiefs of Staff opposed the use of poison gas, fearing that the Germans might retaliate in kind.[28] 'Not at all convinced' by this response, Churchill persisted. The military prepared to attack German cities with anthrax bombs, but nothing came of these plans because sufficient numbers of the weapons – ordered from the United States – were not ready in time.[29]

The new weapons of most interest to Churchill in the final stages of the war were nuclear. On 13 June 1944, he and the recently re-elected Roosevelt signed a 'Declaration of Trust' to gain 'combined control of ores and supplies' needed to produce fissile material.[30] Anderson, carrying out what he believed to be Churchill's wishes, had not attempted to take advantage of the strong British position by, for example, monopolising ore in the Congo. Rather, the declaration permitted the United States to take the lion's share of the ores after the war – the British took a small amount, and even that was resented by the

American negotiators. On the day Churchill signed the declaration, he asked Anderson to explain how it differed from the agreement struck with Roosevelt eight months before in Quebec.[31] The Prime Minister had not understood the document he had signed.

By the summer of 1944, Churchill knew that the world would emerge from the war with only two superpowers – the Soviet Union and the United States. Great Britain now had no choice but to play second fiddle to the Americans. His main aim in the closing months of the war was to maximise the strategic advantage for his exhausted country and its dominions, and to rein in Stalin's ambitions to dominate Eastern Europe, including Poland. For him, it was a grim and depressing end to the war. Determined to maximise the dwindling asset of his reputation and charisma, he repeatedly requested meetings – preferably in Britain – with Roosevelt and Stalin, but both were unwilling to indulge him.

With no appetite for the business of rebuilding Britain after the war and other matters he regarded as small fry, Churchill packed his diary with international travel, most trips leaving him refreshed and celebrating rather more than their outcomes warranted. After a long visit to Italy, he set off for his sixth wartime meeting with Roosevelt, to be held in September, once again in Quebec, this time in the afterglow of the Allies' liberation of France and Belgium. Since they had last met in Canada, both leaders had aged, especially Roosevelt, who looked gaunt and frail – 'You could put your fist between his neck and his collar,' Churchill's doctor noted.[32] The meeting was a disappointment and led to few agreements of lasting consequence.

A few weeks before he set off for Quebec, Churchill – unusually in his dealings with Tube Alloys – took the initiative, requesting a brief on progress from Lindemann. 'The extraordinary American ideas on security' made it impossible to give a valid judgement, the Prof replied, doubting whether the Bomb would be ready in time to be used in the war.[33] On the second day of the meeting, Lindemann wrote to ask Churchill, rather pathetically, if he could discover 'where we stand' on post-war collaboration with the Americans, as no one in the London Tube Alloys office had any idea.[34]

The Bomb was not on the leaders' agenda in Quebec, but they discussed it afterwards in Roosevelt's Hyde Park home, where they talked privately with only the sickly Harry Hopkins – now out of favour with the public and the White House – hovering in the background. The conclusions of the Prime Minister and the President emerged in a brief aide-mémoire, edited by Churchill.[35] It looks forward to the end of the war and notes that nuclear weapons 'might perhaps, after mature consideration [the hesitant wording was Churchill's], be used against the Japanese, who should be warned that this bombardment will continue until they surrender'. After Japan had been defeated, 'full collaboration' for military and commercial purposes would continue between Britain and America unless they agree otherwise. There was no mention of the veto Churchill had given Roosevelt in Quebec on any nuclear-power industry Britain might want to develop after the war.

The remainder of the document concerned Niels Bohr and his thinking about nuclear weapons. A month earlier he had secured a private meeting in the White House with the President, who had been his usual affable self. Roosevelt listened attentively for almost an hour and gave the impression that he was sympathetic to ideas about international control of nuclear weapons and, in particular, bringing Russia in on the secret. These sentiments flatly contradicted the view Roosevelt had held all along – that the 'secret' of the bomb should be kept from the Soviets. He and Churchill maintained this line in the aide-mémoire:

> The suggestion that the world should be informed regarding Tube Alloys, with a view to an international agreement regarding its control and use, is not accepted. The matter should continue to be regarded as of the utmost secrecy . . .

Roosevelt compounded his duplicity by signing up to the Churchillian line that Bohr was potentially dangerous and should be watched carefully:

> Enquiries should be made regarding the activities of Professor Bohr and steps taken to ensure that he is responsible for no leakage of information, particularly to the Russians.

On the following evening, the President accompanied the Prime Minister to the nearby railway station in Poughkeepsie, their car surrounded by Cadillacs full of security guards and special agents. Churchill, preoccupied with the threat he believed Bohr posed, was still fulminating. In a letter the next day to Lindemann, recently returned from Washington, Churchill condemned the Danish physicist as a publicity-seeker, complaining that he had leaked information to Frankfurter and that he was in close correspondence about the Bomb with a professorial friend in Russia (this was Peter Kapitza).[36] 'What is this all about?' seethed Churchill, who wrote that he and Roosevelt were 'much worried': 'It seems to me that Bohr ought to be confined or at any rate made to see that he is very near the edge of mortal crimes.'

Roosevelt met soon afterwards with his Chief of Staff Admiral William Leahy, Vannevar Bush and Frederick Lindemann, who sent Churchill a summary of the meeting.[37] The President underlined what his guest wanted to hear: that Britain and America should share all their nuclear discoveries, and should collaborate not only during the war but afterwards. It was a matter for later governments to decide whether they wanted to continue with this arrangement, Roosevelt said – the only thing he could see that would interrupt it 'would be if he and the Prime Minister, Bush and [Lindemann] were all killed in a railway accident since we all saw eye to eye'. Bush, Lindemann noted, 'said nothing'. The outcome was that Bush would 'check up on Bohr', though he thought there was no reason to worry.

Bohr's supporters and friends on Capitol Hill and in Whitehall sprang to his defence, and the case against him was quickly dropped. Lindemann was among his defenders, replying quickly to Churchill's letter, putting the record straight and pointing out that although the Dane had 'some rather woolly ideas' about how nuclear weapons could induce countries that have them to live in peace and confidence, he was someone of unimpeachable integrity and loyalty.[38] A surprisingly curt comment by Lindemann in this letter shows how little confidence he had that Churchill remembered much about the early public discussions of nuclear energy, a debate the Prime Minister had been among the first to stir: 'I do not know whether you realise that the possibilities of a super weapon . . . have been publicly discussed for at least six or seven years.'

Lindemann stayed in the United States for several weeks, touring sites where the American military was carrying out research and development. His hosts transported him around the country in style and even put an aircraft at his disposal.[39] During this trip he visited the sites of the Manhattan Project and, perhaps, appreciated for the first time how wrong he had been to imagine that such an enormous venture could have been carried out in the UK. He had hardly been alone in making this error of judgement, which had now been proved ridiculous: another who had initially shared this delusion was the leader of the British scientists now working at a frantic pace on the Manhattan Project, James Chadwick.

FEBRUARY 1944 TO JULY 1945

Chadwick witnesses the first nuclear explosion

'[For much of my present duties] I am fitted neither by temperament, training nor ability.'

SIR JAMES CHADWICK writing to Robert Oppenheimer, 25 April 1945[1]

Rutherford would have been amused. The first leading nuclear physicist to be obliged to morph into a back-room diplomat was his deputy 'Jimmy Chadwick', known for his drive, terseness and intolerance of foolishness, not for his skills as a tactful negotiator. He was also a strong patriot, so when duty called on him to be the Churchill government's chief representative on the Manhattan Project, he accepted without demur. Although first and foremost a nuclear experimentalist, he had not done much science in the past year – he spent most of his time embroiled in negotiations with dilatory politicians in Whitehall, and with General Groves, who displayed his chauvinism and his indifference to British interests with some pride, like epaulettes. The contrast with the scientific director of the project was painful to behold – whereas Chadwick always looked harassed, everyone could see that Oppenheimer was having the time of his life.

A welcome bonus of the Chadwicks' move to North America was that they would be reunited with their daughters, who had been living in Nova Scotia for three years. In January 1944, when the family was together briefly in Halifax, Aileen was surprised to see that her daughters had matured from demure young girls to sparky teenagers, wearing make-up and speaking in a mid-Atlantic accent.

Having left their daughters to study at a school in Nova Scotia for another three months, the Chadwicks travelled to Site Y, where Groves allocated them a small wooden cottage next to the Oppenheimers. The Chadwicks' home had the luxuries of two bedrooms, an iron bathtub, indoor plumbing and a roomy living room with a handsome stone fireplace. All this made them the envy of Chadwick's junior colleagues, who had to make do with the spartan accommodation provided in the dorms, with their shared bathrooms and hard beds. Yet life on The Hill was too primitive for Mrs Chadwick's genteel tastes. The absence of the tedious nightly blackouts, the abundance of food, the views across to the snow-capped peaks of the Sangre de Cristo mountains, did not compensate for the temporary loss of their roomy house in Liverpool, the company of friends, the studied calm of the BBC's radio announcers. Uncomfortable in the thin mountain air and often feeling under the weather, she chain-smoked her way through the long days and made no secret of her homesickness or of her dislike of life in America. At one of her afternoon tea parties, she complained of 'the primitive nature of life in the United States', a remark some of her guests found less than endearing.[2]

It was no wonder she felt so lonely: during the year she and her family lived at Site Y, most of the time her husband was away, visiting other Manhattan Project sites and talking with officials in Washington. His longest stay on The Hill was the few weeks he spent there recovering from a debilitating attack of shingles. Permanently tired, when he had a free evening at home all he wanted to do was sit in his armchair with his nose in a book.

Chadwick's greatest diplomatic success was to persuade his American colleagues to support the development of a new nuclear facility in Canada. The cards had been stacked against him. By late 1943, the morale of the Montreal project was rock bottom – mainly owing to the chaotic leadership of Hans von Halban – and the Americans were disinclined to support a satellite project not under their direct control. Yet Chadwick won the day, arranging for the Canadian Atomic Energy project to be led by his former Cavendish colleague John Cockcroft, who had one of the safest pairs of hands in British science.[3] At the picturesque site of Chalk River, just over a hundred miles north-west of Ottawa, Cockcroft oversaw the construction of a prototype reactor, moderated by heavy water, giving Britain and Canada experience in developing a large nuclear facility that was certain to prove crucial after the war. The heavy water and uranium were both supplied by the Americans. When Chadwick scored a victory like this, he seemed to relax, a thin smile appearing across his face like an acute-angled slit. On one occasion, he celebrated a success by putting the best possible spin on the Anglo-American project with a phrase he probably picked up during his Lancashire youth – 'It's all jam and kippers.'[4]

Unlike his junior colleagues, Chadwick found himself worse off financially than he had been at home in England. Although he was the highest-paid British scientist on the project (his pay cheque was 142 pounds a month),[5] much of his salary was spent on travel and hotel expenses that he had no time to reclaim. Nor could the British authorities find a simple way to reimburse him. Eventually, a senior civil servant in Washington made a plea to Whitehall for fairness: 'For us, [Chadwick] is as much of a key figure as Groves on the American side . . . He will certainly not be able to approach the task without wearing himself to death, unless he has a very real personal relief from financial worries.'[6] Fairness eventually prevailed, and the Treasury in London increased his salary by five hundred pounds, backdated to November 1943. Other parts of

Whitehall were not always so cooperative, failing to support wholeheartedly the clear vision that Britain's participation in the Manhattan Project was crucial to the country's post-war nuclear programme. In June 1944, when Anderson and Lindemann responded disappointingly to another of his requests for action, Chadwick confided to Akers that the British government was demonstrating a 'complacency, almost amounting to indifference' to the entire project.[7]

The physicists working on the construction of the Bomb made great progress until they appeared to hit a brick wall in the spring of 1944.[8] The problem was not with Frisch and Peierls's original idea of firing two lumps of ^{235}U at each other to form a critical mass. By then, it seemed virtually certain that this method would work – the physicists had only to wait for the chemists and engineers at Oak Ridge to produce enough ^{235}U. Much more serious was the problem of making a nuclear bomb from plutonium, as the scientists found out when substantial amounts of it began to arrive, also from Oak Ridge. The element turned out to have some unexpected properties, several of them inconvenient for the purpose of making a nuclear bomb. The most serious problem was that plutonium fissions quite naturally at an extraordinarily high rate – with no need for the judicious prod of a neutron – so high that the samples of the element loaded in a Frisch–Peierls 'gun' would fizzle out before there was any chance of an explosion. Nature was making life extremely difficult for Oppenheimer and his team. His response was decisive: after consulting with his experts, in July he reorganised the work of his entire lab, terminating all projects on the gun-type plutonium bomb and redeploying staff to develop a different, more complex way of using the element to build an explosive. This method used an 'implosion' to create the critical mass, by employing a blanket of ordinary explosives to crush a hollow shell of plutonium into the critical mass so quickly that it would explode instantly. Never in the history of human conflict had so many resources been invested, so quickly, to develop a weapon of such power.

One of the leaders of the implosion project was Rudi Peierls, who headed the team of theoreticians investigating the detonation- and shock-waves generated by the plutonium device. One of his colleagues was Lindemann's protégé James Tuck – for the Americans, the great English eccentric of the British mission – who did crucial work on optimising the explosion. Peierls also served as Chadwick's eyes and ears, later compiling for him regular summaries of progress on the entire project, notes that Groves also found useful.[9] Although Peierls's conscience was as well developed as anybody's on the site, he found the atmosphere at Site Y in many ways idyllic, especially after the earlier Anglo-American tensions that had done much to stymie collaboration – 'It is an enormous pleasure', he wrote, 'to be at a place where no distinction is made between members of the American and British organisations and where work is guided by the necessity to get the best answer in the shortest possible time rather than by questions of formal organisation and prestige.'[10]

Most of the credit for creating this agreeable environment was, he knew, due to Oppenheimer, who had persuaded Groves to abandon his attempt to force the scientists to work in compartments, and who understood how to get the best out of everyone in his team, including Peierls's friend Otto Frisch. Later, the American authorities praised Frisch's 'exceptional experimental skill' and 'his ability to see the relation between a laboratory experiment and its practical application to an atomic weapon'.[11] Yet he found plenty of time to play the piano, to draw caricatures of his colleagues, and to go to many of the parties on the site, including one thrown by Robert Oppenheimer (on behalf of 'Insecurity') and Oppenheimer's wife Kitty (on behalf of 'Unintelligence'), their invitation marked 'For Orgy Only'.[12] Perhaps the finest pianist on the site, he often gave evening concerts – his repertoire ranged from Bach to Shostakovich via Schubert – some of them broadcast on The Hill's radio station. Whether he liked it or not, Frisch was given hours of personal advice by Peierls's domineering but kind-hearted wife Genia. She was one of The Hill's great characters, a fortissimo diva, generous host and always the life and soul of every gathering. When the news of D-Day arrived on 6 June she climbed on to the nearest table and danced with a fetching abandon.

One of the waifs and strays that the Peierlses took under their wing was their former lodger Klaus Fuchs, an introspective theoretician whom Genia was gradually coaxing out of his shell. He called her 'Mother Peierls'.[13] Fuchs shared an office with Rudi, working with him on the mathematical physics of the imploding plutonium weapon and becoming one of the British mission's most powerful calculators. Although Fuchs was for many of his colleagues a closed book, he was popular – a willing babysitter, an enthusiast for games of chess and charades, and always ready to relax over a beer or two in the evening, though he could hold his drink.[14] He was a sought-after partner on the dance floor, having proved himself to be the nearest The Hill had to its own Fred Astaire. Like every other member of the British mission, Fuchs's movements were monitored on site, though the security officers left him alone when he ventured outside (for the American scientists, the security services did the opposite). One of the few British physicists to own a car, he went on long drives during the weekends, unhindered by any police presence.[15]

On Sundays, Oppenheimer virtually shut down the lab, urging his colleagues to have at least one day each week away from their punishing schedule. Although the Chadwicks were not the most enthusiastic participants in the week-night activities – the outdoor movies, the dances, the impromptu parties on the staircases of the dorms – they did unwind with everyone else on the lab's days off. Chadwick was fond of trout fishing in local streams and occasionally went on long hikes in the wooded mountainsides. Other colleagues took horse-riding lessons, drove into Santa Fe and read books from the library, which Szilárd had ensured was stocked with a copy of Harold Nicolson's *Public Faces* and H. G. Wells's *The World Set Free.*[16] In the end, though, many of the inhabitants of The Hill found themselves with plenty of time on their hands, which may explain the soaring birth rate on the site – one of the medical staff told Groves that the number of babies born to the scientific personnel at Site Y broke all records.[17]

It was said that identifying the members of the British mission was easy – they had Germanic accents. This was an exaggeration. Some of the American team were refugees from Europe, too, including the head of the theory division Hans Bethe and the protean Enrico Fermi, who had arrived from Chicago in September and quickly been given an entire division of his own, to tackle multi-disciplinary problems. European refugees constituted only about a third of the British mission, but they were among its most prominent members. Niels Bohr – regarded as an honorary Brit – returned to Site Y for several short stays in 1944 to do technical work on the project and to talk about the implications of the Bomb with the scientists, especially with Chadwick and with Oppenheimer, who came to think of him as a demigod ('the noblest man I ever knew').[18] After his foray into power politics, Bohr seemed unperturbed either by the treatment he and Lindemann had been given by Churchill ('He scolded us like two schoolboys!')[19] or by the duplicitous sympathy he had been shown by Roosevelt. Instead, Bohr flouted the British government's directives to steer clear of political discussions and preached the need for statesmanship and openness in order to avoid a long-term disaster in international relations.[20] His influence was growing steadily among some of the scientists on The Hill, as their political masters would soon find out.

Of all the British scientists working on the Manhattan Project, Joseph Rotblat was the most ill at ease. When he arrived in March 1944, he was billeted with the Chadwicks, sleeping in their spare bedroom until their daughters arrived from Canada. With newspapers and radio broadcasts supplying daily reports of the collapse of the Germans' war effort, it seemed increasingly implausible to him that Hitler could be matching the Allies' Bomb programme. Why, then, was the weapon being built? Chadwick listened respectfully, countering that such decisions should be left to political leaders. Over dinner one evening at the Chadwicks', in the company of General Groves, Rotblat's unease turned to disillusion when the General remarked that the real purpose of the Bomb was to subdue the Soviets.[21] Rotblat was flabbergasted. Until then, he later wrote, he had thought the Bomb was being built 'to prevent a Nazi victory'. But he was now being told that the weapon was intended for use against Britain and America's allies – 'the people who were making extreme sacrifices for that very aim'.[22]

A few months later, when it was commonly accepted on Site Y that the Germans had all but abandoned their Bomb project, the disillusioned Rotblat asked for permission to quit. When Chadwick was considering the request, the authorities disclosed that they had evidence that Rotblat had Communist associates outside the site, and Rotblat made it clear that he wanted to return to Europe to search for his missing family.[23] After Chadwick decided it was best to have his colleague out of the way, Groves quickly agreed that Rotblat could return to Britain, provided that he told no one why he had left.

Shortly before Christmas, Rotblat left The Hill to return to Liverpool. He paused to spend a few days with the Chadwicks, who by then had moved to Washington, where Aileen was much happier – the family's conveniently located home on Brandywine Street had much more room than their cottage on The Hill and was also close to Chadwick's office near the White House.[24] At Union Station, Chadwick saw Rotblat off on his journey, helping him lug a box of correspondence and research notes on to the train to New York. When Rotblat arrived, the box had disappeared. The security services had, he was convinced, been at work.

Chadwick was still responsible for recruiting to Site Y British scientists who had skills unavailable in the United States. From the spring of 1944, Oppenheimer needed no more nuclear scientists but urgently required mathematical physicists who were experts in the shock-waves that pass through the Bomb as it explodes, and in the blast it creates. Chadwick brought in Britain's two leading experts in these fields, beginning with G. I. Taylor, from the University of Cambridge. ('Anything short of kidnapping would be justified', Chadwick wrote to London.[25]) A brilliant applied mathematician and unrivalled expert in the motion of fluids, Taylor soon became a popular colleague on The Hill.[26] The second expert, from Imperial College, was Bill Penney, who had accrued years of experience of studying the impact of bombs dropped by the Luftwaffe on British cities. A genial man, Penney's mouth seemed to be fixed in a grin that he maintained even when he was talking in seminars about the number of people who would be killed by the weapons they were working flat-out to build. The American physicist Viki Weisskopf nicknamed him 'the smiling killer'.

The move to Washington made Chadwick's work no easier. A few months before, as the government in France was getting back on its feet, French nuclear scientists in Montreal were beginning to assert themselves over what would today be called their 'intellectual property rights'. The scientists wanted to be free to communicate with their colleagues at home about the progress being made on the nuclear projects – after all, some of it was based on patents that the French had filed years before. Groves was outraged by all this pettifogging talk of patents and was adamant that knowledge of the Bomb and nuclear reactors must, for the time being, remain entirely in North America. Sir John Anderson, however, sympathised with the French, thereby annoying Churchill, who told a colleague that if Frédéric Joliot-Curie had more information than he was entitled to, he should be 'forcibly but comfortably detained for some months'.[27] This was the lowest point of Chadwick's stay in North America – tossing and turning at night,[28] he spent his days wearily triangulating between the indignant scientists and the Anglo-American authorities who showed them little sympathy. The uproar ended only when the scientists were discreetly muzzled.

With the end of the Manhattan Project now in sight, Chadwick and his colleagues began to discuss Britain's post-war nuclear initiatives – if the planning were done well, then all the troubles of the past three years would soon be forgotten.[29] He, Cockcroft, Peierls, Oliphant and others met in Washington in late 1944 to get the discussion started, while General Groves grumbled privately of 'British rascality', resenting any nuclear initiative beyond American control.[30]

The British authorities gave Chadwick, desperately tired and tortured by chronic back-ache, a much-needed fillip in the 1945 New Year's honours list, when he was knighted. Somehow, he summoned the energy to begin his campaign to press the government to capitalise on the modest investment it had made in the Manhattan Project. In a masterly memo to Sir

John Anderson, Chadwick set out his thoughts on future Tube Alloys policy in seven pages of taut but conversational prose.[31] The British policy of joining the project as junior partners had been proved right, he argued, adding that relations with their American colleagues were steadily improving: 'I think we are establishing a sound foundation for a post-war understanding.' The best way forward was to continue to 'throw ourselves heart and soul into the US effort', though 'We should proceed at once to prepare for the post-war development of Tube Alloys.' It was now time for Britain politely to set aside Groves's opposition and to create 'a Government-controlled Experimental [nuclear] establishment'. By urging that Britain should support the Americans until the project's conclusion before focusing on developments at home, Chadwick knew he laid himself open to what he described as the 'superficial and short-sighted view' that he was 'neglecting or forgetting the interests of my country'. He was probably thinking here of Mark Oliphant, who was scathing about Chadwick's stewardship of British interests in the Manhattan Project. In late May, Oliphant told him that Wallace Akers was complaining that the British Tube Alloys office was 'almost impossible to run' because of Chadwick's 'attitude of distrust.'[32] In a letter shortly afterwards to A. V. Hill, Oliphant was even more vitriolic:[33]

> In the [Tube Alloys] field there will be a doleful story to tell of appeasement and of a most undignified servility, dictated by Anderson and [Lindemann] under orders from Churchill, and fostered and carried out by Chadwick and [Ambassador] Halifax in Washington. I believe we have been sold down the river as a nation.

Whitehall was, as usual, slow to respond to Chadwick's suggestions. A probable reason for this was that the most influential of Churchill's nuclear advisers doubted whether the Bomb would work: 'There's many a slip 'twixt cup and lip,' Lindemann told R. V. Jones. 'What fools the Americans will look after spending so much money.'[34] This observation would have neither surprised nor perturbed Oppenheimer – he had commented to Peierls during the Prof's brief visit to The Hill: 'That guy will never understand a thing.'[35]

Late in the afternoon of Thursday 12 April 1945, when Chadwick was working at home, he heard the news that stunned America – Roosevelt was dead. After months of ill-health, the President died suddenly, following an acute cerebral haemorrhage, at his personal retreat in Warm Springs, Georgia. Chadwick and his wife sat in their living room, leaning towards their radio, hanging on every word, shushing away their daughters who couldn't see what all the fuss was about.

The word spread like a panic in the capital. Cab drivers turned their radios up full blast and shouted out the news to incredulous passers-by, some people seeking confirmation of it from complete strangers.[36] It hardly seemed possible that the little-known Truman, who had been Vice-President for eighty-two days, was about to run the country. Chadwick may well have wondered that night what the implications would be for the Anglo-American nuclear partnership, secretly agreed sixteen months before in Quebec by Roosevelt and Churchill. If Truman wanted nothing to do with the policy, the British would soon be in deep trouble.

A common refrain that spring evening was 'What a pity he couldn't have seen Victory in Europe Day,' which came a little over three weeks later, on 8 May. The arrival of the long-expected news on The Hill triggered a noisy party, where Frisch, Peierls and Bethe cast off their European inhibitions, formed an impromptu chorus and belted out some of their favourite German student songs.[37] Their American colleagues were astonished: for them, the focus of the war had long before shifted from Europe to the Far East, as they had been reminded a few days earlier in a circulated message from the Under-Secretary of War: 'Every man-hour of work will help smash Japan and bring our fighting boys home.'[38]

Plans for a test of the first nuclear bomb, built from plutonium, were by then well under way, and a small minority of Manhattan Project scientists were worrying about how the American government might use the 'gadget'. The epicentre of concern was the Metallurgical Lab in Chicago, where work on the project had finished early. A committee chaired by its director of chemistry, the German refugee James Franck, produced in early June a sober but prolix report calling for the government to refrain from making an unannounced nuclear attack on Japan and suggesting a demonstration of the weapon on an uninhabited area.[39] One of the committee members was the irrepressible Leó Szilárd. Already appalled by the carpet-bombing of Tokyo, he was determined to voice his objections to the use of the weapon and his support for international control of nuclear bombs, assuming he was not swatted first by his nemesis General Groves.

Henry Stimson regarded the Bomb as 'a royal straight flush' for the United States in its dealings with the Soviet Union, though he wanted America to play its hand carefully.[40] He had set up and chaired a secret committee of civilians, whose brief was to advise the President on matters relating to the use of nuclear weapons and nuclear energy.[41] Stimson passed the Franck report to a subcommittee of eminent scientists, including Oppenheimer and Fermi, whose views would be respected by the protestors but would be unlikely to rock the boat. The subcommittee dismissed the central recommendation of the report, pointing out that supplies of fissile material were too short to enable the luxury of such a demonstration and – despite knowing almost nothing about the military situation in the Far East – saying that they saw 'no acceptable alternative to direct military use'. In another forum, Vannevar Bush and James Conant agreed.[42] No senior government official in Washington or London showed any interest in consulting Chadwick – or any of his colleagues – about the use of the Bomb. The indifference of the Churchill government to the views of its nuclear scientists was deeply resented by the President of the Royal Society, Sir Henry Dale, who was still ruing the abject failure of Bohr's meeting with Churchill.[43]

At no time had the American government been in any serious doubt that if the nuclear bomb were produced, it would become a legitimate weapon of war. The attacks on Tokyo had strong public support, so it was unlikely that there would be an outcry if the Bomb was dropped – the number of deaths would probably be less than a hundred thousand, the toll of the American raid on Tokyo on 9 March. The nuclear weapon would have an element of surprise, too, and might lead the Japanese to surrender before the planned invasion, certain to cost tens of thousands of lives. Szilárd had no truck with these arguments: he lobbied widely and petitioned the President to think again, but in vain. The political stage was now set for the test of the first nuclear bomb – the Trinity project – described later by Chadwick as 'the boldest and certainly the most expensive experiment in scientific history'.[44]

Chadwick was one of the few hundred civilians allowed by the American army to witness the test, on a hill about twenty miles from the planned explosion. Women had been banned. Shortly before dawn on 16 July 1945, he was sitting in the cold, pre-dawn darkness of Jornada del Muerto desert in New Mexico, listening to the countdown for what was to be, in effect, the first artificial sunrise, half an hour before nature's. This desert, a byword in America for a landscape of almost unimaginable quietness, would for a few moments be the noisiest place in the world. He was on Compañia Hill with a few dozen of the Manhattan Project's leading scientists, including Hans Bethe, Ernest Lawrence, Otto Frisch and Rudi Peierls. Every one of them knew that much could go wrong. The weather, stormy the night before, had already forced a postponement of the ignition to 5.30 a.m. Aware that the radiation blast was expected within the next few minutes, several of the scientists passed the time by applying sun lotion to their skin.

In the control centre, some miles from Chadwick and his colleagues, the nervous Groves – flanked by Bush, Conant and Oppenheimer – was pondering what he would do if the Bomb were a flop. Oppenheimer had not slept a wink and, in the final seconds, his Herculean self-confidence seemed suddenly to drain away: he held on to a post to steady himself and stared straight ahead.

Years of expectation had not prepared Chadwick for what he saw.[45] Suddenly, noiselessly, an intensely bright pinpoint of light appeared and grew quickly into a huge ball of swirling debris, bathing the surrounding hills and desert in dazzlingly bright light, as if someone had flicked a switch and turned on the sun. Chadwick watched it through a piece of welder's glass, sometimes peeping round the edges. As the light faded slightly, the cloud began to resemble a mushroom, the orange-red ball of fire connected to the ground by a short grey stem. Another mushroom cloud grew out of it shortly afterwards, like a mutant offspring. Almost two minutes after the spectacle began the first blast of sound arrived – 'sudden and sharp as if the skies had cracked', as Chadwick later wrote, 'followed by a long rumbling noise', like the sound of a convoy of wagons trundling across the hills.

The crowd on Compañia Hill was ecstatic. According to one report, the buttoned-down Chadwick could not contain his joy: when Lawrence slapped him on the back, he gave one of his trademark grunts before leaping into the air. Several others could not resist mimicking him, before stomping on the ground triumphantly. 'It worked, my God, the damned thing worked!'[46]

More than any of his fellow witnesses, Chadwick appreciated the scientific significance of the spectacle. As a shy undergraduate in Manchester thirty-four years before, he had sat at the back of a lecture theatre and watched Rutherford first present the idea of the atomic nucleus. Chadwick had later proved the existence of the neutron in a bench-top experiment a stone's throw from the centre of Cambridge. But this latest nuclear experiment had to be done in a desert miles from the nearest town and resulted in the destruction of its apparatus, leaving behind a crater 1,200 feet wide and up to six feet deep. Some of the desert sand was fused into glass – Chadwick was later given several samples as souvenirs, which he donated to the Natural History Museum in London.[47] A single one of Chadwick's neutrons had triggered the biggest explosion ever engineered on Earth. Chadwick knew that he and his fellow scientists had been able to do this by working hand in glove with the military and with a government determined to be first to achieve this goal, regardless of the cost, which turned out to be about two billion dollars, a vast sum, though it was easily afforded by the United States, whose Gross Domestic Product had almost doubled in the previous four years.[48] The cost of the Manhattan Project had been about half that of the B-29 bombing campaign against Japan.[49] Many of the world's leading scientists had shown that they could master nature's best-hidden form of energy, and their very first use of this new knowledge was to give politicians an explosive of unprecedented power to kill.

Groves was elated. He had achieved the goal the government had set him – to deliver a usable nuclear weapon – in just over a thousand days after taking on the running of the Manhattan Project. In the nick of time, he had enabled President Truman to attend the conference with Stalin and Churchill as the first world leader to go to the table with nuclear weapons in his arsenal. The conference began in Potsdam, Germany, a few hours later.

1 JULY TO 5 AUGUST 1945

Churchill says yes to dropping the Bomb

> 'If the Russians had got [the Bomb first], it would have been the end of civilisation . . . [The Bomb] has come just in time to save the world.'
>
> WINSTON CHURCHILL, 23 July 1945[1]

The Potsdam Conference was the first gathering of the Allies' leaders at which the Bomb was mentioned, if only briefly and in the wings. Churchill was poorly prepared for the imminent arrival of the nuclear age, having taken little interest in his Tube Alloys briefings, or in the development of the Manhattan Project and its consequences.[2] As Sir John Anderson had told Niels Bohr in March 1945, the problem with Churchill was that his 'mind was so far from being of a scientific nature that he had difficulties in viewing the project in its proper perspective'.[3] Churchill's main concern – the opposite of Bohr's – was to keep the Bomb a secret from the Soviets to maximise the diplomatic advantage over them after the war. He believed that the 1943 Quebec Agreement guaranteed that Britain would share in the Americans' triumph, though it remained to be seen if Truman would endorse Roosevelt's view.

Unlike Stalin, neither Truman nor Churchill was looking forward to the Potsdam Conference. Churchill, wearied by political worries at home and pessimism about the future of Central Europe, predicted to his doctor Lord Moran that the gathering would be of no consequence. Truman was in a more positive frame of mind, but was apprehensive about his debut on the international stage and his first trip to Europe for twenty-seven years. He was also anxiously awaiting news of the outcome of the Trinity test.[4] Churchill and Truman had not yet met, but they had exchanged cables and talked on the phone, confident that the Soviet leader knew nothing about the Manhattan Project.

Truman had begun his presidency ill-briefed. In the eight months since he had won the Vice-Presidential nomination, Roosevelt had told him almost nothing about military, diplomatic or even administrative matters, apart from a vague mention over lunch in August 1944 of the special new weapon the military was developing.[5] The Manhattan Project had been probably Truman's biggest surprise when officials told him about it soon after the Presidency was thrust upon him.[6]

Decent and dedicated, Harry Truman kept Roosevelt's team of advisers and officials almost intact, and strove to continue virtually all of his policies, domestic and foreign, including the use of the nuclear bomb and the fostering of the United Nations. On the train to Norfolk, Virginia, en route to the Potsdam Conference, the journalist Merriman Smith asked him about his enthusiasm for this international forum.[7] Truman bashfully reminded Smith that the organisation was not his idea but was an old one. He took out of his wallet a neatly folded piece of paper on which he had written out Tennyson's prophetic poem 'Locksley Hall'.[8] The President then read the couplets that Churchill had quoted in his 1931 essay 'Fifty Years Hence', the same dark visions of civil and military aviation, and this stirring climax:

> Till the war-drum throbb'd no longer, and the battle-flags were furl'd
> In the Parliament of man, the Federation of the world.

Truman had been carrying this poem – a personal favourite – since he was a high-school student at the turn of the century, and then into the following decade, when the magazines he read often featured stories illustrating the power of American technological genius to change the course of history.[9] Now, almost half a century later, the small-town Missouri kid had taken his place among the new Big Three, and was about to shape the political future of Europe. Still full of enthusiasm for the grave responsibilities of his new job, he had – unlike the exhausted Churchill – prepared himself thoroughly for the Potsdam Conference.

The Prime Minister's morale had been ebbing since early February, when he met Stalin and the desperately sick Roosevelt at Yalta in the Crimea. By then, the Soviets were already in control of much of Eastern Europe, and it was plain that they could not be driven out by anything short of an Alliance-destroying military operation that would have no appeal to Truman and would have unleashed a new and terrible war. Stalin looked every inch the victorious military leader, dressed in the uniform of a marshal, calm and jovial, disdaining to take notes in meetings or even to carry around his papers – there was no need, as all the premises were bugged.[10] As the leaders discussed the war against Japan and the post-war organisation of Europe, he was shown much more sympathy than Churchill would have preferred. It was obvious to the Prime Minister that the Soviet leader wanted hegemony over Eastern Europe and that, helped by Roosevelt's incapacity and apparent indifference, he was going to get it – after the war, Poland would pass from one tyranny to another.

Churchill, as far as we know, had no reproachful words for Roosevelt, who had patently disappointed him. Yet the Prime Minister may have been more hurt than he let on. When the President died two months later, Churchill declined to attend the funeral, even though Truman had suggested they might spend 'two or three days' talking afterwards – a priceless *entrée*.[11] It would have been unimaginable for the energetic Churchill of 1941 to have passed up this opportunity, especially as he was aware that the new President was a foreign-affairs greenhorn and ready to listen to his advice.[12] Six years later, Churchill commented that this was his biggest mistake of the war – saying at the end of the conflict, 'Tremendous decisions were made . . . by a man I did not know.'[13]

One of the Prime Minister's distractions in the weeks before the Potsdam Conference was the crumbling of his coalition government. The trouble had started in late January, when his deputy Clement Attlee sat down at his portable typewriter and bashed out a two-thousand-word letter to Churchill, damning his recent performance as Prime Minister – not reading his papers, maundering in meetings, giving more credence to his cronies' opinions than to his War Cabinet's.[14] Churchill's close friends Brendan Bracken and Lord Beaverbrook nodded their consent, and even Clemmie weighed in to support the 'very brave' Attlee.[15] By the end of May, the Labour ranks had lost patience and Churchill had no choice but to call a General Election.

While he campaigned, he redoubled his efforts to thwart Stalin's intention to pull down an 'iron curtain' across Europe, as he termed it in a top-secret telegram to Truman (H. G. Wells had first coined the phrase in his 1904 novel, *The Food of the Gods*).[16] In an initiative that was not made public for almost fifty years, Churchill instructed his planning staff to consider launching an offensive on the Soviet army in Europe, 'to impose upon Russia the will of the United States and the British Empire', in Operation UNTHINKABLE.[17] His horrified Chiefs of Staff – who felt 'the less that was put on paper on this subject the better' – quickly convinced him that (in his words) 'the Russian bear sprawled over Europe' was too big and too well entrenched to be evicted. Churchill backed off, described his idea as 'a purely hypothetical contingency' and allowed the plan to peter out.

Elsewhere, a Whitehall spat turned nasty when Churchill disappointed eight of Britain's leading scientists – including Patrick Blackett – by vetoing their long-planned visit to the Soviet Union hours before they were due to leave. Embarrassed officials informed the scientists that their request had been turned down because their services might still be needed in the war against Japan – a patent falsehood, as everyone knew. The real reason was that the Prime Minister felt bound ('in conformity with the United States policy') to prevent any scientist who had been within a whiff of the Tube Alloys project to visit the Soviet Union and risk spilling the nuclear beans.[18] On the last day of the coalition Parliament, A. V. Hill and other members of the Commons heard Churchill argue his case: the visits had been banned not because of 'any question of security', but because the scientists were needed 'for the purpose of the Japanese war'.[19] Government officials expected trouble and got it, though only in the form of mildly embarrassing press reports and a brief rumpus among a few Fellows of the Royal Society.[20] Churchill could safely ignore them all, and did.

The Prime Minister decided to forbid the scientists' visit during a difficult week in the election campaign. In a radio address, he undermined his reputation as a consensual leader by slighting his former Labour colleagues, and made the preposterous suggestion that a Labour government 'would have to fall back on some sort of Gestapo'.[21] This was a gift to Attlee, a speaker so dull he could bore for Britain, but who blossomed into a suburban Cicero to give the radio broadcast of his life, praising the Prime Minister's wartime leadership but arguing persuasively that it was now time for new blood.

Britain went to the polls on 6 July, with virtually all the political pundits predicting that the country would not throw out its wartime hero. Many voters were serving overseas in the armed forces, so the parties agreed to a pause of three weeks before the results were announced, midway through the Potsdam Conference. Taking advantage of the gap in his diary, Churchill went on a painting vacation in south-west France, near Biarritz, relaxing in the sun while the prospect of defeat gnawed at his confidence – he would 'be only half a man until the result of the poll', he confided in his doctor.

As Churchill and Truman saw when they arrived in Berlin, the city had been flattened. It was now a black and smouldering ruin, reeking of death and open sewers, with empty-eyed women and men roaming the city in search of food and somewhere to sleep.[22]

The subject of nuclear weapons did not crop up when Churchill and Truman met for the first time, although the matter was on the President's mind – their meeting took place on 16 July, when Groves was in the New Mexico desert, in the final few hours of preparations for the Trinity test. Churchill arrived at Truman's yellow-stucco villa on the Potsdam Ringstrasse at precisely the President's specified time of 11 a.m., having been rebuffed the previous evening, when Truman's staff said the President was 'fully engaged'.[23] They talked for two hours, mosquitoes buzzing around them in the heat. During the conversation, Churchill tried to persuade the President to modify the Allies' demand that Japan surrender unconditionally, but was unsuccessful.[24]

After the meeting, Churchill was relieved. He told his daughter Mary as they walked back to their lakeside residence nearby that he liked the President 'immensely' and was sure he could work with him.[25] Truman was also impressed, writing in his diary that Churchill was charming and very clever, if a little unctuous: 'He gave me a lot of hooey about how great my country is and how much he loved Roosevelt and how he intended to love me etc. etc.' The President was sure that he and the Prime Minister could get along 'if he doesn't try to give me too much soft soap'.

That night, shortly before eight o'clock, Truman first learned of the success of the Trinity test in a coded cable informing him that a child had been born, larger than expected.[26] The news was delivered to him by Henry Stimson, no longer in office and not even an official participant in the conference. Stimson had in effect invited himself, to conclude matters he had been overseeing for years, including the strategic use of the Bomb, which he had called America's 'master card'.[27] The next day, he went to the Prime Minister's villa for lunch, where one of the guests was Clement Attlee, invited by Churchill to Potsdam as an observer. Attlee had only a vague awareness of Tube Alloys, having been excluded from Churchill's inner circle of nuclear confidants for five years.[28]

During the meal, presumably while others were talking, Stimson placed in front of the Prime Minister the cable announcing that a larger-than-expected baby had been born at the Trinity site. This must have been an awkward moment as Churchill had no idea what the note meant; if, as is likely, Stimson told him, Churchill will have quickly snuffed out the subject. At four-thirty, the lunch over, Churchill walked with Stimson to the garden gate and heard the news of Trinity in plain words, although without much detail. Stimson wrote in his diary: 'He was intensely interested and greatly cheered up,' immediately urging that the news should be kept from the Soviets. Within half an hour, Churchill had been driven to the first plenary session of the conference, to join Stalin and Truman, who had met for the first time over a lively lunch.[29]

The conference took place at the mock-Tudor country palace Schloss Cecilienhof, a brick-and-timber-frame building constructed mainly during the First World War, with dark interiors softened by glorious views of the lake. The grassy island in the middle of the central courtyard featured a huge five-point star of bright-red geraniums, planted by the occupying Soviet soldiers who were glumly patrolling the grounds.[30] Churchill's opening oration at the first session was embarrassing, according to his Foreign Secretary Anthony Eden, who made no secret of his views, even to American colleagues: 'He had read no brief and was confused and woolly and verbose.' Eden complained openly that his leader was 'under Stalin's spell, and kept repeating "I like that man"'.[31] Truman, too, was taken with the Soviet leader, who surprised him by looking not so much like a thug and a mass murderer as a softly spoken gnome with a penchant for Chopin.[32]

Stalin was an effective negotiator, subtle and determined. Having lost far more troops and civilians in the war than both his allies combined, he regarded himself as the principal victor of the conflict and was determined to reap a handsome territorial reward. One of his other worries was that Japan would seek a separate peace with the United States, forestalling Soviet entry into that theatre of war and denying him political gains in the Far East.[33] Away from the conference table he suppressed these worries and was a charming host, entertaining his guests while waiters poured champagne and spooned caviar with unstinting generosity. Yet the lavishness of the hospitality could not disguise the horrors of Stalin's regime, at least for Churchill, who was 'rampant' in his private denunciations of Soviet repression, Stimson observed.[34]

President Truman, appointed chair of the meeting, was at first subdued, but four days later he 'was a changed man', telling 'the Russians just where they got on and off and generally bossed the whole meeting', as Churchill later described the President's behaviour.[35] The reason for this was that, shortly before that conference session, Truman had read General Groves's first extensive report on the Trinity test, declaring it to have been 'successful beyond the most optimistic expectations of anyone'. Mid-morning the next day, Stimson took the clutch of typewritten pages to Churchill's villa, where the Prime Minister was meeting with Lindemann. Groves's verbose and repetitive report was the kind of document Churchill normally abhorred, but he was soon won over, as the sheer scale of the event became clear: 'One of [the eyewitnesses] was a blind woman who saw the light . . . an awesome roar [warned] of doomsday and made us feel that we puny things were blasphemous to dare tamper with the forces heretofore reserved to The Almighty . . .' After Churchill had read the report, he turned to his American colleague: 'Stimson, what was gunpowder? Trivial. What was electricity? Meaningless. This atomic bomb is the Second Coming in Wrath.'[36] Perhaps unknowingly, the Prime Minister was echoing phrases he had used in 'Fifty Years Hence' and other articles about nuclear weapons, including one he had published two years before the war began: 'If and when these sources of power become available our whole outlook will be changed.'[37]

The excitement of Groves's report was too much for Churchill – the following morning, he broke his iron rule that the Bomb must be kept 'absolutely secret'.[38] His resistance broke when he was in bed finishing his breakfast. When Lord Moran entered the room, he found his patient desperate to talk, hindered only by his fussing valet. After impatiently dismissing the servant, the Prime Minister turned to Moran and announced, 'I am going to tell you something you must not tell any human being,' adding solemnly: 'We have split the atom.'

The scene a few days before in the New Mexico desert was 'as if seven suns had lit the Earth', Churchill said, before describing the epic scale of the project that made the explosion possible. 'It is H. G. Wells stuff,' Moran commented. 'Exactly,' responded Churchill, perhaps remembering the huge bouquet of flowers he had recently sent to his frail antagonist, who had explained a few months before in a prominent article why 'Churchill Must Go'.[39] The Prime Minister seemed to be desperate to get off his chest the secrets he had been keeping for so many years. Moving up through the gears of indiscretion, he shocked Moran by telling him: 'It is to be used in Japan, on cities, not on armies. We thought it would be indecent to use it in Japan without telling the Russians, so they are to be told today.' Moran was stunned and incredulous, writing in his diary later that day: 'I once slept in a house where there had been a murder. I felt like that here.'

Churchill had still not calmed down when he lunched with his Chiefs of Staff. Their chairman, Lord Alanbrooke, wrote in his diary that he was 'shattered' by the comments of the Prime Minister, who was 'completely carried away':[40]

> The secret of this explosive, and the power to use it, would [Churchill said] completely alter the diplomatic equilibrium . . . Now we had a new value which redressed our position (pushing his chin out and scowling), now we could say that if you insist on doing this or that, well we can just blot out Moscow, then Stalingrad, Sebastopol etc. etc.

Churchill was greeting the advent of nuclear weapons less like a supremely gifted international statesman than a boy playing with his toy armies.

Truman waited until the following day to tell Stalin about the Trinity test. As the delegates were dispersing after meeting, Truman ran after him – without the American interpreter – watched closely by Churchill. The President ensured his disclosure was low-key, informing Stalin via his interpreter Pavlov that America had developed an unusually destructive new weapon.[41] Stalin said nothing, before turning on his heels and departing. Lost for words, Truman stood gazing at the Soviet leader as he hurried out of the room.[42]

Stalin's reaction convinced Churchill and Truman that their Soviet ally had known nothing of the Manhattan Project.[43] Both had been fooled.[44] For years, a network of spies in Britain and the United States had been keeping the Soviet leader and his colleagues well briefed on the Anglo-American plans to build nuclear weapons. The Kremlin had received a copy of the MAUD report in October 1941, when Roosevelt offered Churchill a nuclear collaboration, from the spy John Cairncross, then Lord Hankey's Private Secretary.[45] Five months later, Lavrenti Beria – Stalin's security and espionage supremo – had presented intelligence material to the State Defence Committee.[46] Stalin and his associates had been slow to appreciate the significance of the information, and provided only modest resources to Igor Kurchatov, responsible for the Soviet nuclear venture, who knew that he and his hundred or so colleagues had no hope of competing with the Manhattan Project.[47]

The Quebec Agreement signed in 1943 by Churchill and Roosevelt, enabling British scientists to resume work on the Manhattan Project, was also no secret to Stalin, who had been briefed on its contents.[48] In the year preceding the Potsdam Conference, thousands of pages of top-secret material had been sent to the People's Commissariat for Internal Affairs, keeping the Soviet high command up to date with developments on the American project. Soviet intelligence chiefs expected the Americans to test their nuclear weapon for the first time a week before the Potsdam conference began.[49] It seems likely that, shortly before the Trinity test, Stalin knew roughly as much about it as Churchill.

Stalin also knew that Churchill had contemplated taking on the Soviet army in Europe in Operation UNTHINKABLE.

Planning officials in the Kremlin discussed the operation a few days after Churchill floated the idea to his military chiefs, who quickly persuaded him that it was impracticable.[50] The Soviet leader may have been thinking of these deceptions when he told the Polish Deputy Prime Minister: 'Churchill did not trust us so we could not fully trust him either.'[51] The night the Soviet leader heard about the Bomb, he laughed as he chatted with his colleagues, and arranged to speak with Kurchatov to accelerate their own nuclear-bomb project.[52] Stalin complained to colleagues about their supposed American allies: 'They slay the Japanese, and bully us. Once more everything is done in secret'[53] – a little rich, considering he had shared virtually no technical secrets with the British and Americans. They were about to use the new weapon as a bargaining chip, he suspected – 'They want to force us to accept their plans on . . . Europe and the world,' he said, adding, 'Well, that's not going to happen!' He concluded with one of his ripest curses.[54]

The conference was suspended that evening so that the British delegation could return home to hear the results of the election on 26 July. Churchill was only cautiously optimistic, unlike his doctor, who was so confident that he left his luggage in Potsdam. Around ten in the morning in the Map Room at 10 Downing Street, the Prime Minister climbed into his blue siren suit, lit a cigar, and slumped into a chair to monitor the results as they rolled in.[55] By one o'clock, BBC news reported that the Conservative government had been routed. The British electorate had kicked out the leader they had cheered on the streets three months before – he had, it seems, shown too little interest in rebuilding his country after the war. Churchill stayed in his chair for the rest of the day, watching the disaster unfold and feeling the robes of power slip from his shoulders. In public he was magnanimity itself about the result; in private he was hurt and stunned but impressively candid. He told his son Randolph:[56]

I was received on my [pre-election] tour with public rapture. They then went to the polling booths and voted against me. But the rapture was not insincere. They were not ungrateful. They thanked me for such services as I have rendered in the war. But they did not want me for peace. They may have been right.

On the day Churchill heard of his defeat in the polls, the authorities in Potsdam released the declaration that he, Stalin and Truman had signed, calling on Japan to surrender immediately or face the destruction of its armed forces and 'the utter devastation' of its homeland.[57] The Japanese government's response was *mokatsu* – an icy, dismissive silence that left Truman in no doubt that he should now play his highest card.[58] Churchill had, as British Prime Minister, formally agreed to the use of nuclear weapons on Japan, fifteen days before the Trinity test.[59]

With Chartwell mothballed, the Churchills had nowhere to live. Granted the use of Chequers the following weekend, they threw a party there for their family, wartime colleagues and friends, including Lindemann, Jock Colville, Brendan Bracken and the American Ambassador Gil Winant.[60] This was the kind of party Churchill loved, but even a jeroboam of champagne, abundant good food and a gramophone recording of the mindlessly jolly 'Run Rabbit Run' could not lift his spirits. One of Churchill's few political tasks in these last two days of his wartime premiership was to finalise his message to Britain and its Empire for when the news broke that the Bomb had been dropped.[61] He worked with Colville, Anderson and Lindemann, who liaised with American officials on the texts of the announcements that were to be made after the event. Churchill wanted his to mention the Quebec Agreement and the aide-mémoire he had signed with Roosevelt, but the Prof believed this 'very undesirable – this would only lead to demand for publication, which would be very embarrassing to the Americans and perhaps even to us.'[62] Churchill accepted the advice and backed down.

Most of Churchill's eight-page draft was an outline of the Bomb's history highlighting the early British foundations of the Manhattan Project, naming all the leading British scientists involved except, unaccountably, Otto Frisch.[63] At the end, however, the tone changed suddenly: the author was no longer the project's dutiful chronicler but the romantic who had looked forward sceptically to the nuclear age in 'Fifty Years Hence', albeit with reservations: 'This revelation of the secrets of nature, long mercifully withheld from man, should arouse the most solemn recollections in the mind and conscience of every human being capable of comprehension.' His final words echo the hope H. G. Wells had expressed in his 1914 novel *The World Set Free* that nuclear weapons will be a force for good:

> We must indeed pray that these awful agencies will be made to conduce to peace among the nations, and that, instead of wreaking measureless havoc upon the entire globe, they may become a perennial fountain of world prosperity.

While Churchill and his wife were looking for a more permanent residence in London, they moved into the penthouse suite of Claridge's Hotel, where he contemplated his future and waited for news from Hiroshima.

3

CHURCHILL AS LEADER OF THE OPPOSITION

July 1945 to October 1951

AUGUST 1945 TO JANUARY 1949

Blackett: nuclear heretic

> '. . . the dropping of the [nuclear] bombs was not so much the last military act of the Second World War, as the first act of the cold diplomatic war with Russia now in progress.'
>
> PATRICK BLACKETT, October 1948[1]

The war that had by then killed over fifty million people had one last shock in store. At 9 p.m. on 6 August 1945, a quiet Bank Holiday Monday, millions of radio listeners in Britain heard that nuclear bombs were not – as physicists had repeatedly assured them before the war – merely the pipe dreams of journalists and novelists. The announcer, suffusing his clipped tones with an unusual excitement, began the bulletin with three short sentences, so dense with information that it was difficult for most listeners to take them in:[2] 'Scientists, British and American, have made the atomic bomb at last. The first one was dropped on a Japanese city this morning. It was designed for a detonation of twenty thousand tons of high explosives.'

Patrick Blackett will have read about the news from Hiroshima in the *Manchester Guardian* over his breakfast the next day.[3] He had been bracing himself for the announcement, but years of forewarning had done little to prepare him for the trauma of hearing about what he believed was a disaster, among the most heinous acts of the war.

Blackett was then head of the physics department at the University of Manchester, and he lived with his wife in their apartment in the suburb of Fallowfield.[4] Their two children had recently left home. He had arrived in Manchester eight years earlier, after leaving Rutherford's Cavendish Laboratory and spending four years as a professor at Birkbeck College in London. Although not a national figure, his peers knew him to be a fine nuclear physicist and an outspoken Socialist who was unwilling to kowtow to anyone. Over six feet tall, he had an impressive presence, with the looks of a matinée idol, a tack-sharp mind and the gravitas of an archbishop.

'Man is now well on the way to mastery of the means of destroying himself utterly,' he read in the *Manchester Guardian*'s leader. The newspaper's reports featured no photographs of the damage caused by the Bomb, but had plenty of comments on its significance from Churchill, Truman and others. Henry Stimson, until recently the American Secretary of War, declared the Bomb to be 'the greatest achievement of the combined efforts of science, industry, and the military in all history'. Blackett saw on the back page of the newspaper an entire article on the contribution to 'one of the most remarkable events in history' by his own physics department, beginning with the work of Rutherford. Chadwick's name was mentioned, too, but Blackett's was absent, as he would have wished. Since he joined the MAUD committee in 1940, he had been a law unto himself and had sometimes been wiser than the consensus. His judgement that it would be impossible to build the Bomb in Britain during the war had proved right – the early predictions of Chadwick and Lindemann now looked absurdly optimistic to anyone who could remember them. After Tube Alloys had been folded into the Manhattan Project, Blackett had nothing to do with it beyond opposing Britain's participation.[5]

In the days following the first reports from Japan, journalists struggled to find out how much damage the Bomb had done. At first, all they knew was that Hiroshima – 425 miles from

Tokyo, on the coast of the Seto Inland Sea – was under a vast cloud of smoke and impossible to see from the air. Within two days, however, British newspapers reported that the city had been almost wiped out, with a death toll of about a hundred thousand, and that soon afterwards another Bomb had been dropped on Nagasaki, about 185 miles south-west of Hiroshima. In London, George Orwell noticed that Japan's prompt surrender altered people's perception of the Bomb – after first recoiling from the horror of Hiroshima's annihilation, Londoners were beginning 'to feel that there's something to be said for a weapon that could end the war in two days'.[6] Blackett was less sanguine and doubted that the new weapon would make as much difference to the future of warfare as most people seemed to assume.

Harold Nicolson's phone was ringing off the hook, with journalists asking if he had known about the Bomb when he wrote *Public Faces* thirteen years earlier.[7] He contributed little to the debate and nor did H. G. Wells, who appears to have been too ill to comment. The most prominent public intellectual to set out his views was George Bernard Shaw, now eighty-nine years old. In his *Sunday Express* article 'The Atom Bomb', soon reprinted across America and Europe, he wrote: 'H. G. Wells said all there is to be said, and more, thirty years ago,' before contradicting himself by adding a few ideas of his own.[8] The new weapons might one day be cheap and plentiful, he pointed out, so that wars might easily be waged not only by wealthy countries but also by special-interest groups, such as 'neo-Darwinians and Creative Evolutionists, Fundamentalists and Atheists, Moslems and Hindus'. But he concluded that the Bomb may prove too deadly to deploy with any discrimination – 'It may burn down the house to roast the pig' – not mentioning the possibility that its users might be irrational or even mad.

Blackett would soon become one of the leading public commentators on the Bomb. But in the immediate aftermath of Hiroshima and Nagasaki, he kept his own counsel. He was preparing to take a position as a prominent adviser to the British government on nuclear policy, a role Churchill would never have allowed him to play. Prime Minister Attlee probably knew that he had made a risky appointment – if Blackett did not approve of the policy agreed within government, he would certainly make his dissatisfaction public.

In the United States, by mid-August 1945 the Stars and Stripes had well and truly been wrapped around the Bomb in a report researched and written by the Princeton physicist Henry DeWolf Smyth. Government officials distributed mimeographed copies to the American media after the Nagasaki bombing, making the basic principles of the Bomb's operation available to all-comers. In Washington, a few hours before it was released, Chadwick nervously addressed a press conference, doing his best to promote the British role in the Manhattan Project. But he was on a hiding to nothing.[9] Smyth's presentation set the international agenda for discussions of the project and was a runaway bestseller in the US and elsewhere, going on to nine editions and eventually being translated into forty languages.[10] Attlee's officials, caught on the hop by the report's publication, gave Michael Perrin in the Tube Alloys office twenty-four hours to write the British side of the story. He worked through the night drafting the worthy but dull 'Statements Relating to the Atomic Bomb', rushed into print by the government's stationery office.[11] But it was too little, too late. After the Americans' public-relations coup the British response looked tardy, ineffectual and slightly mean-spirited.

One of Attlee's most pressing challenges was to forge a

nuclear-energy policy for Britain, ideally capitalising on the knowledge and experience his scientists had gained in developing the Bomb. A few days after the end of the war, he set up a select Advisory Committee on Atomic Energy, whose most notable appointment was Blackett.[12] He was the only scientist on the committee to have been trained in military strategy and to have been in the thick of a bloody conflict – in May 1916, he was a cadet in one of the turrets of HMS *Barham* in the Battle of Jutland, the most violent naval conflict of the First World War. Now, thirty years later, he was in by far the most influential post he had held in Whitehall. Chadwick resented the intrusion of the Tube Alloys outsider, remarking later that Blackett's appointment 'was as much political as scientific'.[13] Chadwick believed that Blackett bore a grudge against the Americans after they declined a request he had made in late 1943 to visit their military operations in the Pacific. Blackett would have strongly denied this – he always saw himself as an agent of logic, not someone swayed by mere emotion.

Other core scientific members of the Advisory Committee were G. P. Thomson, Edward Appleton and the Royal Society President Sir Henry Dale, though not ICI's Wallace Akers, who was discreetly sidelined. In a surprising move, Attlee appointed as the committee chairman one of the previous government's most prominent experts on nuclear policy: Sir John Anderson. He was certain to bring experience and fair-mindedness to the role, as well as his fabled caution and some of Bohr's thinking about the new weapons. Churchill's main nuclear adviser, Frederick Lindemann, did not fare so well – after rubbing Attlee up the wrong way for years, the Prof was pointedly excluded.

Among the other committees Attlee set up to deal with nuclear matters, the most powerful was known as GEN 75, staffed entirely by members of the Cabinet. Nineteen days after the Nagasaki bombing, Attlee wrote to his colleagues to set out the stakes. Normally terse and bloodless, the Prime Minister wrote this with a passion and sense of history almost worthy of Churchill: 'We should declare that this invention has made it essential to end wars. The new World Order must start now . . . The Governments of the UK and the USA are responsible as never before for the future of the human race.' He concluded: 'Time is short . . . I believe that only a bold course can save civilisation.'[14] After taking a few weeks to consult his colleagues – and Churchill – Attlee wrote to President Truman, urging that Britain and America work together to deal with the 'entirely new conditions' facing the world.[15]

Blackett's new role in government was the culmination of almost twenty years of political activity. In the first General Election after the First World War, he had voted Conservative, but then turned to the political left and stayed there. During the Wall Street Crash and the slump of the 1930s, he came to believe that planned Socialism was the best basis for running the economy. Although Blackett was impressed by the Soviet Union's apparent economic achievements in the 1930s, he was not a Communist but a principled Socialist, his beliefs underpinned by Marxism and, specifically, the idea that economies should be planned scientifically.

During the war, Blackett had minimal political influence and was mainly deployed as a stellar operational researcher, the boffin's boffin. He had done more than any other British scientist to smarten up the tactics of the air force and navy. One of his most effective contributions had been to the anti-U-boat war of 1943, which he ensured was waged with an unprecedented degree of scientific control. Although publicly loyal to his bosses in Whitehall, he sometimes despaired that the good that he and his colleagues were doing was undermined by incompetent decision-making at the top. 'If I had published the truth of what I have known of parts of our war effort,' he told one of his colleagues in late 1941, 'I would certainly be locked up.'[16]

When Blackett read his first confidential briefings on the British nuclear project during the war, he was scornful of the Churchill government's handling of it. Blackett spelled out his criticisms to Mark Oliphant, who – with a breathtaking lack of discretion – passed them on to the sick and exhausted Chadwick, with comments on the Churchill–Roosevelt Quebec Agreement:[17]

> [Blackett is] appalled by the incompetence and sheer stupidity . . . of the political and administrative history of the British [Tube Alloys] effort and he regards the Quebec Agreement as a degrading document. He is very critical of your views and says that you do not serve as a representative of your country . . . but that you 'side with Groves against [it]'.

Blackett was also scathing about continuing the Tube Alloys project, Oliphant told Chadwick: '[He] expresses amazement at the inadequacy of the organisation . . . and he believes that this reflects the inability of Akers to deal with the situation.' Nor was Blackett any more complimentary about the Advisory Committee on Atomic Energy, which he had joined a couple of weeks earlier: Sir Henry Dale was 'a particularly poor member' and Anderson was 'quite inadequate as chairman'. Oliphant ended by advising poor Chadwick to take a 'real holiday', a break he will have needed after reading that letter.

The 1943 Quebec Agreement did not remain secret for long. It was disclosed in the House of Commons late one evening in October by the nonconformist Raymond Blackburn, a solicitor and former soldier who had recently entered Parliament as a Labour MP. After Hiroshima, Blackburn had become obsessed by the Bomb. Though he was not a scientist, he threw himself into the underlying technicalities, determined to cudgel the government into making its nuclear policy public. In a dramatic statement, he drove a coach and horses through the Official Secrets Act by revealing the agreement's existence, declaring that it 'left the development of the peacetime use of atomic energy by this country very much to the discretion of the President of the United States'.[18] Party officials were furious, as were many others – it was 'a monstrous abuse', Anderson howled.[19] Blackburn refused to name his informants, but the authorities quickly identified one of them as Oliphant, who had spent hours giving the errant MP advice, physics tutorials and a history of Tube Alloys.[20] MI5 could pin no blame on Blackett, whose record in protecting state secrets was exemplary.

Government officials prevented the agreement from becoming widespread knowledge outside Parliament, but the cat was out of the bag. The substance of Churchill and Roosevelt's document was known to most insiders in Britain but not to President Truman, who told the press the day after Blackburn's statement that he doubted the agreement's existence, a remark that did not bode well for Anglo-American nuclear relations.[21] Despite chastisement from Labour Party whips, Blackburn spoke again in the Commons about the agreement, this time in the presence of Churchill, who said that if it were made public he would be only too pleased.

As this controversy swirled in Whitehall, Blackett was looking to Britain's nuclear future, impatient with the plodding pace of the government's Advisory Committee. In early November, he wrote a ten-page paper 'Atomic Energy – An Immediate Policy for Great Britain' and sent it to the Chiefs of Staff. Blackett was probably hoping to influence Attlee, who was shortly to have his first meeting with President Truman and with the Canadian leader Mackenzie King. Blackett challenged virtually every aspect of conventional military thinking on the Bomb. With relations between the US and the USSR already tense, and the near-certainty that the Soviets would have nuclear weapons within five years, he believed it made no sense for the UK to build or acquire them. Britain was, unlike the United States, within striking distance of Soviet aircraft, so if it were to acquire nuclear weapons, the country would be the most likely target in the event of an East–West war. The popular idea of international control of nuclear materials was not viable, in his opinion: it would be better for Britain not to make nuclear weapons but to concentrate entirely on making peaceful use of nuclear energy, as a source of electrical power. The paper was classic Blackett – powerfully argued and more than willing to upset the apple cart of conventional Whitehall thinking.

Attlee, however, was unimpressed. Blackett's paper was the work of 'a distinguished scientist speaking on political and military problems on which he is a layman', he scrawled across his copy.[22] His agenda in Washington bore no sign of Blackett's influence. The most tangible outcome of the meeting was the setting up two months later of the United Nations Atomic Energy Commission, established 'to deal with the problems raised by the discovery of atomic energy' and charged with making proposals for the 'elimination of all atomic weapons'.[23] At the close of the discussions, Truman and Attlee put their names to a few hurriedly drafted lines expressing their desire, in the field of nuclear energy, for 'full and effective cooperation' between the US, the UK and Canada. The spirit of the Quebec Agreement appeared to have endured and Attlee returned home believing that the British and Americans saw eye to eye on nuclear policy. There had been no appetite at the conference for sharing nuclear secrets with the Soviets.

The world was entering what George Orwell had named, within seven weeks of the end of the Second World War, the 'cold war'.[24] Although Blackett was as patriotic as anyone in Westminster, he refused to demonise the Soviets and argued that they had just as much right as the Americans to be concerned about their national security. In thinking about the possibility of nuclear war, he was studiously even-handed – not a popular point of view in Whitehall. Blackett found himself shut out of important meetings and his influence diminished by the week. He sighed to Oliphant: '[the] government doesn't much approve of me'.[25]

Much more congenial to Attlee's government was the stream of advice supplied from Washington by James Chadwick. He ensured a seamless continuity with the thinking that had informed the Churchill government, based on the assumption that Britain should acquire nuclear weapons and begin to set up a nuclear-power industry at the earliest opportunity. As Attlee and his officials knew, this was the national consensus: it was received wisdom in all the mainstream press that Britain must acquire the Bomb if it was to remain a global power and deter a nuclear attack. Winston Churchill said in Parliament that he assumed it was agreed on all sides of the Commons that 'we should make atomic bombs'.[26]

In January 1947, a special Cabinet committee agreed to authorise 'research and development work on atomic weapons'.[27] Lord Portal of Hungerford agreed to lead the project, though reluctantly – he had spent most of the war in the role of Chief of the Air Staff, and was still recovering from exhaustion. The scientists appointed to report to him were all British-born – 'the smiling killer' William Penney was to be head of armaments, the imperturbable John Cockcroft was to lead the research programme, and the hard-driving engineer Christopher Hinton was to be responsible for designing, building and operating the nuclear plants. All these leading players then slid into the shadows – none of their former colleagues and no one outside Attlee's nuclear coterie had any idea what they were doing, except Stalin's spies.[28]

This was a bittersweet time for Blackett. Although disappointed with the minor role he had been given in the new government, he strongly approved of its radical social-democratic agenda. Attlee swiftly began a programme of nationalising industries such as steel and coal, setting up a State-owned railway network, creating the National Health Service and starting to dismantle Britain's Empire. For a few years, rationing was worse than it had been during the war. This made life crushingly drab for a people who expected victory to have a sweeter taste, though the austerities meant nothing to the Gandhian Blackett. He approved of virtually everything the government did, except in his own specialist field of nuclear strategy.[29]

In this area, Blackett advanced his ideas with prodigious energy. While building Manchester University's physics department into one of the best in the country, he became a public figure, Britain's most original and controversial thinker on nuclear strategy. He was also a prime mover in the new organisations promoting open debate on nuclear matters in the scientific community, especially the influential Atomic Scientists' Association, led by Rudi Peierls. Most of the former Tube Alloys scientists had joined the organisation, though not Chadwick, who 'feared they could do something silly and make matters worse . . . I could see no way of controlling them.'[30] Blackett had no wish to be controlled, least of all by conservative thinkers like Chadwick, but wanted to speak freely. He told a meeting of chemists in Manchester that 'the net effect of the discovery of [nuclear] energy up-to-date has been wholly bad'.[31]

*

In the spring of 1946, relations between the United States and the Soviet Union were worsening by the month.[32] America was determined to develop its nuclear arsenal and maintain its position as the world's leading military power, while Stalin was refusing to be intimidated. Scientists warned that it was only a matter of time before the Soviets acquired these weapons, so it was all the more important for America to maintain its nuclear lead. Blackett's friend Niels Bohr, though politically jejune, had been right to predict that a potentially dangerous nuclear-arms race would begin within months of the end of the war.

The breakdown in trust between the Cold War's leading protagonists was perhaps even worse than Bohr had foreseen. The United States, convinced of the moral superiority of its political system, was determined to maintain its military pre-eminence so that it would never suffer another Pearl Harbor. The Soviet Union, equally committed to its own social and political structures, was no less determined to avoid another of the terrible invasions that had left its economy and infrastructure in ruins. The US was fearful that Stalin would gain an irreversible hold on war-torn Europe, especially in Germany, whose future had not been properly resolved at Potsdam.

From the point of view of the American and British governments, Stalin was a bully, imposing puppet regimes in Bulgaria, Poland and Romania, as well as an authoritarian mini-State in the USSR's occupation zone in eastern Germany. Yet in the early post-war period Stalin did allow relatively free elections in Czechoslovakia and Hungary, and he agreed to the formation of representative governments in Austria and Finland. Most people in the US and Western Europe, however, soon became alarmed by the increasing repressiveness of administrations now answering to Moscow, fears that were later given voice by George Orwell in his novel *Nineteen Eighty-Four*. 'If you want a picture of the future,' his hero Winston Smith is told before one of his bouts of torture, 'imagine a boot stamping on a human face – forever.'

Fearful lawmakers on Capitol Hill had no interest in sharing with Britain what most of them regarded as America's secret of the Bomb. The Quebec Agreement had been news to Truman, whose officials could not find it and had to request a copy from London.[33] Precisely the same thing happened in the case of the aide-mémoire signed by Churchill and Roosevelt in September 1944.[34] To have lost not just one of the documents, but both, may attest less to sheer carelessness than to the relatively low importance that Roosevelt placed on them.

Truman cast aside the Quebec Agreement and went along with the consensus in Washington that the United States should work alone on nuclear energy. On 1 August 1946, he signed the McMahon Act, which made it illegal for any American to share nuclear information with any other country, signalling that the United States was now supremely dominant in the nuclear field. Britain was left to its own devices. On Capitol Hill, the British diplomat Roger Makins – the civil servant most familiar with Tube Alloys – brandished copies in Washington of the Quebec Agreement and aide-mémoire signed by Churchill and Roosevelt, but the American lawmakers were unimpressed. As Makins later recalled, 'They were very weak documents, and they had no legislative backing.'[35] Attlee complained about the passing of the Act in a letter to Truman, but did not receive a reply and let the matter drop.

The passing of the McMahon Act was a bitter blow to most of the British scientists who had worked on the Bomb, especially for those still working on the laboratory on The Hill, now known by its place name, Los Alamos.[36] Groves directed them all to leave, though not before some of them were denied access to reports they had written days before. Chadwick, however, looked on the bright side: 'Are we so helpless that we can do nothing without the United States?'[37] Indignation among British scientists and government officials at the Americans' decision did not spill over into the British press – Attlee's government ensured that public discussion of nuclear weapons never caught fire: during the entire five-year life of this Parliament, there was not a single debate on nuclear matters.

It was a different story in the United States, where the media often reported the discussions on the subject between scientists and legislators. At one extreme, the hawkish Secretary of State James Byrnes wanted to make the most of the US's monopoly and play tough with the Soviet Union; at the other, Robert Oppenheimer and the now-retired Henry Stimson emphasised that the monopoly was not permanent and that the US should take the lead in setting up an international framework to control the spread of the weapons. American policy was hammered out by a high-powered committee, whose members included Oppenheimer, Groves, Vannevar Bush and James Conant. They recommended that fissile material be placed under the aegis of a neutral international agency and that the US should abandon its nuclear monopoly and reveal its Bomb-making secrets to the Soviet Union. In exchange, America should secure an agreement limiting the number of weapons both countries would be allowed to make. Truman, unable to swallow whole such a radical proposal, turned to an unlikely authority to develop the proposal for submission to the United Nations: Churchill's friend and admirer Bernard Baruch, seventy-six years old, almost deaf, in poor health but as flamboyant as a Hollywood star on the make. At a UN gathering in the Bronx on 14 June 1946, Baruch unveiled his plan at an event with all the hoopla of a movie premiere.[38]

The Baruch proposal appeared to offer a generous deal to the Soviets, but specifically insisted that 'condign and swift punishment' for control violations would be imposed by the UN Security Council and decided by a majority vote of the Great Powers. The Soviets rejected the plan, having smelt a rat: America could always command a majority in the UN by calling on the support of its allies, so it could get away with whatever it liked.

Blackett predictably disagreed with the proposal and in September 1946 set out his reasons in a pamphlet, for the first time publicly declaring his opposition to Anglo-American control initiatives.[39] That month, during a visit to the US, he was alarmed by the hysterically anti-Soviet mood – politicians were vying with each other during the mid-term hustings to whip up public anxiety about the Communist threat. American scientists, realising that plans for international controls were likely to fail, appeared to be making matters worse, Blackett believed. He told Peierls that Oppenheimer had admitted that 'the attempt to solve the atomic bomb problem in isolation led to one logical solution only – preventative war'.[40] Oppenheimer's former Manhattan Project colleague Harold Urey announced that America might be forced to declare war 'with the frank purpose of conquering the world and ruling it as desired and preventing any other sovereign nation from developing mass weapons of war'.[41]

Returning to Manchester angry and shaken, Blackett decided to challenge the consensus in favour of international control, which he believed was unfair to the Soviets. In early November 1946, after a private meeting with Attlee in 10

Downing Street, Blackett recommended to the Prime Minister that, in the event of the break-up of the UN Atomic Energy Commission, Britain should renounce nuclear weapons, introduce a purely defensive strategy and declare itself politically neutral, like Switzerland.[42] Yet none of Blackett's arguments ultimately had any purchase in Whitehall, nor was he making much impact among his peers, who thought he was overreacting to American extremism. They lobbied him hard to support – or, at least, not to oppose – the consensus that it was crucial to forge some sort of agreement with the Soviets to control the spread of nuclear weapons, but he would not budge.[43] Foreign Office mandarins thought Blackett was spouting 'dangerous and misleading rubbish', while some of his colleagues, weary of his self-righteous expostulations, believed he was becoming a crank.[44]

It is not clear why Attlee agreed to meet Blackett. The Prime Minister had already accepted, in secrecy, almost two weeks before their first meeting, that Britain needed the Bomb, and he appears not to have told Blackett this when they talked. The decision was taken in Downing Street on 25 October 1946, after the Foreign Secretary Ernest Bevin waddled into a meeting of the secret 'Atomic Energy Committee', having fallen asleep after a heavy lunch.[45] Attlee and his colleagues were discussing whether to build a costly gaseous diffusion plant that would enrich uranium-reactor fuel and accelerate the production of plutonium. When Attlee summarised the view of his leading economic ministers that it would be best to take the proposal off the table, Bevin piped up, 'No, Prime Minister, that won't do at all.' Having recently been 'talked at' by the swaggering American Secretary of State James Byrnes, Bevin said he did not want any future Foreign Secretary of Britain to suffer such a humiliation – 'We've got to have this thing over here, whatever it costs . . . We've got to have the bloody Union Jack on top of it.' The meeting swung his way and the decision was finally confirmed three months later, a verdict known only to Attlee and the four ministers he trusted to be on the committee. No one else in Parliament was aware that Britain was about to develop its first nuclear weapons, although – as Attlee knew – few MPs would have doubted that Britain must have the Bomb, to demonstrate its military muscle and to help sustain some of its prestige on the international stage.

When Blackett discussed the decision with his colleagues on the Advisory Committee, he realised that he had been banging his head against a brick wall. It was time to try another way of getting his message across, by writing a book. He set out to demonstrate the futility of area bombing, to explain why current international control initiatives were unfair to the Soviet Union, and to argue for policies that treated nuclear weapons alongside conventional armaments.[46] The American policy on nuclear weapons was misguided, he believed – America's monopoly would not prevent the huge Soviet army in Eastern Europe from invading the west of the continent.

After reading a proof copy of the book, Tizard advised Blackett to avoid giving the impression that 'everything that America has done is wrong and stupid, and that everything Russia has done is right'. Blackett accepted that he 'might have erred slightly in objectivity', but left the substance of his argument unchanged.[47]

On 12 May 1948, when Blackett was finalising the book, the government disclosed that Britain was building its own Bomb, making the announcement in the most downbeat way imaginable. During a quiet afternoon in the Commons, new Labour MP George Jeger put a planted question to the Minister of Defence, A. V. Alexander, asking if he was satisfied that 'adequate progress is being made in the development of the most modern types of weapon'.[48] The news that Britain was making nuclear weapons was delivered in passing, so quickly that it seems most of the MPs did not even notice:

MINISTER: Yes, sir. As was made clear in the statement relating to defence, 1948, research and development continue to receive the highest priority in the defence field, and all types of modern weapons, including atomic weapons, are being developed.
JEGER: Can the Minister give any further information on the development of atomic weapons?
MINISTER: No. I do not think it would be in the public interest to do that.

Seconds later, the Commons' agenda moved on to consider the quality of imported Danish beef. The disclosure was reported in the press so discreetly that only the most diligent of readers would have seen it and appreciated its importance. As a result, the announcement was a non-event, no doubt as Attlee had intended.

Behind the scenes, the government had achieved a measure of success in its negotiations with the Americans over their nuclear policies. Senior members of Congress had been horrified the year before when they heard the details of the Quebec Agreement, especially the British veto on the American use of the Bomb. A few months later, in January 1948, the governments' leading nuclear scientists – including Vannevar Bush and John Cockcroft – secretly agreed a *modus vivendi* in which the veto was revoked, along with America's freedom to halt a nuclear-power industry in Britain, and some other clauses favourable to the US.[49] Most importantly, the new agreement enabled the British to share American nuclear information that would help them build a new weapon of their own design. Unknown to the British Parliament, the Quebec Agreement was now dead.

*

Blackett's short book *Military and Political Consequences of Atomic Energy*, so dense that much of it is barely readable, went on sale in early October 1948. By then, the temperature of the Cold War had fallen further, making it unlikely that Blackett's ideas would have much appeal. The Americans, looking to win influence in Europe, had pumped huge amounts of money and resources into the continent to alleviate economic hardship through the Marshall Plan, which the Soviets regarded as part of an attempt to forge an alliance hostile to the Kremlin and its friends. The plan would also reduce the chances of Communist regimes taking power in Eastern Europe. On 24 June, Stalin snapped – the Soviets blockaded West Berlin, aiming to take control of it by cutting off its Western supply lines, threatening hundreds of thousands of Berliners with starvation. A tragedy was eventually averted by Allied airlifts, but only after Europe had spent weeks on the brink of another war. Soon afterwards, the Western powers created the Federal Republic of Germany and the Soviets established the German Democratic Republic, a split that mirrored the division of Europe into American-led and Soviet-led spheres of influence. The creation of the North Atlantic Treaty Organization (NATO) a few months later cemented the trans-Atlantic security agreement that sought to deter Soviet expansionism and quell a revival of nationalist militarism in Europe. This new security arrangement was, the Soviet news agency TASS protested, 'openly aggressive'.[50]

In normal circumstances, *Military and Political Consequences of Atomic Energy* would have been ignored. But Blackett was in luck – only weeks after the book was published, he heard that he had won the Nobel Prize for physics (the literature prize went to T. S. Eliot). The award gave his tract a special credence and made it an improbable bestseller, soon translated into eleven languages.[51] Blackett and his wife celebrated the award of the prize by throwing a party, defying departmental protocol by inviting not just his fellow academics but everyone who worked with him, including secretaries, technicians and cleaners.[52] His idea of splashing out usually involved nothing more than buying a new pipe, but his Nobel winnings enabled him to buy a second-hand yacht, twenty-five feet long, with rust-red sails. It had been named *Red Witch*.

The book shook up the nuclear debate on both sides of the Atlantic, and made Blackett several influential enemies. One of them was George Orwell, who included him among thirty-eight pro-Communist writers and intellectuals on a list he sent a few months later to a friend at the Foreign Office, to help with its clandestine anti-Soviet propaganda initiative.[53] Blackett was in good company on this list, which also included J. B. Priestley and Charlie Chaplin.

Most commentators were more level-headed about Blackett's intentions. Many of his scientist colleagues – even those who were not Socialists – were warm in their praise of his presentation. The steadfastly moderate Rudi Peierls found himself 'mostly in the position of taking your side'.[54] The Conservative G. P. Thomson regretted Blackett's suggestion that the Americans had 'Machiavellian motives' when they dropped nuclear bombs on Japan. Thomson, however, accepted Blackett's criticisms of the Baruch Plan and praised him for being 'almost certainly right in deprecating the extremer claims for the effectiveness of atomic bombs'. The root of the arms-control problem, Thomson agreed, was the 'Russian fear of being in a permanent minority position on any control body'.[55]

Some of Blackett's American colleagues were less complimentary. In *The Atlantic Monthly*, the eminent atomic physicist Isidor Rabi dismissed the British scientist's 'ostentatious display of scientific objectivity' as merely 'a thin veneer which covers an extraordinary piece of special pleading'.[56] Blackett was hopelessly confused and in denial that nuclear weapons had revolutionised modern warfare, Rabi wrote: 'On international affairs, [Blackett] writes like the amateur which he is . . .'

Sir John Anderson and Frederick Lindemann agreed. In an uncharacteristically forthright BBC radio talk, Anderson attacked Blackett's 'warped' judgement, dubbing him 'a conscience-smitten atomic scientist' who had entered 'quixotically into the unfamiliar field of politics'. A few weeks later, in a *Sunday Dispatch* article entitled 'Britain's Red Scientists', Lindemann was equally disobliging, complaining about the disproportionate public prominence of a few lefties among Britain's scientific elite. What pests they are, with 'their unlimited time, unbounded energy . . . [as well as] insatiable itch and unerring nose for publicity'. He was almost certainly thinking of Blackett. The two men could scarcely bring themselves to give each other the time of day, each believing in his own ability to bring rationality to other people's confusions about political and military matters.[57]

Although Lindemann was the better mathematician and writer, Blackett was much the finer scientist. In Oxford, Lindemann's interfering stewardship of its nuclear research had proved to be expensive and unproductive.[58] In Manchester, however, Blackett had shown himself to be the best research leader to have graduated from Rutherford's school. Among his achievements, he helped to set up the Jodrell Bank Telescope and formed the group of cosmic-ray physicists that discovered elementary particles whose bizarre behaviour indicated the need for a new property of sub-atomic matter, later called 'strangeness'.[59]

Blackett's Nobel Prize and the publicity given to his views appear to have inflamed Lindemann's jealousy. The Prof sent his denunciations of Blackett to several of his friends, and supplied Churchill with a copy of 'Britain's Red Scientists'.[60] Over the past four years, Lindemann had remained a close adviser to his hero, who had developed firm opinions on nuclear weapons – views that Blackett abominated.

AUGUST 1945 TO AUGUST 1949

Churchill the Cold Warrior

> 'Mr Churchill honestly desires peace; but he is convinced that Stalin is waiting for a favourable opportunity to launch an aggressive war . . . and that he thinks of nothing else by day or night. Therefore, says Mr Churchill, the Western Powers must arm to the teeth.'
>
> GEORGE BERNARD SHAW, 1950[1]

Churchill believed that, after the war, it was vital for the United States to make the most of its nuclear monopoly while it lasted. Horrified to see millions of Europeans living under Stalin's thumb, he believed it crucial to keep the Soviets in check to stop the cancer of Communism from spreading any further.

Just as depressing for him was the sight of Britain emerging from the war exhausted, impoverished and – for the first time in centuries – playing a secondary role in the world. The only sensible way forward for Britain, in his view, was to line up behind the Americans and maintain his country's role in the 'special relationship' by acquiring nuclear bombs. This is why he was disturbed that his country appeared to be making such slow progress in developing them. In the meantime, he wanted America to capitalise on its nuclear advantage, and even suggested – at the height of his frustration with Stalin's bullying – that the US make a pre-emptive strike on the Soviet Union.

Churchill's views were already in place on the day after Hiroshima, when he visited his old friend Lord Camrose, the proprietor of the *Daily Telegraph*, at an off-the-record meeting in the newspaper's Fleet Street headquarters.[2] Afterwards, Camrose wrote in his notes:

Churchill is of the opinion that, with the manufacture of this bomb in their hands, America can dominate the world for the next five years. If he had continued in office he is of the opinion that he could have persuaded the American Government to use this power to restrain the Russians. He would have had a show-down with Stalin and told him he had got to behave reasonably and decently in Europe, and would have gone so far as to be brusque and angry with him if needs be.

After the war, freed of the burdens of high office, Churchill had plenty of time to write a multi-volume account of the conflict that enshrined his version of the epic story. Camrose pressed him to get on with the job and strike while the iron was hot – if the books were not written by Christmas 1947, their commercial value would be diminished.[3]

The new government's parliamentary majority was so large that, barring disasters, it would be in office for at least one complete term (five years) and probably two – a disheartening prospect for an elderly opposition leader. In his spasmodic appearances in the Commons, he fulminated against the government's social-democratic legislative programme and its bureaucratic controls. But his opposition was weak and ineffectual. Most upsetting of all for Churchill was the government's decision to give India independence, destroying the Empire 'by a hideous act of self-mutilation'.[4]

Churchill was now approaching his seventy-first birthday. On the day after he met Camrose in the *Telegraph* offices, he was miserable, confiding to his doctor:[5]

It's no use, Charles, pretending I'm not hard hit. I can't school myself to do nothing for the rest of my life . . . After I left Potsdam, Joe did what he liked. The Russians' western frontier was allowed to advance, displacing another eight million poor devils . . . I get fits of depression.

The autumn of his life promised to be one long, gloomy November afternoon, made bearable only by the company of friends and family, the challenge of writing his memoir, and plenty of vacations in the sun. With luck, he would live long enough to have another crack at the premiership.

Although Churchill had excluded Attlee from years of discussions about the Bomb, the new British Prime Minister showed no resentment. He often consulted his former boss on nuclear policy, though he did not always accept his advice.

Churchill was angry when he heard that the government's new Advisory Committee on Atomic Energy included Blackett but not Lindemann.[6] Blackett's views and activities had long attracted the suspicion of Churchill, who had ordered MI5 four years before to 'see if they had anything against' him, but had been told that he was 'entirely harmless'.[7] That did not put Churchill's mind at rest, however. Soon after he heard of Blackett's appointment, he lobbied Sir John Anderson, putting in writing for the first time what many had suspected – after Tube Alloys, Blackett had been excluded from the British nuclear project:[8]

> As you know, [Blackett] had been kept carefully away from all this business, and I think it very likely that the inclusion of his name will have the effect of drying up . . . contacts in the United States.

Anderson's reply was masterly. He pointed out that Lindemann's name had not even been mentioned during a discussion with Attlee of possible advisers.[9] It was important, Anderson added – perhaps with his tongue in his cheek – that Churchill was advised by someone not involved with the committee so that he was not inhibited when he was criticising the government's proposals. 'As regards Blackett,' Anderson concluded crisply, 'I think he is probably safer on the committee than off it.' When Churchill replied a week later, he was beginning a painting vacation on the shores of Lake Como in northern Italy. A few days later, he told Attlee that it was 'not so much what [Blackett] would do as what the Americans will do . . . I apprehend that they will be increasingly shy of imparting the further developments'. Churchill backed down, but he had marked Blackett's card.[10] A few months later, when Blackett was nominated for an honorary degree at the University of Bristol, Churchill used his power as its Chancellor to veto the nomination.[11]

Churchill's intervention on Lindemann's behalf was, however, not in vain: Attlee gave the Prof a place on a technical committee that put him in the stalls of the nuclear debate, but not close to the main actors.[12] This kept both Lindemann and Churchill well informed but far enough from the decision-making to prevent them mounting guerrilla attacks like the ones they had waged on the Tizard Committee a decade before.

Attlee also sought Churchill's comments on a draft note to President Truman, which proposed that the United States make a Bohr-like 'Act of Faith' by sharing its knowledge of nuclear weapons through the United Nations.[13] Churchill's response was like a controlled explosion on the shores of Lake Como: such talk would 'raise immediate suspicion in American breasts', he said. He then argued – convincingly, in Attlee's view – that Britain and America should try to achieve security based on 'a solemn covenant, backed by . . . the force of the atomic bomb'. Churchill concluded with impassioned comments on the 1943 Quebec Agreement, which he said 'almost amounts to a military understanding between us and the mightiest power in the world'. The operative word was 'almost'.

*

In the early spring of 1946, Churchill spoke out on what he believed to be the growing threat of Soviet power. To make maximum impact, he chose to make a big speech not in Britain, where he was a controversial figure and apparently on the wane politically, but in the United States, at the invitation of President Truman. 'The Iron Curtain speech', as it is now remembered, was certain to receive the blanket press coverage no longer given routinely to his pronouncements in the UK.

Churchill gave the speech in Missouri, the President's home state, in the tiny college town of Fulton. He rose to the occasion, talking with all the eloquence and aphoristic brilliance his audience hoped for. The speech, titled at the last minute 'The Sinews of Peace', focused on a grand theme, redolent of his 1930s campaign on the dangers of growing militarism in Europe, with the Nazis now replaced by the Soviets. No one knew the limits to the 'expansive and proselytising tendencies' of the Soviets and their international Communist organisation, he warned.[14] Condemning any thoughts of appeasement, he urged that the only hope of countering the Soviet threat was through 'a special relationship between the British Commonwealth and Empire and the United States'. God was on their side, he believed, as He had willed that only America had nuclear weapons for the time being, providing a crucial means of containing the Soviets. It would be 'criminal madness' to give away the secret of the Bomb in the current climate, though Churchill knew that the Soviet Union would have the Bomb within the next few years – by that time, he hoped nuclear weapons would be overseen by the United Nations. His message was not, as is often depicted, a call to begin a Cold War but an argument that the UK and the US must stand together and negotiate with the Soviet Union from a position of strength, to avoid another global conflict. This was to become his clarion call.

The speech was coolly received by the British press and went down badly in America. The *Nation* objected that Churchill had poisoned 'the already deteriorating relations between Russia and the Western powers', while the *Chicago Sun* complained that he was seeking 'world domination' by America and the British Empire, though the United States wanted no such alliance.[15] Truman and Attlee quickly distanced themselves from the speech, though Churchill had been careful to consult them both when he was drafting it.[16]

Churchill had gone out of his way to express his 'strong admiration and regard [for] the valiant Russian people and for my wartime comrade, Marshal Stalin'. But this cut no ice with the Soviet leader, who made the most of the controversy by giving an interview to *Pravda* (soon reprinted in the *New York Times*) in which he pointed out that Churchill's call for English-speaking peoples to control the fate of the world was 'strikingly reminiscent of [the rhetoric] of Hitler and his friends'.[17] Not only did Churchill brush off these and other criticisms, he also did little to correct the misinterpretations of his text, commenting to an admirer that he had made the most important speech of his career.[18]

In Fulton, he had proved beyond doubt that he was still a figure to be reckoned with on the global stage. Six months later, he reinforced this impression in another powerful speech at the University of Zurich, where he spoke on 'The Tragedy of Europe'.[19] He urged that a 'United States of Europe' should be formed as another bulwark against tyranny and once more stressed that the West should use its narrow nuclear lead over the Soviet Union: 'We dwell strangely and precariously under the shield and protection of the atomic bomb.'

A few weeks later, Churchill wrote to Attlee, who had yet to be persuaded that the Soviets were bent on global domination:[20] 'It is clear to me that only two reasons prevent the westward movement of the Russian armies to the North Sea and the Atlantic,' Churchill wrote. 'The first is their virtue and self-restraint; the second, America's possession of the atomic bomb.' If the United States was not prepared to use the Bomb, then Europe could soon be overrun by Stalin's troops, he believed.[21]

In private, Churchill was as robust about using the Bomb as he was in public, as Augustus John heard over lunch at Chartwell.[22] The painter said little during the meal, but a mention of Hiroshima made him explode: the dropping of the Bomb was 'the most monstrous crime in all history'. Churchill disagreed: 'I have many things worse than that on my conscience.' The conversation then moved on. Churchill was rather more forthcoming at another lunch, this time with Lord Mountbatten, soon to oversee Britain's withdrawal from India, and Sir John Anderson. When Mountbatten suggested that the use of the Bomb had allowed the Japanese to save face after spending years as aggressors, Churchill replied that it was fair to question the decision to use the weapon:[23]

> I may even be asked by my Maker why I used it but I shall defend myself vigorously and shall say – 'Why did you release this knowledge to us when mankind was raging in furious battles?'

Anderson retorted, 'You cannot accuse your judges.'

Churchill's spirits were lifted later that summer by his last correspondence with George Bernard Shaw, in which they discussed nuclear weapons. After Churchill sent greetings to the writer on his ninetieth birthday, Shaw replied with a valedictory testimonial: 'You have never been a real Tory,' he wrote. Churchill was 'a phenomenon that the Blimps and Philistines and Stick-in-the-Muds have never understood and always dreaded', Shaw wrote.[24] Churchill replied, in a warm note. He seemed despondent about the state of the world and, once again, decided that some of the blame ought to be shouldered by God:

Do you think that the atomic bomb means that the architect of the universe has got tired of writing his non-stop scenario? There was a lot to be said for stopping at the Panda. The release of the bomb appears to be his next turning point.

Churchill wrote those words on 18 August 1946, five days after the death of his old frenemy H. G. Wells, whose original predictions thirty-five years before about the post-Bomb world had proved so wide of the mark. In his final creative spurt, Wells had worked on the scenario for a movie, *The Way the World Is Going*, aiming to show that mankind should 'face his culminating destiny with dignity and mutual aid and charity, without hysteria, meanness and idiotic misrepresentation of each other's motives'.[25] The work never reached his publishers and appears to have been lost or destroyed.

Churchill's frustration with what he perceived to be Soviet expansionism apparently drove him often to strike a more hawkish tone, as Lord Moran found when he visited him at Chartwell.[26] Over lunch, his tongue perhaps loosened by champagne and claret, Churchill was gloomy:

MORAN: You think there will be another war?
CHURCHILL: Yes.
MORAN: You mean in ten years' time?
CHURCHILL: Sooner. Seven or eight years. I shan't be there.

When Moran questioned how a country as small as Britain could take part in a nuclear war, Churchill cut loose: 'We ought not to wait until Russia is ready.' His face brightening, he went on:

America knows that fifty-two per cent of Russia's motor industry is in Moscow and could be wiped out by a single bomb. It might mean wiping out three million people, but they would think nothing of that.

Just over a year later, he repeated the same sentiment, still in private but this time when talking with the Canadian Prime Minister Mackenzie King.[27] The West should make it clear that the Soviet Union must not extend its regime any further in Western Europe, Churchill argued. He added that if the Soviets did not accept the ultimatum, a Western leader should tell them straight: 'We will attack Moscow and your other cities and destroy them with atomic bombs from the air.' This was the zenith of Churchill's nuclear bellicosity.

He soon softened his line. In the House of Commons, he went no further than the words he used after British relations with the Soviet Union deteriorated again, in January 1948:[28] the best chance of avoiding war was 'to bring matters to a head with the Soviet Government . . . to arrive at a lasting settlement'. His enthusiasm for a prompt showdown diminished during that year. He told his colleague Anthony Eden in September that he wanted to give the Americans time to stockpile more nuclear weapons and to improve their ability to deliver them – then, in a year's time, a demand for a showdown 'would certainly have to be made'.[29] He dreaded the moment when the Soviets acquired the Bomb, for then 'nothing can stop the greatest of all world catastrophes'.

When Churchill was writing his war memoirs, he made no bones about his approach: 'This is not history, this is my case.'[30] He had a terrific story to tell and the literary talent to make it sing; all he needed was a crack team of researchers, administrators and several years of hard labour to bring it to fruition. The arrangements for the research, writing and checking were in place by spring 1946, along with lucrative publishing contracts, including the all-important one in the United States, arranged by his friend Emery Reves. If it went well, the venture would enable Churchill's account to set the agenda for public discussions about the war in the English-speaking world.

He quickly set up an impressive team, often called 'the Syndicate'. Its de facto chief executive, Churchill's former research assistant Bill Deakin, was supported by Churchill loyalists Pug Ismay (formerly military Chief of Staff to the Minister of Defence), Henry Pownall (formerly Vice-Chief of the Imperial General Staff) and the Cabinet Secretary Norman Brook, who read and commented on the entire text and even wrote some of it. Among dozens of other specialists, R. V. Jones and Lindemann spent months drafting and checking the sections concerning science and technology.

Churchill persuaded the government to give the Syndicate colleagues unprecedented access to files and papers that custom and practice decreed should normally remain under lock and key for decades.

The headquarters of the writing project was Chartwell. In the summer of 1946, the unsustainable burden of running the property was removed by Lord Camrose and other philanthropists who clubbed together to finance a scheme that enabled Churchill to live there for the rest of his life, at a peppercorn rent.[31] At the Churchills' London home near the Royal Albert Hall, the Syndicate pored over thousands of documents retrieved after months of rummaging through Whitehall files, marshalling facts, composing summaries and text for Churchill to rework, and commenting on his drafts. The first volume, *The Gathering Storm*, included coverage of the period between the end of the First World War and the beginning of the Second, telling of Hitler's rise, Germany's rearmament and the feeble response of the English-speaking peoples.[32] The story – a rich, character-driven drama – had a resonant subtext: 1948 may be like 1938 all over again, with different actors.

The genesis of nuclear weapons was one of Churchill's themes.[33] He makes no mention of the articles on the subject he wrote before the war, remarking only that 'from time to time Professor Lindemann had talked about atomic energy'. When Churchill introduces the subject, he quotes in full a letter he claimed to have sent to the Air Ministry in early August 1939 – after a briefing by Lindemann to pour cold water on scare stories that the Germans had a nuclear weapon on the eve of war. This recollection was inaccurate – he was actually consulting the Ministry about whether to send the letter, drafted by Lindemann, to the *Daily Telegraph*.[34] Churchill had begun his account of the development of nuclear weapons as he would continue, with a misremembered anecdote, unqualified praise for the Prof, and no personal regrets.

Although the completion of the volume, in late 1947 and early 1948, cost Churchill a huge amount of work, it was a welcome respite from the grinding frustration of opposition. Burrowing away on the book project, he was a young journalist reborn, driving his publishers to distraction with his quest for perfection. He demanded ever more sets of proofs, tweaking previously finalised passages, agonising over difficult sections and adding new perspectives right up to the wire.

The Gathering Storm, published in the United States in June 1948 and four months later in the UK, was an instant hit, selling by the truckload and winning laudatory reviews.[35] Among the book's admirers was Noël Coward. He later wrote to Churchill to thank him for writing it, praising his 'impeccable sense of theatre', 'sublime use of words' and for telling the story with 'so little bitterness'.[36] Many critics pointed to Churchill's brazen self-centredness and faulted his account of the 1920s and '30s – there was no mention of the defence cuts he had made at the Treasury – and of his relaxed attitude to the aggression of General Franco, Mussolini and the militaristic leaders in Japan. However, none of the reviewers in the leading publications commented on Churchill's distortions of the history of the British radar programme, especially his false implication that he had always strongly supported it. He minimised Tizard's role while building Lindemann's to the sky.[37] Yet these few shameful pages and other shortcomings did nothing to mar the book's appeal, and its publication made 1948 a banner year for Churchill.

The business of parliamentary opposition bored him, and it showed. In early March 1949, his wife was so upset by the rumblings of discontent in the Tory ranks that she told him in a letter: 'I have felt chilled & discouraged by the deepening knowledge that you do only just as much as will keep you in power. But that . . . is not enough in these hard anxious times.'[38] He took her advice, cutting short his next trip to the US, where he was to speak at the Massachusetts Institute of Technology (MIT). Uneasy about addressing an audience certain to contain a preponderance of scientists, including Sir Henry Tizard and G. P. Thomson, Churchill turned to Lindemann for advice. The Prof took the opportunity to promote one of his favourite causes – the standard of engineering teaching in British universities was in his view now so poor that the UK's status as a military and economic power was under threat.[39] Britain needed its own MIT, Lindemann believed.

Churchill's arrival in New York on 23 March 1949 afforded a welcome opportunity to draw attention to the second volume of his war memoir, *Their Finest Hour*, due to be published six days later. At the New York quayside, he posed for the flotilla of photographers, one hand holding a fat cigar and huge hat, while the other made his now-trademark V-sign.[40] 'Nodding like a beaming pink cherub', as the *New York Times* put it, he was given a stirring welcome by the crowds.

His main reason for making the journey was not to give the well-paid talk, which turned out to be boilerplate Churchill – sparklingly delivered but bereft of new ideas. Rather, he wanted to talk with President Truman, recently re-elected against all the odds, and the former Secretary of State George Marshall, whose Recovery Plan (warmly endorsed by Churchill) was assisting several of the crippled economies in Europe.[41] In their meetings, Churchill urged the President to make clear to Stalin that the US was prepared to deploy the Bomb against the Soviets. Truman obliged soon afterwards, and in late June Churchill wrote to him with all the obsequiousness the President had come to expect:[42]

> I was deeply impressed by your statement about not fearing to use the atomic bomb if the need arose . . . Complete unity, superior force and the undoubted readiness to use it, give us the only hope of escape. Without you nothing can be done.

Yet Truman was more cautious about using the Bomb, as Churchill realised when he read the President's reply a few days later:[43] 'I am not quite so pessimistic as you are about the prospects for a third world war. I rather think that eventually we are going to forget that idea, and get a real world peace. I don't believe even the Russians can stand it to face complete destruction . . .'.

Churchill's memories of the Bomb project were hazy, probably reflecting the fitful attention he paid it during the war. He began to reconstruct the story in early 1949, tracking down facts and documents to clarify his confused recollections. By the late summer, he had written a brief semi-technical introduction to the Tube Alloys story and, after lunching with Lindemann, asked him to comment on a draft.[44] Although Churchill did not use the passage in the published book, his text has survived, along with annotations by Lindemann and R. V. Jones.

The passage displays Churchill's ability as a vivid writer about basic science. When setting the scene for the physicists' foray into the heart of the atom, he used what he described as 'simple metaphors and [linking] to a more scientific basis in the reader's mind':[45]

> Until recent years all chemical processes – whether it be digesting our food, the burning of coal in our grates, or the detonation of high explosives, like TNT – entailed merely the rearrangement of the [electrons]. No one had interfered with the nuclei. These powerful bodies dwelt unperturbed each with their own cluster of gnats (electrons) buzzing about them.

He gives a clear account of nuclear chain reactions, though he makes no explicit mention of Frisch and Peierls's memorandum or of any other contribution by refugee scientists. Churchill does, however, outline the work of the MAUD committee – 'Their battlefield was the laboratory' – and comments that he relied on Lindemann 'to give me a prod if any action were needed [as] I had a lot of other things to do at the time'. No extant document or diary entry from the spring of 1941 makes any mention of his emotional reaction to the committee's report, though he writes in his war memoir that he was profoundly concerned:

> I had a very deep fear of a new explosive and was much alarmed to learn that it was approaching the threshold of a war which already seemed bad enough. If we had found out so much, what about the enemy? We knew they were groping in this direction . . .

American scientists were kept apprised of the MAUD proceedings, Churchill writes, so he was 'not surprised to receive a note from President Roosevelt' in the late autumn of 1941. This was the offer of equal partnership that Churchill did not answer for seven weeks. Making no mention of this, he simply remarks: 'I naturally replied I would be delighted to receive his nominee and to put him into the picture . . .' The draft passage ends there. After he wrote it, he continued his struggle to get his nuclear story straight and to present it as another vindication of his policy of working hand in glove with America.

Two days after Lindemann began working on Churchill's draft, President Truman announced that his military had determined that the Soviets had successfully tested a nuclear weapon, at 6 a.m. on 29 August 1949. Stalin at first said nothing and confirmed only later that it was true. His acquisition of the Bomb so soon shook the American and British governments to the core – virtually every expert in the West was taken by surprise.

Churchill now had to think afresh about nuclear strategy and Britain's bond with the United States. It was not long before the relationship was again under strain, when it was revealed why the Soviets had been able to build the Bomb with such astonishing speed: they had been receiving inside information on the Manhattan Project. As Lindemann later remarked: 'Despite their elaborate spying system, it is incredible that in 1945 the Russians should have known more about the production of atomic bombs than we did . . .'[46] It soon emerged that the principal mole was one of the scientists sent to America seven years before by Churchill's officials.

FEBRUARY AND MARCH 1950

Peierls and 'the spy of the century'

> 'My Russian childhood and youth taught me not to trust anybody else, and to expect anyone and everyone to be a Communist agent.'
>
> GENIA PEIERLS to Klaus Fuchs, 4 February 1950[1]

Rudi Peierls never forgot the events of Friday 3 February 1950, especially the phone call he received shortly after noon. He was working in Birmingham University's department of theoretical physics, housed in a former army hut, the site of what was now one of Britain's leading research centres. His office looked every inch the workplace of an academic – a bicycle in a corner, a blackboard covered with quantum hieroglyphics, faculty papers stacked on his desk and piles of books covered in chalk dust.

A few minutes later, rain beating against the windows, he took the call that would scar him for life. It was a reporter on the line. This was no surprise – the press had been hounding him since President Truman, three days before, had ordered the Atomic Energy Commission to produce the hydrogen bomb.[2] The prospect of building much more powerful nuclear weapons and another escalation in the arms race had prompted Peierls – along with G. P. Thomson, Jo Rotblat and several other scientists – to sign a petition urging that 'utmost attempts' should be made now 'to eliminate atomic warfare'.[3] But this journalist, from London's *Evening Standard*, wanted Peierls's views on something quite different, news that had just broken in the capital: his old friend and colleague Klaus Fuchs had been arrested and charged with being a spy.[4] Shattered, Peierls refused to comment.

Like many anti-Nazi students in the early 1930s, Fuchs had associated with Communists, but Peierls had never detected any sign that his friend was on the far left, still less that he was a Soviet sympathiser. Nor, in Peierls's view, had Fuchs ever seemed capable of such disloyalty to the country that had given him a home and a good living when he was a desperate refugee.[5] Fuchs was now the UK government's leading nuclear theorist, and had brought dozens of secrets from Los Alamos that were extremely valuable to the British Bomb project. The two men had spoken on the phone only the day before and, to Peierls, nothing seemed amiss.[6]

As usual when Peierls had to deal with difficult news, he phoned his wife and soul mate Genia at their home. She prided herself on her ability to take the measure of anyone and give them wise advice on everything under the sun ('I am the cleverest person in Birmingham,' she once told a friend).[7] Yet even Genia was bewildered. Fuchs had been a close friend of the family for nine years, a lodger for eighteen months. He was now a regular guest, good with the children and helpful around the house.[8] The notion that he had been lying to them all for years was inconceivable. As they struggled to put words to their anguish, Rudi and Genia's conversation was fragmented and barely coherent:[9]

GENIA: But, my dear, you are in the same danger yourself.
RUDI (replying in Russian): No.
GENIA: How?
RUDI: I don't know, but I couldn't care less now, anyway.

This exchange was recorded by the British security service MI5. Some of its officers had installed a phone tap at the Peierlses' home that morning, just in time to catch their reactions to Fuchs's arrest (Peierls's university correspondence had been tapped since his return from Los Alamos). A few hours later, Peierls was still reeling. Desperate to talk to his wife, he cycled through the rain to their home, a huge Victorian house in Edgbaston, a fifteen-minute ride from his office. Waiting until their two babies had been put to bed and their two teenage children were out of earshot, the Peierlses mulled over the news again. Neither could believe that Fuchs was a traitor – he was simply not capable of that level of pretence.[10] Peierls decided that he must go and see Fuchs to get things straight and make sure he had a solicitor.[11] At 9.15 that evening, he telephoned the Metropolitan Police to ask if he could visit Fuchs in London the next day. His request was granted.

In Scotland Yard the next morning, Peierls heard that his friend had confessed eight days earlier. In a four-thousand-word statement, Fuchs had explained to the security services why he had decided to become a Soviet spy soon after joining Peierls in Birmingham to work on the Bomb:[12]

> When I learned the purpose of [Peierls's] work I decided to inform Russia and I established contact through another member of the Communist Party. Since that time I have had continuous contact with [intermediaries]. At this time, I had complete faith in Russian policy and I believed that the Western Allies deliberately allowed Russia and Germany to fight each other to the death. I had, therefore, no hesitation in giving all the information I had . . .

Peierls – barely able to believe what he had heard – made his way to Brixton Prison. There, he was ushered into the Deputy Governor's office to meet Fuchs, who was brought from his cell shortly after lunch. Their conversation never left first gear, each of them speaking slowly and so softly that the police observers could barely hear what they were saying. Fuchs said that he regretted his disloyalty now, having come to appreciate the virtues of Western life. Now that he had been unmasked, he sought no mercy.[13] Peierls, glum and dispirited, asked him if he needed any help, but Fuchs was indifferent to everything except the offer to send him a few books and some clean underwear. After talking for little over a quarter of an hour, the two men exchanged pleasantries and Peierls, shaken by Fuchs's unutterable naivety and foolishness, headed back to Birmingham.

At home that evening, Peierls poured out his anger and frustration to his wife. He also wrote a statement for the Metropolitan Police, detailing his recollections of the conversation with Fuchs, and putting himself at their disposal.[14] Genia, weeping in an armchair close to the fire in the sitting room, reached for her fountain pen and wrote Fuchs a long letter, as excoriating as it was far-sighted:[15]

> Do you realise what will be the effect of your trial on scientists here and in America? Specially in America where many of them are in difficulty already? Do you realise that they will be suspected not only by officials but by their own friends . . .?

Fuchs could now never be happy, she wrote angrily, but he should at least 'Try to save as much as you can of this decent and warm and tolerant [and] free community of international science which gave you so much . . .' She signed off 'God help you!' and handed Peierls the letter to type up and send to Fuchs. The pages, barely legible, were soaked in tears.

Fuchs quickly replied, pleading that he had not thought about what he was doing.[16] He had suffered from 'controlled schizophrenia', he said, trying unconvincingly to explain his actions. 'I didn't control the control; it controlled me.' Now that he had decided to come clean and get everything off his chest, he said, his greatest fear was that the prison authorities 'would discover the safety pins which held my pants together'. Fuchs thanked Genia for being kind enough not to mince her words, writing that it was 'funny that women see things so much clearer than men'. In a postscript, he added that at least he had now learned again how to love and cry – tears stained his letter, too.

In Britain and America, Fuchs's face stared expressionlessly from dozens of newspapers, his ample forehead shining above circular-framed glasses.[17] His unmasking occurred at a time when fears of a Cold War were growing alarmingly. On 1 October 1949, eight days after President Truman's announcement that the Soviets had exploded a nuclear device, Mao Zedong formed the Marxist-Leninist People's Republic of China, after besting his nationalist enemies. When the Fuchs story broke, Mao was in Moscow negotiating what became the Sino-Soviet Treaty of Friendship, Alliance and Mutual Assistance, concluded on 14 February. According to Wisconsin Senator Joe McCarthy, Communism was on the march. Five days before, he had launched his campaign to expose hundreds of spies that he alleged were working in the US government and military. It was these people, he insisted, who must have enabled the Soviets to acquire the Bomb so soon. His claims resonated with the public mood of fear and panic.

Other nuclear spies had been unmasked, but none of them had been given Fuchs's access to top-secret information. Day after day, the British newspapers castigated MI5, imagining the secrets that Fuchs might have divulged to the Kremlin and the effect this embarrassment was having on Britain's relationship with America. There was no choice for the British security services but to sit in the stocks, mute and red-faced, while they were pelted with richly deserved abuse.

On the day the Fuchs story broke, General Groves testified for two and a half hours in Washington before a closed session of the Joint Committee on Atomic Energy. Its chairman Brien McMahon called in the press and summarised the General's testimony. He gave the *Washington Post* the impression that Fuchs had handed to Russia 'data involving the super-secret hydrogen bomb', fanning worries that the Soviets might beat the Americans to an even more powerful type of nuclear weapon.[18] Republican Senator John Bricker remarked: 'I've always opposed the use of foreign scientists on [nuclear] projects [and] the arrest of Fuchs makes me even more certain that I am right.'[19] This argument, Peierls observed, missed the fundamental point that without foreign-born scientists America would have no nuclear weapons at all.[20]

The vortex of the Fuchs scandal swirled over the United States for weeks but it dissipated within a few days in Britain. There, most people were more interested in the imminent election, in which Churchill was apparently poised to give Attlee a bloody nose. Peierls kept up with the Fuchs story in radio news bulletins and reports in *The Times*. He regarded himself as politically centrist and a thoroughgoing democrat, content to respect the decisions of Parliament, which he believed were almost always reasonable.

It was in this spirit that he led the UK Atomic Scientists' Association. His politically neutral leadership confounded MI5's suspicions both of him and of his organisation. One report concluded that he believed the Association was needed only as a watchdog that needs to do 'very little, so long as politicians and others are sensible about atomic energy'.[21] The organisation concentrated mainly on promoting discussions between scientists about nuclear matters and on public education initiatives, including a drab Atom Train exhibition about nuclear physics organised by Chadwick's colleague Jo Rotblat. It chugged its way round Britain in 1947–8.[22] Of Peierls himself, the security services found nothing more incriminating than the fact that he was married to a Russian, and nothing more unpleasant than a comment by one of its agents that he was 'a shifty and rather oily individual'.[23]

Peierls's colleagues would not have recognised that description. He was popular and admired for his integrity, his ability both as a theoretical physicist and a cultivator of fresh academic talent. Having turned down professorships from Oxford, Cambridge and other leading universities,[24] he built the reputation of theoretical physics in Birmingham. He worked on an impressive variety of problems, concentrating on trying to understand the behaviour of the electron and other sub-atomic particles using quantum theory and relativity. After the war, his experimental colleagues had been led by Mark Oliphant with a good deal of sound and fury, but with precious little success. Oliphant had returned to settle permanently in Australia during the previous summer, having lost patience with what he regarded as the Attlee government's poorly coordinated and parsimonious support for basic science. Peierls had proved to be a much more effective leader: his theoretical physics group was now the best in Britain, according to his protégé Freeman Dyson, who was lodging with the Peierlses when they heard of Fuchs's treachery.[25]

Eleven days after the news broke, shortly after Fuchs's first court hearing, Peierls still found his colleague's behaviour incomprehensible,[26] and was in no doubt about its consequences. Unburdening himself in a letter to Bohr, Peierls concluded that spying was inevitable in open societies, but not in the Soviet Union:[27]

Russia has found how to stop leakages very effectively. If this is the only effective solution do we want to go that way ourselves or should we not say that at that price security is not worth having [?]

Peierls visited Fuchs again in late February, this time accompanied by Genia, but they learned nothing new. By that time, Peierls had convinced himself that his friend's contrition was genuine and told Fuchs's defence lawyer to say that his client understood how much he had let down the scientific community. 'This is his worst punishment,' Peierls wrote, 'worse than whatever the judge may impose.'[28] The lawyer, alarmed by the deep sympathy evinced in these words, immediately alerted the security services.[29]

Fuchs came to trial at the Old Bailey on 1 March, less than a month after his arrest. The public gallery was crammed with reporters from all over the world, as well as celebrities including H. G. Wells's former mistress Rebecca West. Fuchs pleaded guilty on all counts, and the proceedings were over in ninety minutes, after which he thanked the court for a fair trial.[30] Ill-acquainted with English law, he had earlier thought he would be executed, but that was not permissible as his crime of passing secret information to allies during wartime was relatively minor. The judge, Lord Chief Justice Goddard, was nonetheless in no mood to be lenient. He concluded that 'the maximum sentence Parliament has ordained is fourteen years: that is the sentence I pass upon you'. He banged his gavel and security guards promptly whisked Fuchs off to Wormwood Scrubs.

The story of his imprisonment livened up the post-election bathos in Britain. Attlee's Labour government had won a second term, though its majority had collapsed to only six, putting Churchill within sniffing distance of a return to power. On 14 February 1950, nine days before polling day, he had been the first leading politician to talk openly about the possibility of nuclear war. As the *Manchester Guardian* reported, he introduced the Bomb into the election campaign 'with explosive effect'.[31]

FEBRUARY 1950 TO SPRING 1951

Churchill softens his line on the Bomb

> 'I really do not know why it is that when we were so far advanced in this new, mysterious region of atomic war we should have fallen so completely behind in these last four years.'
>
> WINSTON CHURCHILL, Edinburgh, 14 February 1950[1]

Churchill made his explosive remarks about nuclear policy towards the end of a long speech in the Usher Hall, Edinburgh, to an audience of almost three thousand. Having complained that the British government had not yet been able to equip its military with the Bomb – 'one of the most extraordinary administrative lapses that have ever taken place' – he moved on to the Cold War.[2] Things were now so grave, he intoned, that it was 'not easy to see how matters could be worsened by a parley at the summit'. This was the first use of the word 'summit' in this context – perhaps he chose it to chime with the national obsession with conquering Mount Everest.

Although he made the best of the United States' lead in the nuclear race – it had 'almost a monopoly' – he implicitly accepted that the time for making veiled threats to the Soviet Union was over. Labour politicians dismissed all this as a gimmick, but Churchill had assuredly changed his views. He never again spoke of showdowns and attacks on Moscow, but switched to the need to talk with the Soviets to avoid a nuclear Armageddon. This new stance was to become the leitmotif of his final years as a political leader.

Before he could press this new line in Parliament, he had to deal with the matter of Klaus Fuchs, who had been recruited on his watch. After the Fuchs arrest, Whitehall officials had closed ranks and declared that no one was at fault, although the security investigations unearthed material that would have made General Groves blanch. As soon as news of the arrest broke, Attlee agreed with Sir John Anderson that 'there was no case against the Security Services' and promised to have a quiet word with Churchill, probably to keep him on side.[3] Attlee addressed the 'deplorable and unfortunate incident' near the beginning of the new Parliament's first full session, on 6 March, exonerating both his own government and Churchill's wartime administration.[4] The Commons listened to him in silence, Churchill showing little interest.

This was to be a tetchy, unpleasant Parliament, permanently on the verge of dissolution. Attlee tried to prolong his previous government's programme of nationalisations, high taxes and austerity measures, but his slender majority forced him to steer a moderate course and to avoid any danger of a rebellion.

Though sometimes taunted by Labour backbenchers, Churchill remained a revered figure in the House, feted on the fiftieth anniversary of his election to Parliament, and honoured when the Members' entrance was named after him. But in the Commons he sometimes stooped to behaviour that was below the standards his admirers expected of him, repeatedly taunting Attlee and trying to put him off his stride.[5] Yet, over a whisky in the Commons' Smoking Room, Churchill could turn the sourest of resentments into the sweetest of amities. Despite his sometimes frail health and increasing deafness, he was impatient to win back what he regarded as his job, whatever it took.

Early in the new Parliament, in June 1950, the Cold War entered an exceptionally dangerous phase when war erupted in Korea after a hundred thousand North Koreans, with Sino-

Soviet backing, invaded their southern neighbour.[6] To the Americans, this was almost as great a shock as the Pearl Harbor attack. Two days after the invasion, President Truman issued a statement: 'it is plain beyond all doubt that Communism has passed beyond the use of subversion to conquer independent nations and will use armed invasion and war.' Under the auspices of the United Nations, the Americans went to South Korea's aid, and Churchill supported the Attlee government's decision to provide an auxiliary fighting force. It remained to see how many casualties the Americans would tolerate before they considered playing a nuclear card.

Although Churchill almost always backed initiatives that promoted Anglo-American military collaboration, he was concerned about Attlee's decision to allow American bombers to be based on British soil, in East Anglia. He wrote privately to Attlee that the base was 'the bull's eye of any Soviet target, should they decide to make war'.[7] The Baruch plan for international nuclear controls also appealed to Churchill, partly because he believed its inevitable rejection meant that America would not be obliged to share any of its nuclear know-how with the Soviets. Churchill wrote to his colleague Anthony Eden:[8]

> The Baruch Commission had to find some way of refusing to disclose or share the Atomic weapon with the Soviets. They therefore insisted upon inspection[,] feeling sure that this would never be accepted by the Russians.

Lindemann's support for Churchill was as strong as ever. In the Common Room at Christ Church, the ailing Prof railed against Attlee's defence strategy, which he thought left Britain dangerously exposed to a sudden Soviet attack, and about the red tape that he believed was hindering the production of the first British nuclear weapons.[9]

Lindemann was also helping Churchill to complete the fourth volume of his war memoirs. It was proving difficult to tell the story of the Bomb coherently, Churchill found, partly because the paper trail was incomplete but mainly because his recollections were so muddled. At weekends with his family at Chartwell, he spent as much time working on his book as he did painting, entertaining friends and playing with his grandchildren. Much as he did for most of the 1930s, he was now spending more time writing than on politics, though now at least he had the incentive that his return to power could be imminent.

In early October 1950, he and his wife flew to Copenhagen to receive Denmark's highest honour, the Order of the Elephant. It was to be presented by the Danish King in the Palace of Fredensborg, an eighteenth-century baroque building a short drive from the capital.[10] As Churchill knew, he could scarcely avoid meeting Niels Bohr, one of the few other living civilians to have been given the same honour. A few months before, the great physicist made another ineffectual foray into international politics with an 'Open Letter to the United Nations', calling for greater openness in dealing with the challenges of the nuclear age.[11]

The three-day trip was one of the highlights of Churchill's year. It was such an important event in Denmark that its government brought forward the date of the scheduled General Election to avoid a clash with 'Churchill week'. Many of the former Prime Minister's Danish admirers remembered huddling around their radios during the war, listening to his defiant speeches. He and his wife were driven to the city centre at dusk in an open-topped limousine, the route lined with cheering crowds waving Union flags, with dozens of shop windows displaying lighted candles – a unique gesture of welcome to a foreigner.

Churchill had heard himself eulogised hundreds of times since the war, but this praise was especially sincere and particularly appreciative of his intellect. At the University of Copenhagen, where he was awarded an honorary PhD, the Rector went so far as to compare their visitor's 'unique concentration of mental power' to 'the energy concentration in an atomic nucleus'.[12]

After the ceremony, Churchill spoke engagingly about the role of universities, pointing out that the number of exams he had passed was now exceeded by his tally of honorary degrees. In a passage alluding to his early spats with H. G. Wells, Churchill praised his host university's science, but with a notable qualification:[13]

> The first duty of a university is to teach wisdom, not to train, and to confirm character and not impart technicalities. We want a lot of engineers, but we do not want a world of engineers. We want some scientists, but we must make sure that science is our servant and not our master. It may be that the human race has already found out more than its present imperfect and incomplete structure will enable it to digest.

This was one of the few speeches he gave in Copenhagen that did not call for a united Europe or even for 'an effective world super-government'.[14] He had floated this idea before and it had played well with his audiences in Denmark, and all over the continent.

Churchill talked with Bohr at the farewell lunch in the Palace of Fredensborg. Like Einstein, Bohr was fond of breaking the ice with new acquaintances by playing with favourite toys. On this occasion, he brought a spinning top that unpredictably and repeatedly flipped over, while still in motion on the floor. Smiling broadly, he demonstrated it, prompting a brief exchange as they leant over the rotating toy: when Churchill commented, 'I don't understand that,' Bohr replied, 'Neither do I.'[15] Thus ended their final conversation – more equable than their first, though no more consequential.

Two months after Churchill's visit to Denmark, the world again appeared to be in danger of slipping into nuclear war. The crisis began in the White House on the last day of November, after President Truman was quizzed in a press conference about the war in Korea, which had taken a turn for the worse – the Chinese soldiers had openly joined the North Koreans in pushing back American forces. Under pressure from the journalists the President, usually cautious about the prospect of using nuclear weapons, said rashly – and inaccurately – that his most senior military commander in the Korean War had the authority to use them against China.[16] Within hours, there was global panic. With the British Labour Party close to uproar, Attlee rushed to Washington, his first visit for five years. The result was a calming announcement from the White House, expressing the wish that 'world conditions would never call for the use of the atomic bomb'.

In the Commons, Churchill declared in a measured speech that the Washington visit had 'done nothing but good'.[17] He seemed to be in fine form, but beneath his composed demeanour, he was seething – a few hours before Attlee boarded his flight to America, he told Churchill for the first time that the Quebec Agreement had proved unsustainable and was no longer in force. Britain now had no power to veto an American decision to drop the Bomb. Churchill liked Attlee and admired many of his qualities, but never forgave him for surrendering the hard-won agreement, for being so 'weak and incompetent'.[18]

Towards the end of his speech, Churchill turned his attention to the Bomb, taking a swipe at the modish suggestion that it should be used only in retaliation for another nuclear attack. This was tantamount, he said, to saying that 'you must never fire until you have been shot dead'. Again he prodded Attlee on the government's nuclear policy, which still had not been set out in the Commons. What, Churchill wondered, was the state of Anglo-American nuclear collaboration now that the Quebec Agreement was defunct?

During the next two months, Churchill piled pressure on the government to make the text of the agreement public. As it was now no longer in force, it would do no harm to release it, he argued in the Commons.[19] This was a lonely battle – even Lindemann refused to join him. 'Frankly I had never thought of your agreement with the President as binding in perpetuity,' the Prof wrote to Churchill. 'It was, I thought, a provisional agreement.'[20] But Churchill did find an ally in the new Labour MP Raymond Blackburn, who, in August 1945, had scandalised Parliament by revealing the existence of the secret agreement. Blackburn was an occasional spokesman for the UK Atomic Scientists' Association and a regular weekend visitor at Chartwell.[21]

In Whitehall, civil servants found the clamorous Blackburn 'clearly impossible to satisfy'. Nor did Attlee find it much easier to deal with Churchill, who refused to accept the Prime Minister's argument that while the signing of the agreement was a 'great achievement' in August 1943, the document was not a treaty and so had to be renegotiated after the war.[22] Publishing it now would embarrass the American government, at a time when Britain was winning back trust after the Fuchs affair. In what appears to have been a tense tête-à-tête in Downing Street on 31 January, Churchill seemed to be all but clueless about recent developments in nuclear politics.[23] Most worrisome for Churchill was that the new arrangements with the US allowed America to launch a nuclear attack from one of its air bases in East Anglia without even consulting the British.[24] If Blackett had been with them, he would have been cheering Churchill on.

After the meeting, Attlee reluctantly agreed to sound out the American authorities. But Churchill was in too much of a hurry to wait for the response and wrote to Truman asking him to release the agreement. Two weeks later, a brief handwritten reply reached Churchill via special courier:[25]

> I hope you won't press me in this matter. It will cause unfortunate repercussions both here and in your country, as well as embarrassment to me and to your government. The reopening of this discussion may ruin my whole defense program . . . Your country's welfare and mine are at stake in that program.

The message had the desired effect – Churchill quietly dropped the matter, at least for the time being.

Churchill had an opportunity to shed light on the origins of the Anglo-American nuclear partnership in *The Hinge of Fate*, the fourth volume of his Second World War memoir, published in the US in November 1950. The critics were almost unanimous in their praise. Edward Murrow, the distinguished American broadcaster, was especially complimentary, declaring the book to be 'the most revealing document to come from the pen of this great craftsman of the English language'.[26] Given that the Bomb was one of the hottest topics of the day, none of the leading critics remarked on the sketchiness of Churchill's description of its origins. He mentions them only in two passages, an account of the project's development that amounts to a whitewash.[27] The first passage begins with Roosevelt's suggestion in October 1941 that the British and American nuclear projects might 'be coordinated or even jointly conducted'. Then follows a description of the leaders' meeting eight months later in Hyde Park, at which Churchill falsely claims that Tube Alloys was high on their agenda. It was during this meeting that Roosevelt first told him that the US 'would have to do it'. Churchill also writes implausibly that if the Americans had declined to go ahead with the project, the British would have done it 'in Canada [or] . . . in some other part of the Empire'.[28]

Nowhere in *The Hinge of Fate* does Churchill mention – or even allude to – the breakdown of the Anglo-American relationship, or the political machinations that followed, after he and his colleagues failed to respond quickly to Roosevelt's offer of collaboration. Tube Alloys resurfaces only much later in the book, when Churchill disrupts his narrative by quoting a brief message of good news about the project written in May 1943. In this note, written to Sir John Anderson, Churchill reports that Roosevelt agreed that the exchange of information on the project should resume and that 'the enterprise should be considered a joint one'. Careful readers will have been perplexed by this, as Churchill had said nothing earlier about the Americans freezing the British out of the Manhattan Project in late 1942. More seriously, the quoted message gives no sense of the junior role that Britain had accepted.[29]

Churchill had the opportunity to put this right in the next volume of his memoir, *Closing the Ring*, which covered the period between June 1943 and June 1944, including the signing of the Quebec Agreement. His problem was that in July 1945 he had gone along with the US government's request to keep the agreement secret, a policy Attlee had been unable to overturn. The result was that although Churchill regarded the Quebec Agreement as the linchpin of the Anglo-American nuclear relationship, his memoir did not even mention it.

Around this time, the nuclear agendas of Churchill and Lindemann began to differ in emphasis. Whereas Churchill focused on Attlee's surrendering of the Quebec Agreement, Lindemann was concerned only with what he believed to be the slow progress of the British nuclear project. For the Prof, the government had made a disastrous error in placing the venture – at Patrick Blackett's insistence – under the aegis of the Ministry of Supply, a languorous bureaucracy capable of grinding down even the liveliest innovator. It was vital, the Prof believed, to free the scientists and technologists contributing to the programme by allowing them to work in a fleet-footed organisation that operated at arm's length from government.[30] Without this change, Lindemann believed, Britain's capacity to build bombs would suffer and 'The possibility of achieving full collaboration concerning plutonium and hydrogen bombs with the US will vanish unless we have something of our own to show.'

Lindemann was keeping a close eye not only on the government's nuclear plans but also on the deliberations of his fellow scientists in the Atomic Scientists' Association. With an irony that always eluded him, he condemned every attempt by any scientist to influence politics. Peierls bent over backwards to keep the Association free of party-political bias and in particular to ensure that Lindemann was not provoked to resign. The two men had rubbed along for the past decade, but their superficial amity ended on a conference platform in Chicago. Peierls joked that Lindemann's defeatist attitude to arms control demonstrated that, in his attitude to history, he was at root a Marxist.[31] That did it. Peierls was now beyond the pale for the Prof, who briefed the security services against him, telling them that he neither liked nor trusted him, and that he 'frequently behaves like a silly ass in matters of security'.[32] In MI5's sheaf of security records on Peierls, these are the only words a fellow scientist spoke against him.

Churchill showed little interest in either the organisation of the government's nuclear scientists or their attempts to lobby politicians. He was more perturbed by the time they were taking to build the Bomb. But he had no idea of the huge nuclear enterprise that Attlee had funded, led by a scientist Churchill had scarcely heard of, soon to be nicknamed 'the British Oppenheimer'.

AUGUST 1945 TO OCTOBER 1951

Penney delivers the British Bomb

> 'The discriminative test for a first-class power is whether it has made an atomic bomb and we have either got to pass the test or suffer a serious loss in prestige . . .'
>
> WILLIAM PENNEY, 1951[1]

Bill Penney's sobriquet 'the smiling killer' had not stuck, rightly: he was so peaceable that he would think twice about crushing an ant. Yet underneath his beatific calm he was absolutely determined to provide his country with the most advanced nuclear weapons. During the British Bomb project – a 'comic opera', as he later described it – he did more than anyone else to deliver the weapons, and won the admiration of the leader who later commissioned him to deliver the H-bomb, Winston Churchill.[2]

Penney was remarkably self-possessed. At Los Alamos, he had been as stoical as a martyr after his wife died of post-natal depression in the spring of 1945, three and a half years after the birth of their second child. None of Penney's colleagues saw any sign of grief.[3] He also had a singular ability to separate the strategic imperative of making the Bomb from the awful consequences of dropping it on civilians. Alone among the British nuclear scientists, he had witnessed the obliteration of Nagasaki, on board one of the aircraft shadowing the plane that dropped the Americans' second Bomb. A few years later, he recalled the distress of everyone on the flight: 'We realised that a new age had begun and that possibly we had all made some contribution to raising a monster that would consume us all.'[4] A few days after he saw Nagasaki destroyed, he and two

American colleagues walked the city, examining everything that remained – parts of buildings, the odd telegraph pole and gravestone. Back in his hotel room, he analysed the data with the detachment of a pathologist and the rigour of a mathematician. He eventually concluded that the explosion was equivalent to twenty-two kilotons of TNT, ten per cent higher than President Truman had announced.[5]

Penney was without peer as a curator of nuclear carnage, which is why he was the only British scientist the Americans regarded as indispensable. He was valued equally highly by the British government – unusually for a top-flight scientist, he worked as effectively with ministers and officials as he did with his academic colleagues. Penney was popular, too, for his accessibility – everyone who knew him well called him not Professor Penney, but 'Bill'.

As a young researcher, he had collaborated productively with the renowned American quantum theoretician John Van Vleck, who thought highly of him. Although Penney was not among the cream of scientists, he was a fine mathematical physicist, writing admired papers and giving exquisitely well-crafted lectures.[6] He was a popular leader of the British nuclear project, especially with his young colleagues. Sometimes he would surprise them by showing up at their desks, enthusiastic and encouraging, dressed in a jumper two sizes too big, and switching easily from the subtleties of mathematical ballistics to the latest fortunes of Arsenal Football Club. As a lad, he had been a gifted sportsman, on the pitch, on the track and in the ring. Now in his late thirties and with an ampler waist bulging over his belt, he was content to spend his weekends tending the roses in his garden and playing a few gentle rounds of golf. It was easy to believe that he had once been a boxer, however, and he brought his sometimes intimidating presence to dozens of inquisitions by government officials. Swathed in the jacket of his baggy double-breasted suit, he spent hours fielding one tough question after another from worried ministers, batting each delivery to the outfield with an artistry no other scientist of his day could match.[7]

Penney later said that he had been dragged into the British nuclear project. It was C. P. Snow, supported by James Chadwick, who twisted his arm soon after the war ended. Even though the job of leading the Armaments Research Department was unattractive – a knacker's yard for scientists of Penney's ambition – Snow argued that Britain was undoubtedly going to build its own nuclear weapons and the job had to be done by someone with Penney's knowledge and aptitude. 'Will you do it?' Snow pleaded.[8] Motivated mainly by patriotism, Penney accepted, but no information was forthcoming about the prospect of a British Bomb.[9] He had other things to attend to: he remarried, to his children's nurse, and worked almost full-time as an adviser to the American military. As part of his close involvement with the first peacetime nuclear tests on the Bikini Atoll in the Pacific, he was responsible for gauging the power of the blasts and the toxicity of the lingering radioactivity.

Having heard nothing for fifteen months about the British government's intention to build a Bomb, Penney was eventually summoned to Whitehall in May 1947. Lord Portal did not beat about the bush:

> We're going to make an atomic bomb. The Prime Minister has asked me to coordinate the work. They want you to lead it. I am not going to worry about all the details. But I'm prepared to use my influence, with the Prime Minister if necessary, to get all the resources you need. In a few days, bring me a plan saying what you need.

Penney typed out his requirements on two pages, including 'one hundred scientists and engineers', new facilities at the

Royal Arsenal Research and Development Establishment in Woolwich, and new fences to go round it. The Ministry of Supply must grant all Penney's requests, Portal told his officials, who quickly found ways round their instructions. As Penney later recalled: 'I got the fences!' The comedy had begun.

The entire project was to be administered in London by a dozen Ministry staff working on the fourth floor of Shell Mex House on the Strand.[10] The office, protected round the clock by armed guards, was known to insiders as 'the Cage' because it was located behind a network of bars, vertical and horizontal. The Ministry was notorious for following the government's regulations to the letter on every detail, from monitoring orders for the canteen's cooking fat to the delivery of nuclear warheads. What the project needed was a leader of General Groves's drive and gumption, but it was never given one. Instead, Penney was expected to deliver with his hands tied by blinkered and uncooperative civil servants. It is hardly surprising that he was, though outwardly as cheerful as ever, privately disillusioned.

He quickly set up his teams at Woolwich and at Fort Halstead, a campus-sized outpost of the Ministry of Supply hidden behind trees on the main road linking London to Sevenoaks. He did not attempt to replicate the hothouse atmosphere of Los Alamos, with its galaxy of research stars and its bountiful resources, nor was he given the opportunity. Instead, he worked within extremely tight budgets and with mostly inexperienced researchers, although he recruited a few alumni of the British mission at Los Alamos, including Ernest Titterton, James Tuck and Klaus Fuchs, at that time assumed to be entirely loyal. Using the notes and memorised details that Fuchs and Penney brought back from The Hill, the team decided to base the design of their weapon on the Americans' plutonium Bomb.

By the time the British Bomb project had been made public

in May 1948, Penney realised that he had seriously underestimated the number of staff he needed to deliver it: not one hundred but five times as many. His target delivery date of 1951 was beginning to look precarious. Most frustrating of all, although Attlee had committed his government to the project, Penney had to spend month after enervating month pleading with unsympathetic officials for funds.[11] Sir Henry Tizard, whose 'cake-cutting committee' apportioned funds to British defence projects, had agreed that Penney's Bomb should be given overriding priority,[12] though by the autumn of 1949 his enthusiasm had waned. The rude shock of the Soviets' first nuclear test, the apparently high cost of Penney's project and quite probably his reading of Blackett's book convinced Tizard that Britain should rethink its defence priorities. In a bold, top-secret strategy paper for the Chiefs of Staff, he reasoned that Britain should focus on working with its military allies to defend mainland Europe against Soviet attack using conventional means. This would be at the expense of acquiring the Bomb, which would not make much difference in any scenario they were likely to encounter, Tizard believed.[13] He concluded that Britain 'should cease for the time being to manufacture [nuclear weapons], but should continue research on a scale necessary to keep in the forefront of knowledge'. The message was clear: Tizard wanted Penney's budget slashed.

The Chiefs of Staff, eager to have the Bomb, were having none of it. Tizard was humiliated into backing down and overseeing cuts to conventional weapons programmes to help fund the Bomb project, which belatedly gathered momentum. Reinvigorated, Penney waded through the sludge of regulation, but had no luck at all. No sooner was his project back on schedule than he was dealing with the loss of his finest theorist, Klaus Fuchs. Soon afterwards, Churchill piled on the agony with his complaint about the time it was taking Britain to develop the

Bomb.[14] Maintaining his poise, Penney sustained his campaign for adequate resources by exploiting the urgency that Churchill had helped to promote.

In the spring, Penney took possession of a bleak, disused airfield on the outskirts of Aldermaston, a picturesque village forty-five miles west of London. This was soon to be the headquarters of British nuclear weapons research and development, as Aldermaston's unsuspecting residents learned from local reports of nearby 'atomic devilries'.[15]

There was still no sign of an end to the infighting at the Ministry of Supply. After Attlee had deemed the project 'super-secret',[16] Penney could share his frustrations with no one – he seemed destined to spend the next few years starved of resources, and to have to suffer in silence. However, in the House of Lords, one well-placed expert did choose to speak out – Frederick Lindemann. Briefed by Penney's boss Lord Portal,[17] Lindemann tabled a motion in the House of Lords regretting the slow progress of Britain's nuclear programme. In a speech almost as powerfully argued as it was soporific, he praised the scale of the government's project ('they have not grudged men, money or materials')[18] but argued that it should never have been delegated to the clodhopping bureaucrats at the Ministry of Supply.

The British nuclear venture, Lindemann insisted, should be run by a quasi-autonomous organisation, fleeter of foot and untrammelled by outdated customs and practices. The government's spokesman Viscount Alexander of Hillsborough found himself on the defensive. He offered only a weak defence of the status quo and conceded, rather more readily than the Ministry of Supply would have liked, that Attlee and his colleagues might consider running the project differently.

The Prof had drawn blood. The Lords passed his motion with a handsome majority, an outcome Penney probably found gratifying, though he was too discreet to say so.[19] Yet he and his Aldermaston colleagues knew the implications of the Westminster debate: with the government's hold on power looking precarious and with Lindemann so close to the leader of the opposition, fundamental changes to the British nuclear project might well be afoot.

By the late summer of 1950, Penney was confident enough to inform Lord Portal that the Bomb should be ready to test in about a year, on schedule.[20] With the project still a well-kept secret from all but a few of the most senior ministers in Parliament, in May 1951 Attlee's nuclear confidants began to plan Britain's first nuclear test, dubbed 'Operation Hurricane'. By this time, Penney's patience with the sclerotic Ministry of Supply and his paltry salary had worn desperately thin. He considered quitting his post and taking a well-paid professorship,[21] a move that would have put his project's timetable in jeopardy. This concentrated the minds of ministers wonderfully – they persuaded him to stay and gave him the funds he needed for the project.

In spite of Attlee's strictures about the need for security, news of the planned British test soon leaked, in the *New York Herald Tribune*.[22] While Penney's team worked frantically to finalise the Bomb's design and assemble it, he and the Chiefs of Staff and other officials deliberated over the options for the location of the first test, preferring to keep it on a site in the Commonwealth. The Monte Bello Islands, off the north-west coast of mainland Australia, were the most popular option, followed by sites in the Canadian wilderness. The American government at first appeared to be disinclined to host the test, but in September changed its mind, offering to discuss with Penney and his colleagues the possibility of a site in the Nevada desert.[23]

In September 1951, with the back-room debate about the

nuclear test site in full swing, Attlee announced that there would be a General Election on 25 October. As well as putting the quarrelsome Parliament out of its misery, he hoped that voters would increase his majority and strengthen his mandate. Labour MPs swiftly went on the offensive, repeatedly calling Churchill a 'warmonger', a charge he angrily rebutted as 'cruel and ungrateful' while maintaining his composure and running a surprisingly moderate campaign that concentrated on domestic policy.[24] In his final election speech, given at a football ground in Plymouth two days before the vote, Churchill unfurled his true colours, focusing not on dreary domestic details, but on saving the world. He wanted to be Prime Minister so that he could persuade the Commissars in the Kremlin to have a 'friendly talk' and 'make an important contribution to the prevention of a Third World War'.[25]

The speech made a splash in the next morning's newspapers. Penney and Tizard probably read Churchill's words en route to another Chiefs of Staff committee meeting to discuss nuclear weapons. Tizard had recently set aside his usual caution, and warned in a confidential policy review that Britain's military strategy was deluded:[26]

> We persist in regarding ourselves as a Great Power, capable of everything, and only temporarily handicapped by economic difficulties. We are not a Great Power and never will be again . . . A wholly exceptional set of circumstances enabled us to become a Great Power for a century; a small period of time compared with the Romans, the Spaniards and the French. That set of circumstances will never recur.

This clear-headed *cri de cœur* left no trace in Whitehall, and it was certain that a Churchill administration would regard such views as defeatist nonsense.[27] Sir Henry plodded on, implementing defence policies he regarded as wrong-headed and delusional, including the high-priority funding of nuclear weapons at the expense of more rational investments.

Penney's team was now finalising the design of its Bomb, which was similar to the implosive device that the Americans had detonated over Nagasaki.[28] All the details of the design of the first British nuclear weapon were, however, finalised on home soil, and the plutonium for the core was to be supplied by Christopher Hinton's reactors, eked out if necessary with additional fissile material from Chalk River in Canada. If the Bomb worked, it would demonstrate to the Americans that the British were quite capable of building a nuclear weapon on their own.

Penney had recently been in Washington, where he had been impressed by the Americans' willingness to help test the British Bomb.[29] Nevertheless, it seemed to both him and Tizard that Monte Bello was the best place to detonate the weapon. Most important, the site allowed them to plan and deliver Operation Hurricane without having to consult the American military and risk a sudden change in their attitude.[30]

With the politicians back in their constituencies, the civil servants and other officials in Whitehall were kicking their heels, waiting to hear the names of their new bosses. At the Ministry of Supply, officials in 'the Cage' had good reason to be jittery – the scientists were now poised to deliver the Bomb, but a change in government could put an end to all their plans.[31] If Churchill, famously most interested in matters of defence and international relations, were returned to Downing Street, he would probably have Lindemann at his side. The Prof had been making trouble for the Ministry and, though friendly with Penney and the other leaders of Britain's nuclear project, was sometimes woefully ignorant of its progress. Only a few weeks before, he had told the House of Lords that the civilian use of nuclear energy was 'several decades' away.[32]

Officials were more concerned about how Churchill would react when he learned that Attlee had hidden from him – and almost everyone in Parliament – the budget for building nuclear weapons. Penney would soon find out the fate of his plans to detonate the first British Bomb: two days later, Churchill was Prime Minister again, elected to the office for the first time.

4

CHURCHILL'S SECOND PREMIERSHIP

OCTOBER 1951 TO DECEMBER 1952

Churchill – Britain's first nuclear Premier

> '[Churchill] is imaginative, unpredictable, firm in [his] belief in his own genius, and apparently determined to attempt one last crowning act on [the] world stage.'
>
> WINTHROP ALDRICH, US Ambassador to the UK, in a secret briefing to the American Secretary of State, 27 October 1953[1]

In his first speech to the new Parliament on 6 November 1951, Churchill struck a moderate tone: 'What the nation needs is several years of quiet, steady administration.'[2] In private, he was more graphic about the Tory Party's top priorities – they should be 'houses and red meat and not being scuppered'.[3] According to Jock Colville, Churchill at first believed that his second term in Downing Street would be only brief. He remarked soon after the election that he intended to be Prime Minister for only a year, after which he would hand over to Anthony Eden.[4] Yet it was unlikely that Churchill, having been out of power for over six years, would leave it after twelve months, having focused on the comforts of home and hearth. Consciously or not, he wanted to make another appearance in the pageant of world history.

Churchill was soon to build on the work of Attlee's government, becoming the first British leader to be armed with nuclear weapons and to put in place plans to set up a nuclear-power industry. But the new Prime Minister was ill-prepared to take over the stewardship of a country with its own nuclear capability – as his colleagues would soon see, his views on nuclear matters had scarcely changed since he left Downing

Street in July 1945. As far as he was concerned, what mattered most was the restoration of Anglo-American relations, which he believed Attlee had grievously neglected. With that achieved, he believed, Britain's nuclear projects could flourish under the wing of American power.

Within hours of his election, Churchill walked into 10 Downing Street. The event was like the return of an exiled monarch, with choruses of 'Auld Lang Syne' ringing out around Whitehall. After six grey years under Attlee, government officials were glad see 'The Old Man' back.[5] Probably expecting a return to the hyperactivity of the war years, the Office Keeper in Downing Street laid before the Prime Minister's seat at the Cabinet table a sheaf of labels screaming ACTION THIS DAY.[6] In some ways, it was like May 1940 all over again: Churchill appointed himself Minister of Defence, attempted to set up a coalition (with the Liberals, this time without success) and reappointed many of his cronies and wartime colleagues. Among them was, predictably, Frederick Lindemann. 'I must have Prof,' Churchill told his Cabinet colleagues. 'He is my adder. No. I can add – he is my taker away.'[7]

Lindemann was reluctant to return to government. Ailing with heart disease and diabetes, he was out of breath after climbing a single flight of stairs and, according to his fellow don A. J. P. Taylor, 'looked as though he banged his head against a stone wall for half an hour each day to keep it in trim'.[8] For three days the Prof resisted Churchill's pressure, but then accepted the Cabinet post of Paymaster General, which came with the perk of a two-room apartment at the top of 11 Downing Street.[9] He asked for no salary. Lindemann joined the government wanting only to serve Churchill and to extricate the UK's nuclear programme from the Ministry of Supply, ambitions that would soon come into conflict and cause the Prof a good deal of grief. Yet he was to achieve several notable successes, standing up to Churchill and establishing himself as Whitehall's most valued supporter of the government's leading nuclear scientists.

Lindemann quickly learned how much opposition there was to his plans for the government's nuclear programme. The Treasury, the Ministry of Supply and sundry Whitehall mandarins heard him out and offered warm words of support, while doing everything they could to thwart his intentions. But he appeared to hold the trump card – Churchill had pledged to support his plans as a condition of returning to government. Or so the Prof believed.[10] Whether out of insensitivity or mischief, Churchill had installed as his Minister of Supply his son-in-law Duncan Sandys, long one of Lindemann's arch-enemies in Whitehall. A bloody battle was in prospect.

The first clash between Churchill and his favourite scientist took place within two weeks of the government's taking office. In an uncharacteristically wordy message, the Prof asked the Prime Minister to approve the choice of the Monte Bello Islands as the site of the first British nuclear test.[11] The reply left Lindemann stunned.

Churchill wrote that he had 'never wished . . . that England should start the manufacture of bombs', contradicting almost everything he had ever said on the subject.[12] He now believed that Britain needed only to be an expert in the science of the Bomb, not the weapons themselves – 'the art rather than the article', as he put it. He was sure, he said, that when he met Truman in a few weeks' time, the President would gladly hand over 'a reasonable share' of the Americans' own nuclear weapons. For the Prof, worse was to come. Churchill told Lindemann that decisions about the first British nuclear test were not urgent and could surely wait until he had met the President. 'When we produce the [Quebec Agreement] and demand that it shall be published', Churchill wrote, 'we shall get very decent treatment' from the Americans.

It took the Prof six days to recover his composure. In a robust reply, he reminded the Prime Minister of the story of Anglo-American nuclear relations since the war, and ended by rebutting all Churchill's points. Lindemann strongly supported Attlee's nuclear policy, noting that it had so far cost 100 million pounds, and emphasising that it was vital to take a decision on the Monte Bello test immediately. If they delayed, the Prof pointed out, Penney and his colleagues would be too late to take advantage of optimal weather conditions, and would have to postpone by a year. If that happened, Lindemann knew they risked 'being jockeyed [by the Americans] into a test in Nevada where we should reveal our secrets without getting any return'. Lindemann ended his memo on his knees: 'I beg you to accept the advice of the Chiefs of Staff, the Foreign Office and all the other Departments concerned and definitely to decide in favour of the test being made in Australia.'

Churchill's comments on Lindemann's memo are revealing. Obviously unconvinced by Lindemann's case, he ringed the project's cost of 100 million pounds, a figure that amazed him.[13] When Lindemann noted that it was 'inconceivable' that the Americans might hand over bombs or details of their construction, Churchill circled the word and wrote in the margin 'No'.

Two weeks later, while Lindemann was still pondering his next move, Churchill fired off a short minute to a Treasury official about one of the details he had read in Lindemann's brief: 'How was it that the £100 million for atomic research and manufacture was provided without Parliament being informed?'[14] A few days later, with all the figures in front of him, his predecessor's accountancy on the Bomb project became clear for the first time. By manipulating the numbers, Attlee and his colleagues had kept the cost of producing the

Bomb from his country's legislature, exactly as Roosevelt had done.[15] After the Prime Minister's enquiry had been answered, Lindemann pressed again for a decision on the test. Churchill replied, 'Proceed as you propose.'[16] Thanks to Lindemann, Britain's nuclear policy was back on track, for the time being.

No one else in government appears to have been aware of this spat. To most of his colleagues, Churchill was making a confident start to his second premiership, pulling his Cabinet together, declaring that Whitehall was now 'drenched in Socialism' and ordering a cull of Attlee's sprawling committees.[17] He wanted to put an end to 'party brawling' and concentrate on pushing through his domestic policies, overseen by the forward-looking ministers he had appointed, including Rab Butler at the Treasury and Harold Macmillan in the Housing Ministry. Consensus and continuity were to be the watchwords of this administration. Churchill's main priority, however, was to restore the warm Anglo-American relationship he believed he had bequeathed to Attlee, and in particular to persuade the Americans to repeal the McMahon Act.

As Christmas 1951 approached, Churchill was looking forward to his first Prime Ministerial visit to the United States for more than seven years. At home, he savoured the trappings of power he had long missed, a cadre of loyal assistants always at his beck and call. After spending a morning working on his papers in bed, he still liked to take an afternoon nap and round off the day with a good dinner, sometimes with his old friends in The Other Club.[18] At weekends, he and Clemmie decamped to Chequers or Chartwell – Lindemann was still his most frequent visitor.

Long breaks in the parliamentary calendar afforded Churchill plenty of opportunities for restorative vacations. Now seventy-seven, the infirmities of ageing were among his most serious problems. His eyes were giving him trouble, he worried that he would have another stroke or heart attack, and, most inconveniently, he had gone seriously deaf. 'A bloody nuisance,' he sulked.[19]

The purpose of the Washington visit, Churchill insisted, was not to 'transact business' but to 'establish intimate relations' with President Truman.[20] No matter how hard his officials tried to push him to prepare for the trip, he was determined to wing it. After officials in both the White House and Downing Street pestered him for a clear statement of what he wanted to discuss at the meeting, he grudgingly offered only the bones of an agenda. At the top was 'The "Cold War" – policy of the West towards Russia', with Anglo-American cooperation on 'Atomic Energy' second to bottom.[21] None of these was among Truman's top priorities, and he took care to leak his opposition to the idea of another 'Big Three Summit' to the *New York Times* well before Churchill's arrival.[22]

Truman was at that time mired in problems – perpetually under siege for being soft on Communism, plagued with scandals and under pressure to put an end to the unpopular Korean War. Although he regarded Churchill as the greatest public figure of the age, he had been looking forward to the Prime Minister's visit with mixed feelings, worried that he was a potential trouble-maker. The 'old lion would have to be watched by the White House's gamekeepers'.[23]

When the two leaders met at the National Airport in Washington on 4 January, Truman saw that Churchill was a shadow of the figure he had first met – white-haired, slower in step and with a pronounced stoop, though still with charisma to spare. That evening, cruising down the Potomac on the Presidential yacht, they chatted with each other and their colleagues at the reception party, Churchill beaming with pleasure at his return to the Presidential court. Drinking tomato juice to demonstrate that he could do without alcohol, Churchill treated his hosts to a well-practised entertainment, in which he and Lindemann performed like an aristocratic Laurel and Hardy.[24] After Churchill had loudly demanded an estimate of the total volume of alcohol he had drunk in his life, the Prof theatrically pulled out his slide rule. He calculated that if all the alcohol the Prime Minister had consumed were poured into the room, it would come up to their knees. Feigning great disappointment, the Prime Minister delivered his punch line, saying that he had hoped everyone would drown in champagne and brandy. The double act was a highlight of the party.

In the coming days, the two leaders discussed everything from the military challenges in East and South-East Asia to the possible threat from Mao's government in China, as well as meat-and-potatoes matters of strategy and shared intelligence. Before they had exchanged a word on policy, Churchill opened his heart, recalling with surprising candour his impression of Truman at their meeting in Potsdam:[25]

> I must confess, sir, I held you in very low regard then. I loathed your taking the place of Franklin Roosevelt. [Pause] I misjudged you badly. Since that time, you more than any other man, have saved Western civilisation.

Impatient as ever with the Prime Minister's fawning, Truman often cut him off as he began another of his paeans to Anglo-American goodwill.

For all his presence and dominance over his ministers, especially Eden, Churchill struggled at the conference table. Bereft of much of his former energy and clarity of mind, he struggled to grasp detail and repeatedly cupped his hands over his ears to hear the President's words. He failed entirely to make any headway with his passionate belief in the need for another 'Big Three' summit. Truman explained that, while relations with the Soviets were so poor and with so little common ground, this was not the right time to begin talks, especially as their failure might well increase tensions still further. Besides, he had already invited Stalin to Washington, but did not expect him to accept. Truman deflated Churchill's concern about an imminent nuclear showdown by stating categorically that he had no interest in precipitating one.

Churchill did, however, have some diplomatic successes, persuading the Americans to agree that the use of their military bases during an emergency 'would be a matter for joint decision'.[26] Away from the conference table, he was determined to get to the bottom of how the Quebec Agreement had been abandoned. Four days after he arrived in Washington, his only American guest for lunch at the British Embassy was Senator Brien McMahon, who had done more than anyone else to terminate Anglo-American nuclear cooperation. Churchill had corresponded with the Senator a few years before but, out of ignorance or discretion, had not raised the subject of his Act.[27] With the culprit now in front of him, the Prime Minister examined him like a prosecutor: he sent for a copy of the Quebec Agreement and handed it to McMahon, who read it.[28] According to fellow guest Lord Moran, Churchill declared that Britain 'has been grossly deceived' and that there had been 'a breach of faith'. McMahon replied with the words the Prime Minister wanted to hear: 'If we had known this, the Act would not have been passed. Attlee never said a word.' Churchill's suspicions had been vindicated – he could now blame the rupture in Anglo-American nuclear relations entirely on Attlee.

Yet Churchill did not pursue the matter with the President. Two weeks later in the White House, at their final meeting of the visit, the Prime Minister was sheepish about pressing for the Quebec Agreement to be released.[29] According to the British record, 'This was not a matter which he wished to bring up formally but he wanted to mention it to the President' (Truman was content for it to be published, but not for the time being). When Lindemann was similarly circumspect about the possibility of revising the McMahon Act, Truman brushed his concerns aside, commenting that there was 'no prospect of amending this law in the present session of Congress'. Senator McMahon certainly made no effort to reverse it before he died of cancer later that summer.

At the end of the visit, when asked by a journalist how he felt about the imbalance between British and American power, Churchill replied, 'I don't feel on uneven terms, with your might and my prestige.'[30] Privately, he had spoken of talking to Truman 'as equals', but it was a delusion. Before the visit, American officials had agreed that 'we do not want to reconstruct the Roosevelt–Churchill relationship'. They had achieved their goal and Churchill's hopes of setting up a summit with the Soviets rested with the election of a new President in November.

In the thirty years of friendship between Churchill and his favourite scientist, 1952 was to be the stormiest year of all. They first clashed a few weeks after their return from Washington, when Churchill withdrew his half-hearted support for Lindemann's scheme to take the British nuclear project out of direct government control. The policy was, in Churchill's view, causing more political trouble than it was worth. At Chequers one Sunday in early March, the Prime Minister dictated a minute approving the transfer of the British nuclear project to the Ministry of Defence, with Lindemann in the room to make quite sure he did it. Only a few hours later, the Prof heard that the Prime Minister had changed his mind. By the end of the month, Churchill was angry with Lindemann, who was so frustrated that he was ready to throw in the towel.

The Prof's doctor had counselled him to give up ministerial office, he told Churchill, whose own physician had dispensed the same advice a few months before.[31] They both stayed put, Lindemann perhaps hoping that his political fortunes would be buoyed in the early autumn, when Britain's first nuclear weapon was to be tested.

Churchill was staying with the Queen at Balmoral Castle when he heard, on 3 October 1952, that the Bomb had been tested successfully. It had been detonated inside the hull of a 1,450-ton frigate, HMS *Plym*, which was all but vaporised, as Churchill reported to the Commons a few weeks later. The explosion had gone almost exactly as predicted and he paid handsome tribute to Penney and even complimented Attlee, sitting opposite, for initiating the Bomb programme.[32] Penney had given his political masters good value for money: they had acquired a nuclear weapon for only a tiny fraction of Britain's total defence budget over the past seven years – less than one per cent.[33]

Eight days later, the Americans raised the stakes again by detonating an even more powerful weapon, the first hydrogen bomb, obliterating an entire island in the Enewetak Atoll, some three thousand miles west of Honolulu. The device had an explosive power equivalent to 10.4 megatons of TNT, twice the total power of all the explosives used in the Second World War. Advised by Lindemann, Churchill knew that it was only a matter of time before the H-bomb was in the hands of the Soviets, who by then also had a substantial stockpile of fission weapons. In a top-secret report, the Atomic Energy Intelligence Estimates Unit advised Churchill that, partly because Soviet bombers were not yet able to reach the US, 'The primary target will be the UK.'[34] The havoc that would follow such a nuclear attack had been spelled out a few months before in another secret report that, for the first time, made Churchill think seriously about the consequences of a nuclear assault on Britain.

His response to these challenges was weak. A decade before, he would have given his colleagues a vigorous lead, but he was now bereft of ideas and energy, dominating his Cabinet 'more like Buddha than Achilles', as Jock Colville noted.[35] Deafness was now making it hard for Churchill to talk with his colleagues, especially those who spoke quietly. Microphones were installed around the Cabinet table, along with loudspeakers to project the amplified voices.[36] At home, Churchill avoided hard work and retreated into novels, including *Nineteen Eighty-Four*, which he regarded as 'a very remarkable book'.[37]

Lindemann was less disturbed by his hero's hearing than by his erratic judgement. The Prof began to run out of patience with Churchill after he denied that he had ever pledged support for Lindemann's plans for the British nuclear industry, 'still less that I made a bargain with you about it when I took office'.[38] The Prime Minister was interested only in developing a nuclear policy that aligned Britain ever more closely to the United States, even though he was 'greatly disturbed' by the election to the Presidency of General Eisenhower: 'I think this makes war much more probable,' Churchill told Colville.[39] Shortly after Christmas, Lindemann was at the end of his tether. He sent Churchill two memos on the case that Britain should continue to make its own nuclear weapons and on its relationship with the United States.[40] The Prof argued forcefully that it was now time for Britain to go it alone in nuclear research, fostering close relations with uranium-rich Commonwealth countries and to stop chasing 'the will-o'-the-wisp of full American collaboration' – working with Truman's officials had been fruitless and there was no reason to believe that the new administration would be any more accommodating. Churchill's response showed that he had changed his mind again about the wisdom of acquiring nuclear weapons.

Would it be possible for Britain to acquire H-bombs? Churchill asked him. Lindemann replied:[41] 'We think we know how to make an H-bomb' – a statement Penney would have regarded as cavalier, to say the least – 'but at this stage we can only make very rough estimates of the various materials required.' Such explosives did far more damage than ordinary nuclear weapons, the Prof wrote, underlining the fact that – unlike the type of Bomb dropped on Hiroshima – there is no theoretical limit to the explosive power of an H-bomb. All this seemed to wash over Churchill. It would be almost another year before he got the point.

Lindemann was no less disappointed with Churchill's attitude to developing nuclear power for peaceful purposes. At a Defence Committee meeting earlier in December, the Prime Minister had refused to sanction the full British nuclear programme until he had talked with the new American administration, which he still believed would supply Britain with 'atom bombs for our use'.[42] This prompted Lindemann to compose the most withering memo he ever wrote to Churchill.[43] In the unlikely event that the US gave Britain nuclear weapons, it was crucial that the British nuclear programme went ahead, the Prof wrote, otherwise the initiative would be delayed. Indigenous supplies of fossil fuels in Britain were dwindling so rapidly that it was heading for a shortfall of coal production of twenty million tons in the next thirteen years.[44] 'To give up serious work on the development of atomic energy' would be, he argued, 'a disastrous line which might well in the long run spell national suicide'. If Churchill wrote a reply to this note, it has not survived.

On the question of whether to develop nuclear power, Lindemann himself had made a U-turn, though he apparently never acknowledged it. Soon after the war, he had dismissed plans by Blackett and others to develop this new industry, sneering that there was no hope of running a heat engine off a nuclear reactor – it would be 'as efficient as Stephenson's Rocket'.[45] One of the experts who changed the Prof's mind was Christopher Hinton, who was pushing hard to start building nuclear plants in Britain, to make his country a world leader in the field – if Churchill would let him.

1953

Hinton engineers nuclear power

> 'All pioneering engineering is very much like poker; there are some things that you know for certain, some things that you think you know and some things that you don't know and know that you can't know.'
>
> SIR CHRISTOPHER HINTON, 1970[1]

When his colleagues nicknamed him 'Sir Christ', they were alluding to his virtual near-omniscience as a nuclear engineer, not to his spirituality. Although Sir Christopher Hinton was difficult, short-tempered and sometimes unreasonably demanding, he was as dependable as an atomic clock. In early 1953, he and his team had completed the huge task of delivering the low-enrichment part of Britain's first gaseous diffusion plant, within a week of its scheduled opening date and well below its 14-million-pound budget.[2] Yet he received little thanks from Whitehall. He had been so fierce and unpleasant to deal with that his boss, Churchill's son-in-law Duncan Sandys, did not even send him the traditional note of ministerial congratulation.

Five months before, Churchill had been more generous. After the Monte Bello trial, the Prime Minister had sent a telegram to Hinton, Penney and Cockcroft, known in Whitehall as 'the atomic knights'.[3] Churchill thanked Hinton and his colleagues for 'the devoted efforts they have made and the brilliant engineering skill they have shown' in producing plutonium for the first British nuclear weapon. The telegrams were drafted by Lindemann, who had warmer relations with the atomic knights than they had with each other. The Prof's admiration was reciprocated, especially by Hinton. From where he stood, Lindemann was the best friend that the government's nuclear project had in Whitehall – it was certainly not Duncan Sandys or any of his officials, always on hand with more reels of red tape.

Of the three knights, Hinton had arguably the most demanding task – to set up a national nuclear industry from scratch. He joined the government's project in late 1945, knowing next to nothing about nuclear science. Officials told him that his job was to build the necessary factories to purify uranium to make reactor fuel, to enrich uranium with fissile ^{235}U and to produce plutonium. The *raison d'être* of the entire project was to build weapons, even though Parliament had not approved the programme. The overwhelming secrecy made life extremely difficult for Hinton and his colleagues, who harassed contractors to meet barely possible deadlines but were forbidden from explaining the urgency.

Hinton spent most of his time working in the north-west of England. His headquarters were in the small Lancashire town of Risley, on a bleak site that had been occupied during the war by the government's munitions factories. The offices he inherited were poorly equipped and some were squalid – an environment unappealing to many potential recruits, one of the reasons why Hinton was dogged with staffing problems. He and his colleagues selected five main locations for their nuclear project. A small prototype nuclear reactor would be built at Harwell in Oxfordshire, where the physics, chemistry and metallurgy research would be done. The other sites were in the north-west of England. At Springfields in Lancashire, uranium would be extracted from its ores and turned into reactor fuel. A huge plant at Capenhurst in Cheshire would be able to enrich this fuel, if necessary. At Windscale (later renamed Sellafield) in Cumbria, the main nuclear reactors would be constructed, along with the facilities for extracting plutonium from their spent fuel.

To complete these projects Hinton and his colleagues spent 74 million pounds, almost twice the combined budget of Penney and Cockcroft.[4] It was perhaps for this reason that Hinton was the highest paid of the three and, by 1953, was indisputably Britain's leading nuclear engineer, on a par with the best in the world. Although the least familiar of the atomic knights to the public, he was a valuable asset to Churchill, enabling him to preside over a new and successful power industry. Hinton and his colleagues were, however, not the first engineers to deliver nuclear energy to the national grid – the Soviets achieved that goal at the Obninsk reactor in June 1954, six months after the date predicted by H. G. Wells in *The World Set Free*, four decades earlier.[5]

Hinton's exceptional talent as an engineer had shone since he was a teenager. At the age of sixteen, in 1917, he left school to spend six years being trained at the Great Western Railway and Metro-Vickers – 'the best craft apprentice I have ever had', one fitter later remembered.[6] After winning a place at Cambridge University to study mechanical sciences, Hinton powered his way through the degree in two years, leaving the third free for research. By the time he was in his early thirties, he was one of ICI's top engineers, responsible for building some of the biggest and most complex chemical plants in Europe, a decade ahead of his peers on the career ladder. He stood out in other ways, too: handsome, well-spoken, forceful and six feet six inches tall, this sequoia of a man could dominate a crowded room merely by walking into it. Marriage and children did nothing to mellow him. During the Second World War he worked for the Ministry of Supply, as a senior manager at the huge armaments plants in Risley, directing the construction of a new cordite factory that broke production records. He emerged from the conflict frazzled but with a reputation so strong that no one was surprised when the Ministry asked him to help found the British nuclear industry. He quickly assembled a team of five former colleagues from the armaments factory, none of whom knew the first thing about nuclear science.[7]

Above all, Hinton was an autocrat. He thought nothing of lambasting an errant colleague in public and was slow to compliment even his highest achievers. Praise was unnecessary, he said, because 'They know that they are good or they would not be working for me.'[8] Yet he was a popular leader with a reputation for straightforwardness – everyone on his staff always knew precisely where they stood with him and were rarely on the wrong end of a tongue-lashing that was not deserved. When ideas were being thrashed out, he welcomed open debate but would stick to agreed decisions and policies, though he could never quite bring himself to admit that he had argued a discredited case. Underneath his titanium hide, however, was an unspoken craving for friendship his junior colleagues saw only when his guard was down.

Hinton had no friends among his peers at the top of the nuclear industry. One difficulty was that he liked to tackle problems head-on, bringing every challenge out in the open – an approach that appeared congenial to William Penney but not to Cockcroft, who irritated Hinton by shying away from every confrontation. Cockcroft's scientists at Harwell were doing almost all the fundamental research that underpinned the design and operation of Hinton's factories. This included the techniques to handle industrial quantities of uranium, to isolate every last microgram of plutonium from the reactors' spent fuel rods and then to transport it safely to its destination. If Cockcroft's team took longer than expected to solve a problem Hinton's team had set them, the inevitable consequence at

Risley was missed deadlines, redrawn production charts and frayed tempers.

Involved in every decision taken about every site under his command, Hinton was as knowledgeable about the minutiae of local geology as he was about the choice of the suppliers of his office stationery. At the same time, he oversaw the finances of the entire operation with the fastidiousness of an actuary. He played his hand like the accomplished player of poker he had once been, though the game had lost its appeal for him now that he was gambling not merely with cards but with millions of pounds.[9]

As a negotiator, Hinton was powerful and often intimidating, qualities that helped to lay the foundations of the British nuclear-power industry in early 1953. After over a year of discussion and argument, Churchill's government finally approved a programme in which nuclear weapons and power would be developed together, using a group of new reactors with the dual functions of making plutonium and delivering energy to the national grid. Initially, the plan had been to build reactors that produced plutonium as a by-product of generating energy, but the outcome was the opposite: the top priority was to make nuclear weapons, not to solve a national fuel crisis. As Hinton repeatedly pointed out, Britain's coal industry was doomed and nuclear power might solve the problem, but with the important proviso that it could be delivered safely and economically.[10]

Cockcroft had wanted to take the lead in designing these new reactors, but Hinton outmanoeuvred him. On the day after Churchill's officials had given the go-ahead, in February 1953, Hinton was at his desk in Risley drawing up the master charts for the construction of four new reactors.[11] He had undertaken to deliver the first of them at Calder Hall, near the Cumbrian coast in the north-west of England, within three and a half years.

With Britain's policy on nuclear energy now clarified and operational, Hinton accepted an invitation to present it publicly for the first time, at a conference in New York, in front of hundreds of representatives of the embryonic international nuclear-power industry. Before the gathering, he joined Lindemann and Cockcroft to lobby American officials to relax the restrictions on nuclear cooperation imposed by the McMahon Act. Hinton had long believed that the Act had been good for Britain, forcing it to develop its own ideas rather than depend on American expertise. The prospects for improved Anglo-American collaboration were now much brighter. Soon after taking up the Presidency, Eisenhower had announced that he wanted to see a freer exchange of nuclear information with Britain – Churchill's negotiators, armed with clear evidence of success in the nuclear field, were no longer approaching the Americans as humiliated supplicants.[12]

The conference was held at the Waldorf Astoria Hotel in the heart of Manhattan. After lunch on the final day, 30 October, as the representatives sipped their coffee, Hinton approached the lectern, where a huge microphone awaited him.[13] He began bullishly, declaring: 'It would be a mistake to imagine that, on peacetime standards, you in the US had only three years' lead over us.' He pointed out that the huge investment by American industry in the Manhattan Project – far greater than would be conceivable in peacetime – meant that by the end of 1945 the American lead over Britain in nuclear power was not three years of normal activity, but 'something like six years'. Making no mention of nuclear weapons or the McMahon Act, he outlined the British nuclear-energy undertaking. Government-overseen organisations would get the project off the ground, with the intention of eventually handing it over to private industry, which would run it at a profit.

Hinton spoke for almost an hour, making eye contact with

all parts of the audience and holding his glasses at his waist with both hands. Safety was the most pressing challenge for nuclear engineers, he said. They had to find ways to dispose of radioactive waste without polluting the environment and to run their reactors without risking a disaster: 'We have no practical experience of what happens if a reactor runs away.' The United States might benefit handsomely in the long run, he concluded, if its engineers investigated what happened when a nuclear reactor melts down, an experiment that 'would cost no more than the trial of a single atomic bomb'. His advice was declined by the American authorities, a decision that did not look wise in the aftermath of the accident at Three Mile Island twenty-six years later.

Hinton's talk appears to have gone down well, attracting media attention on both sides of the Atlantic. But compared with the H-bomb, nuclear power was low on the political agenda. The weapon's development had divided nuclear scientists in America, where Senator McCarthy's campaign to purge alleged Communists from the government's employment was in full swing. Edward Teller and Ernest Lawrence were among the scientists who argued that America should acquire ever more powerful nuclear artillery to stay ahead of the Soviet foe. Others wanted to rely much less on the H-bomb but more on arms control, conventional forces and less powerful nuclear weapons. This cohort included James Conant (now US Ambassador to Germany) and his former boss Vannevar Bush, then out of favour in Washington after campaigning to halt the first H-bomb test on the grounds that it would cause a dangerous escalation of the arms race.[14] The most prominent opponent of the unfettered development of thermonuclear weapons was Robert Oppenheimer, who had made well-publicised calls for debate on the American nuclear strategy. He was now in the cross-hairs of the business executive Lewis Strauss – appointed chair of Eisenhower's Atomic Energy Commission early in the summer – who told the FBI chief J. Edgar Hoover that he would 'purge Oppenheimer'.[15] Strauss meant business, as he quickly demonstrated.

An edition of *Time* magazine with Strauss on the cover caught Hinton's eye soon after he arrived in North America.[16] The article, for the most part a gushing portrait of the 'US Atom Boss', was shot through with anxiety over the inevitable moment when the Soviets would have H-bombs of their own. 'Does the armed free world just sit back and wait for the clock to strike in 1956 or 1957 or 1958?', the article concluded ominously. After Stalin's death in March 1953, Eisenhower and his officials had assumed that the new Soviet leadership would not be interested in establishing more constructive relations with the West, so tried to put them under unremitting psychological pressure to reform.[17] When Hinton returned home, desperately in need of a long vacation after an enervating year, he saw that his Prime Minister was approaching the Cold War very differently.[18]

MARCH 1953 TO FEBRUARY 1954

Churchill the nuclear missionary

> 'The PM feels that Stalin's death may lead to a relaxation in tension. It is an opportunity that will not recur . . . He seems to think of little else.'
>
> LORD MORAN, 7 March 1953[1]

By March 1953, Churchill's plans to set up another 'Big Three' summit had gone nowhere. President Eisenhower – a high-ranking underling, in Churchill's opinion – refused to cooperate, believing the British leader too old to be Prime Minister and stuck in the mindset of the recent world war.[2] Churchill had always been confident that he could do business with Stalin, and when the Soviet leader died, the Prime Minister was no less convinced that at least a few of the new leaders in the Kremlin would be equally susceptible to his eloquence. Churchill could be accused of naivety and of being a political anachronism, but his buoyancy was undeniable.

To realise his dream of another summit and follow on from where he left off at Potsdam, Churchill needed a theme to justify the urgency. It arrived in an epiphany eleven months after Stalin's death, when he first fully understood the potential impact of thermonuclear conflict, which would make even the Second World War look like a fracas. Avoiding such a catastrophe by easing the tensions of the Cold War was to be his final great cause, the nuclear successor to his campaign for rearmament and air-defence research in the 1930s.

In the first few weeks after Stalin's death, Churchill trod cautiously. The politician wielding the most power in the Soviets' new 'collective leadership' appeared to be Georgy Malenkov, who struck an encouraging note at Stalin's funeral. During a tearless eulogy, he declared that the Soviet Union wanted a 'prolonged coexistence and peaceful competition of two different systems, capitalist and socialist'.[3] This, together with a few signs that the Soviets wanted to ease tensions with the West, encouraged Churchill to believe that it would do no harm to meet them halfway. His Foreign Secretary and deputy, Anthony Eden, was initially supportive, but officials soon talked him out of it.[4] For Churchill – now Sir Winston – this was going to be another lonely struggle.

In the next two months, he wrote thirteen times to Eisenhower, gently leaning on him to begin talks with the Soviet leaders.[5] The President's replies were courteous, but privately he found Churchill's messages 'tiresome'.[6] For Eisenhower, concerned that the West was being lulled into a false sense of security, this was no time to extend the hand of friendship to the Soviet Union, a leopard that was not going to change its spots.[7] The opposition from the White House took the Prime Minister by surprise, as did the cool response from the Kremlin. Having failed to persuade the President, and with his Foreign Secretary on long-term sick leave following a botched operation, Churchill cut loose. If necessary, he decided, he would travel alone to Moscow. On 11 May, he delivered the most emotional speech he had given to the Commons since he returned to office, and became the first Western leader to suggest publicly a new approach to relations with the Soviet Union.[8] It was time for the West to focus not only on the threat posed by the Red Army, he argued, but also to take into account the Soviets' security interests. He even suggested the possibility of a neutral, reunified Germany. Most importantly, 'the leading powers' should meet, at the earliest opportunity:[9]

This conference should not be overhung by a ponderous or rigid agenda, or led into mazes and jungles of technical details, zealously contested by hoards [*sic*] of experts and officials drawn up in vast, cumbrous array. The conference should be confined to the smallest number of Powers and persons possible. It should meet with a measure of informality and a still greater measure of privacy and seclusion.

Churchill's speech went down badly with Eisenhower, with the American Secretary of State John Foster Dulles, with the West German government, with the British Foreign Office and the Cabinet. It was of little help to Churchill that the Soviet government tentatively welcomed his proposals or that Attlee's Labour opposition was encouraging. To win widespread support for this first proposal of an East–West détente was going to be exceedingly difficult.

Undeterred by the welter of opposition, Churchill believed he was making good progress, until his initiative was curtailed by a medical trauma. It happened in Downing Street late in the evening of 23 June, at a dinner in honour of Italian Prime Minister Alcide De Gasperi, one of the architects of the European Community.[10] Churchill had appeared to be on fine form, giving a witty after-dinner speech on the Roman conquests of Britain. Afterwards, as the diners were leaving, he could take only a few faltering steps from the table before flopping into the nearest chair. 'I want a friend,' he said to his guest Lady Clark, taking her hand as his voice trailed off. 'They put too much on me. Foreign affairs . . .' Officials called for help and hustled the guests away.

Churchill had suffered a serious stroke. Against the advice of his doctor Lord Moran, he attended Cabinet the next day, somehow managing to chair it and make several pertinent interventions ('We thought he was rather quiet,' Rab Butler later said).[11] Kept carefully away from photographers, the Prime Minister was soon whisked off to Chartwell. There, he lay in bed for days, reading Trollope's political novels, blinking when he wanted an assistant to turn a page. 'I am a hulk,' he told his doctor, 'only breathing and excreting.'[12] The loyalty of his closest colleagues and his friends among the newspaper barons ensured that his indisposition was a well-kept secret. He spent several weeks convalescing at Chartwell, and was soon back in Downing Street, chairing Cabinet meetings.[13]

In mid-August, Churchill heard the news he had long known was inevitable: the Soviet Union had detonated an H-bomb. With the US and the Soviets now able to blow each other to bits, Churchill saw no alternative but to pursue peace, to reduce the chances of a nuclear catastrophe. News of a political upheaval in Moscow did not make his task any easier. Although Malenkov still seemed to be the boss in the Kremlin, its 'collective leadership' appeared to be crumbling. Churchill believed that 'there has been no change of heart in Russia, but she wants peace' and that Malenkov might just deliver, given the opportunity.[14] The consequences of failure were almost too appalling to contemplate, a depressed Churchill told Lord Moran: 'That hydrogen bomb can destroy two million people. It is so awful that I have a feeling it will not happen.'

Churchill's wife and his confidants were at one in recommending him to stand down, but he was not going to budge 'at the present crisis of the world'.[15] First, he set himself the task of asserting his authority at the Tory Party conference in the autumn, vowing to resign if he failed. Once again he triumphed: on 10 October, in front of a huge crowd of supporters in a Margate auditorium, he gave a masterful speech, albeit with help from performance-enhancing amphetamines regularly supplied by his doctor.[16] A week later, he had even more to be pleased about. After a summer of ministerial in-fighting, and helped by a committee chaired by Sir John Anderson (now Lord Waverley), Lindemann finally won the Cabinet's approval for his plan to reorganise the British nuclear project.[17] The creation of the UK Atomic Energy Authority, as it became known, was an enduring triumph for him – it still exists today. For Churchill, the main benefit of this was that it soothed the running sore in his relationship with the Prof, who was now debilitated by angina, diabetes and a list of other complaints.[18] Soon afterwards, Lindemann left the Cabinet to return to Oxford, but remained a close Prime Ministerial adviser and emissary on nuclear affairs. He had left ministerial office on a high note that several of his fellow scientists a decade before would scarcely have believed possible.

The Cabinet, most of them now itching for Churchill to leave, saw him take on a new lease of life in the autumn. In mid-October he was buoyed by the news that he had won a Nobel Prize, though for literature and not for peace, as he had hoped. Preferred by the appointing committee over his fellow nominees – the novelists Graham Greene and Ernest Hemingway along with the poet Robert Frost – the Prime Minister was awarded the prize for his 'brilliant oratory in defence of exalted human values' and for his biographical work.[19] His life of the Duke of Marlborough and his memoir *My Early Life* had impressed the Nobel committee rather more than his account of the Second World War, whose final volume was published in the US shortly afterwards. It included several passages that he had toned down for fear that they might jeopardise the possibility of a 'Big Three' summit by offending Soviet and American leaders.[20]

Two weeks later in the Commons, Churchill spoke on the need to ease Cold War tensions. He concluded with a bravura passage about the threat of the H-bomb and his forthcoming summit in Bermuda with President Eisenhower.[21] The most powerful explosive the world had ever seen may possibly bring 'an unforeseeable security' to the world, Churchill argued.

During the preparatory talks for the meeting, the British negotiators discovered that the Americans were now prepared to consider relaxing the McMahon Act. Almost three weeks before Churchill set off from London on 1 December, Lindemann informed him that 'the Americans have now definitely agreed to exchange [nuclear] information with us (and the Canadians)' and that the new arrangement would soon be officially ratified. Churchill was delighted – 'Many congratulations on this great achievement,' he wrote to the Prof.[22]

To Churchill's annoyance, the UK and the US were joined at the summit by France. Making no attempt to hide his irritation, on the flight to Bermuda he voraciously read C. S. Forester's *Death to the French*, and soon after arriving at the airport avoided talking to the French Prime Minister by making a clumsy detour, ostensibly to stroke the honour guard's mascot, a goat.[23]

The gathering had only limited success. It ended in the routing of Churchill's attempts to improve relations with the new Soviet regime, by taking every opportunity to promote trade and better cultural relations, but giving them no quarter in military negotiations.[24] After he made his case at the first plenary meeting on 4 December, Eisenhower peremptorily dismissed it, declaring that there had been no change at all in the Soviet policy 'to destroy the capitalist system and the free world by force, lies and corruption' since the days of Lenin.[25] It made no sense, the President argued, to foster relations with the Soviets now.

Churchill did, however, make some headway with his nuclear policy. On the following morning, he met to discuss it in Eisenhower's quarters at the Mid Ocean Club with Admiral Strauss, Lindemann and the President himself (Churchill refused to allow the 'bloody frogs' to attend this session). Despite the success of the Prof's preparatory talks with the Admiral,[26] Churchill seemed to want from Eisenhower nothing less than an unconditional promise that the Anglo-American cooperation first spelled out in the 1943 Quebec Agreement should resume. The one-and-a-half-hour conversation appears to have been amicable, and was later notable mainly for Lindemann's premature assurance that Britain did not intend to do any work on the hydrogen bomb.[27]

Shortly before the meeting broke up at lunchtime, the discussion turned to the desirability of sharing intelligence information about Soviet weapons. Sensing that the Americans were being evasive, Churchill decided to tackle head-on what he believed to be the fundamental problem – he pulled out a photocopy of the Quebec Agreement and read it to Eisenhower. The official record does not mention the President's response, but does note Strauss's surprise – he claimed never to have seen the document before. Churchill moved swiftly on, proposing that it was now time to publish the agreement and to delegate the matter to Strauss and Lindemann. Later, apparently after the gathering, the President remarked to officials that he was ashamed of the McMahon Act, which he described as 'one of the most deplorable incidents in American history'.[28]

During the remainder of the summit, Churchill was inconsistent about nuclear policy. After declaring that he 'quite accepted' Eisenhower's intention to use nuclear weapons on military targets if the Communists broke the truce recently negotiated in Korea, Churchill backtracked.[29] When he read a draft of the 'Atoms for Peace' speech the President was to give at the United Nations directly after the Bermuda meeting, Churchill asked Eisenhower to replace a phrase about the US being 'free to use the atomic bomb' with a more emollient reference to 'reserving the right to use' the Bomb.[30] The requested change illustrates a difference between the leaders' attitudes to nuclear arms, as Jock Colville learned when he talked with the President, who believed they were simply 'just the latest improvement in military weapons', and would soon be regarded as merely conventional. Churchill initially agreed but then changed his mind, insisting that 'the atomic weapon [is] something entirely new and terrible'.[31]

Eisenhower's speech at the United Nations contained an initiative that gave Churchill pause for thought. The President suggested that governments should begin to make 'joint contributions from their stockpiles of normal uranium and fissionable materials to an international agency under the aegis of the UN'.[32] This went against Churchill's previous line on international control, but he now accepted the proposal and went out of his way to support it. After returning to London, he told the Soviet Ambassador that the American initiative 'was not a mere propaganda move' and that the Soviets should take it seriously.[33] He was doing his best to bring the superpowers together, but was preaching into a void.

With his détente agenda all but played out, Churchill was tired, bored and contemplating leaving office. He repeatedly promised to leave, only to think better of it soon afterwards, leaving his colleagues – especially Anthony Eden – close to despair. What Churchill needed was a reason to carry on, preferably a grand cause that only he could pursue. He found it in his Downing Street bedroom on the morning of Thursday 18 February 1954, while flicking through the daily newspapers. On the front page of the *Manchester Guardian*, he read a five-hundred-word article that brought home to him the extent of the H-bomb's destructiveness. The piece reported the after-lunch speech given in Chicago the day before by no less an authority than Sterling Cole, chair of the US Joint Committee on Atomic Energy. Cole described the aftermath of America's first H-bomb test, which had wiped out an entire island, he said, hinting that even more destructive weapons were on the way.

Jock Colville walked into the room to find an agitated Churchill with his copy of the newspaper open on his bed-table.[34] The Prime Minister's halitotic poodle was probably, as usual, yapping at the foot of the bed, his budgerigar chirruping in its cage or flying round the room.[35] Churchill insisted on reading the article aloud:

> . . . the heat and blast generated in the 1952 hydrogen test would cause absolute destruction over an area extending three miles in all directions . . . The area of severe-to-moderate damage would stretch in all directions to several miles . . . the Russians would be able to deliver [such an attack on the US] in 'one or two years from now'.

Churchill said – 'with a mixture of triumph and indignation', Colville later wrote – that he had called his Foreign Secretary, the Cabinet Secretary and all three Chiefs of Staff to see if they knew what had happened. None of them knew anything about it. Colville later recalled Churchill's conclusion: 'It was lucky that at least one person in Whitehall read the newspapers.'

From that morning, Churchill was obsessed with the H-bomb. He told Colville that the world was now almost as far from the original nuclear weapon as the Bomb had been from the bow and arrow. When the Soviets caught up with the US, the consequences did not bear thinking about: the UK was extremely vulnerable to a Soviet attack, which could lay waste to London, Manchester, Edinburgh, Cardiff and Belfast in seconds.

He was now more determined than ever that the 'Big Three' leaders should meet to reduce the likelihood of nuclear war. Hinting that a few felicitous phrases – no doubt his own –

might put an end to the threat, he told Eisenhower: 'I can even imagine how a few simple words, spoken with the awe which may at once oppress and inspire the speakers, might lift this nuclear monster from our world.'[36] Two days after Churchill wrote those words, he told Rab Butler: 'I feel like an aeroplane at the end of its flight, in the dusk, with the petrol running out, in search of a safe landing.'[37] He now had a grand cause to power him smoothly to the tarmac.

First, however, he needed to be briefed on the H-bomb. He did not – as he would have done during the Second World War – turn only to Lindemann, but also to Bill Penney, and to the most laconic and enigmatic of Rutherford's 'boys', Sir John Cockcroft.

MARCH TO DECEMBER 1954

Cockcroft becomes a confidant of the Prime Minister

> 'Dear Mother, I was sorry not to see you this week but to lunch with Winston Churchill does not come often in a lifetime.'
>
> JOHN COCKCROFT, 17 December 1954[1]

It was no surprise that Cockcroft was called in to advise Whitehall on the hydrogen bomb. Quiet, determined, resilient and straightforward, he was the Clement Attlee of nuclear science.

Like Attlee, Cockcroft was often underrated, especially by those who preferred talent to be flaunted rather than quietly made plain. 'John is more a manager than a physicist,' Mark Oliphant once commented – a remark that contains more than a grain of truth.[2] Such aspersions were common in nuclear circles, but they did not trouble Cockcroft, especially after he and Ernest Walton shared the 1951 Nobel Prize in physics for being first to split the atom artificially. The award burnished Cockcroft's reputation in Whitehall as one of the most dependable scientist-administrators, blessed with a rare combination of specialist knowledge and common sense. Rutherford aside, no other British Nobel laureate had ever been more effective in what C. P. Snow later described as 'the corridors of power'.

In 1954, Cockcroft achieved a distinction unequalled by any of his former Cavendish colleagues – he got to know Churchill, advised him personally on nuclear policy, and even became a valued lunch guest. It is perhaps surprising that the Prime Minister's confidence had been won by someone so reserved, so short of fine words and so lacking in allure. At gatherings,

Cockcroft was content to be a wallflower, but would chat amiably with all-comers, who could easily mistake him for a modestly prosperous businessman.

He had accepted the job of running the government's nuclear-research establishment in January 1946, when he was still directing the Chalk River facility in Canada. Chadwick had supported him for the post, but noted with rather unbecoming candour that Cockcroft's 'knowledge is wide but it is not at all profound' and that 'his views are of rather a dull, every-day hue'.[3] It did not take Cockcroft long to prove that, pedestrian thinker or not, he was exceptionally well qualified for the job. Drawing on his experience in North America, he led the conversion of a bleak airfield at Harwell into what was, in effect, a new university of nuclear science.

Although Harwell produced some useful blue-sky research, its four thousand scientists and engineers focused on providing technical advice and data for the British government's nuclear projects, as well as radioactive materials for medical researchers. Dearest to Cockcroft's heart was the project to supply nuclear energy to the national grid – he was determined to make the project a reality and tended to underplay weapons research in his public talks.[4] Equipped with state-of-the-art nuclear reactors and sub-atomic particle accelerators, his experimenters worked alongside theoreticians and mathematicians, enthusiastically collaborating with nuclear physicists overseas. Cockcroft stressed that the toughest challenges were chemical, for example, isolating minute quantities of rare and extremely toxic substances, while maintaining the highest standards of health and safety. The chemists were not allowed to leave their laboratories until they were certified sufficiently clean 'to go out and mix with the physicists'.[5]

Like many of Rutherford's most successful protégés, Cockcroft was a cross between a physicist and an engineer, though he also had another skill – after studying mathematics at Cambridge, he had graduated as a Wrangler. Young colleagues at Harwell teased him that his heyday in the early 1930s was 'the Stone Age of nuclear physics',[6] but they knew their leader was no Neanderthal. A conscientious reader of the technical literature in the nuclear field, he had a good feel for the best lines of research to pursue and was quick to detect attempts to pull the wool over his eyes. He even brought to Harwell some of the brio of his hero Rutherford, going so far as to license a 'Crazy Committee' to abandon temporarily the constraints of scientific conservatism.[7] The committee amply justified its name.

Cockcroft's Zen-like calm deserted him only occasionally. He later said he panicked only when his Chalk River colleague Alan Nunn May was exposed in 1946 as a Soviet spy, though others occasionally saw him lose his equilibrium during rows with Christopher Hinton over the design and operation of nuclear plants.[8] Cockcroft never raised his voice during even the most tense exchanges, though his family remember him yelling encouragement to his five children as they learned to ski during their winters in Canada. At home, he was a caring and involved father, though exceptionally reserved: the children had a rule that Daddy was not allowed to leave the dinner table until he had uttered two complete sentences.[9] Physics was never far from his mind. When working at home in the afternoon, he would sit down with a cup of tea and a plate of biscuits and mull over his problems, a gramophone record playing soothingly in the background, perhaps Tchaikovsky's *Swan Lake* or Copland's *Appalachian Spring*.

Cockcroft's reticence sometimes drove his Harwell staff to distraction. In meetings, he spoke no more than was absolutely necessary, often leaving staff with little sense of being led. His economy with words sometimes led to serious misunderstandings – he had appeared to be sympathetic to his staff's opposition to Lindemann's plan to transfer the British nuclear industry out of the Ministry of Supply, though he supported the plan in Whitehall. As a result, several of his colleagues felt betrayed.[10]

Although Cockcroft had amicable relationships with all the leading politicians and his scientist peers, he was in no one's pocket. Of the government's top nuclear administrators, he was the only one active in the Atomic Scientists' Association, giving upbeat lectures to thinly attended gatherings in village halls. He knew that most people regarded scientists as 'a race apart – long-haired gentlemen without any normal human desires or weaknesses who go coldly about our esoteric researches without any contact with the community'.[11] The onus was on scientists, he believed, to engage effectively with the public – 'Many of us are learning, by painful experience, the techniques of making ourselves understood.'

In the privacy of the Athenaeum club in Pall Mall, and at home with his wife, Cockcroft was sometimes cutting about the government's nuclear policies, past and present. For him, official accounts of the British nuclear project during the war failed to give 'a real picture of events' – for example, he believed that Churchill had been foolish to hand over Tube Alloys to ICI. Cockcroft never regretted leaving the project when the industrialists took over as 'They did nothing very effective.'[12] He deprecated the sidelining of Tizard and Blackett from policy-making circles during the conflict, and the disproportionate influence of Lindemann, whom Cockcroft and his wife described as Churchill's 'henchman'.[13]

At the end of the war, Cockcroft could not understand why the Americans did not demonstrate the first nuclear weapon on an uninhabited island. In Washington a few days before the attack on Hiroshima, he had been disturbed to hear that a senior member of the Manhattan Project was afraid that the war might be over before they could drop the Bomb.[14] When a journalist invited him a few months later to express regret for the Allied scientists' involvement in the project, Cockcroft refused and wearily explained himself:[15]

> We had to do it because we did not know whether the Germans would do it. We had no choice. There was a war on and we couldn't sit back and let the other fellow do it first . . . At the same time, we felt if we found it impossible to produce, it would be better for the whole world.

After nuclear weapons had proved viable, Cockcroft believed his country needed them in order to defend itself and preserve its international influence, which in his view was almost entirely benign. Like Bill Penney, he could be relied upon not to rock the government's nuclear boat and to give wholly accurate and constructive advice to support Britain's emergence as a nuclear power. This was another reason why Churchill's Cabinet Secretary Sir Norman Brook invited these two experts to his office on 12 March 1954 for a super-secret briefing about the implications of the hydrogen bomb. Also present were Sir Edwin Plowden – recently appointed chair of the UK Atomic Energy Authority, responsible for overseeing all the country's nuclear programmes – and four senior Ministry of Defence officials, who awaited new information like goldfish expecting their feed.

Cockcroft was chauffeured that morning to Whitehall in his official limousine, a Mark 6 Bentley. At 10.30, Brook opened the meeting with his usual directness, pointing out that the scientists were there to help him brief the Chiefs of Staff about the weapon's implications. Brook believed the advent of the H-bomb obliged the government to rethink 'our foreign policy and general strategy and, thereafter, the "size and shape" of the Armed Forces, our civil defence policy and our atomic weapons policy'.[16] Bill Penney was first to speak, summarising his understanding of the Americans' and Soviets' progress on the weapons. He pointed out that the Soviets had not detonated a real H-bomb, but a 'hybrid' weapon, something like a standard nuclear bomb but much more powerful, though it was only a matter of time before they had the real thing.[17]

The British government had a lot of thinking to do. A few days later, Cockcroft and Penney agreed to prepare a statement for ministers about the ability of the Soviet Union to manufacture the new weapons. Their brief began with two arresting sentences:[18]

> Thermonuclear weapons are undoubtedly simpler to make than scientists thought. Great skill may be required, and precise engineering, but a vast industrial effort beyond that required for producing plutonium and ^{235}U is not needed.

For the Chiefs of Staff, and for their Commander-in-Chief Churchill, none of them known for declining opportunities to acquire new weapons, the crux of the message was that the H-bomb would not be ruinously expensive. Cockcroft and Penney probably guessed that they would soon be told to make the new weapons – and quickly.

One of Cockcroft's projects at Harwell was to help set up a European nuclear laboratory, later named CERN (an acronym for the French name of its original governing body, Conseil Européen pour la Recherche Nucléaire).[19] The idea of such a venture, first mooted after the Second World War, took wing in the closing weeks of the 1940s and was backed in Britain by Cockcroft and several other leading physicists. Many of their colleagues believed, however, that it would be better to invest the UK's modest resources in research at home rather than risk an unwieldy European collaboration. In the early planning stages, Britain appeared to be leery of the initiative and never formally joined the European organisation. Cockcroft then quietly demonstrated leadership. Backed by his young Harwell physicists, he was among the leading advocates of the idea that European countries should combine their resources to build a high-energy particle accelerator to rival the one at Brookhaven National Laboratory in the United States. This enthusiasm had been kindled in part during his visit in October 1954 to Brookhaven, where the laboratory's physicists were still reeling from the decision four months earlier to strip Robert Oppenheimer of his security clearance, after a long hearing in Washington.[20]

Part of the appeal of a European laboratory for Cockcroft was that it would be the fruition of Rutherford's dream of bringing together leading scientists from all over the world to enquire into the innermost workings of nature. CERN was planning to pool the resources of its dozen contributing countries to build a machine that could accelerate protons to a hundred thousand times the energy Cockcroft and Walton had achieved a quarter of a century before.

Lindemann was the most influential of the CERN sceptics, dismissing Niels Bohr's suggestion that the new laboratory might be located in Denmark ('too vulnerable to Russia', the Prof told Chadwick).[21] As usual, Lindemann's bark was worse than his bite: in November 1952, members of the British physics community – including Blackett and Peierls – set aside their differences and backed the proposal to set up the laboratory near Geneva. Lindemann supported it in the Cabinet and, at the end of the year, Chancellor of the Exchequer Rab Butler approved the long-term plan for Britain to contribute to the running of CERN. Seven months later, the British representative on the venture signed a document expressing his country's willingness to contribute to the project's long-term funding and Parliament ratified the agreement soon afterwards. So it was on Churchill's watch that Britain began to play its part in the organisation that now operates the Large Hadron Collider.

Cockcroft had much to thank Churchill for when they met on Friday 3 September, the first of their three meetings in 1954. That morning, Cockcroft had been surprised to receive a phone call from the Prime Minister's office and even more taken aback to be summoned to lunch at Chartwell, along with Bill Penney and their boss Edwin Plowden. With no idea why they were needed, Cockcroft cancelled his other appointments and instructed his chauffeur to drive him to Kent. When he arrived, at 1.15 p.m., he was met by one of the servants, escorted upstairs and introduced to Churchill, who was dressed in a blue striped boiler suit with an open collar. The Prime Minister was sipping whisky and holding court with the other guests in his study. Twenty-eight years before, in this room, Churchill had taken time off from his budget preparations to dictate his understanding of the quantum theory of the atom, demonstrating that he understood the idea of the nucleus. On the shelves were the complete novels of H. G. Wells, including *The World Set Free*, where Churchill had almost certainly first read of the possibility of harnessing nuclear energy.

Churchill looked younger than Cockcroft had expected. Only when the Prime Minister stuck out his jaw – drawing attention to his neck's scrotal complexion – did he look his age. Cockcroft was not easily star-struck, but now he was powerless to resist, as is clear from the account of the afternoon he sent his mother.[22] Although written in his staccato prose, his words radiate pleasure:

. . . we moved to the drawing room, a cheerful room with flowers and a fine view over the park. Then Lady Churchill and a daughter came in, both very lively and talkative, and we partook of sherry or tomato juice. Then we had lunch – dressed crab and salad, chicken, fruit and cheese, and then good black coffee. There was much talk on both sides – then the ladies left and the PM held forth on World politics till 4 o'clock – most entrancing . . .

Cockcroft ended his letter, 'I must write a record of what was said for posterity,' though he never did. It is, however, safe to assume that Churchill rammed home the crucial importance of standing alongside the US, and of the need for Britain to acquire the H-bomb. He probably wanted to know when the weapons would be ready, assuming Parliament approved their acquisition. At the end of the afternoon, following a long walk round the gardens, Cockcroft and his colleagues departed. Churchill then went to sit in an adjacent room for Graham Sutherland, who, at the request of Churchill's parliamentary colleagues, was painting his portrait as an eightieth birthday present.

Three months later, eight days before Christmas, Cockcroft received another invitation to lunch with the Prime Minister, this time at 10 Downing Street. Again, Cockcroft had no idea why he had been invited, but he found out soon enough, when he was joined by Penney, Plowden, Lindemann and the Cabinet minister Lord Salisbury in the eighteenth-century drawing room upstairs.[23] Uncharacteristically punctual, Churchill entered looking smart in a black suit, well-ironed white shirt and spotted bow tie. He was on fine form, apparently still basking in the afterglow of his eightieth-birthday celebrations. However, he had not liked the Sutherland portrait, which seemed to show him as an elderly, almost frangible leader clinging on to power, rather than a veteran statesman resolutely staying in office to oppose tyranny. The painting upset him so much that Clemmie later had it cut up and burnt.[24]

Cockcroft was impressed to see the Prime Minister propounding his views on world affairs with the vim of a young man. Churchill declared himself 'pretty concerned' with the discussions at the NATO meeting then taking place in Paris. That gathering was chaired by its Secretary General Pug Ismay and attended by the US Secretary of State John Foster Dulles, a politician Churchill loathed ('a terrible handicap . . . this bastard'[25]). NATO was contemplating the use of tactical nuclear weapons to supplement its conventional forces, and delegating the authority to deploy nuclear arsenals to its military commanders.[26] The use of battlefield nuclear weapons was in line with President Eisenhower's 'New Look' defence policy, summarised in the popular slogan 'More bang for the buck'. The discussion was driven in part by the need to reduce military budgets and to deter the Soviets in case they were tempted to invade Europe. Churchill appears to have been concerned that the new policy might cause a mere skirmish to escalate into a nuclear conflict.

After the Prime Minister had fed his poodle, the party sat down to a four-course lunch. Over the Stilton and fruit, Churchill discoursed on the changing landscape of geopolitics, occasionally probing his guests on nuclear matters. He commented that he would like to visit the Aldermaston weapons establishment and Harwell. Scientists there were already working on the H-bomb with American help – after the renewed exchanges that followed Eisenhower's election to the Presidency – though without authorisation from Parliament.[27] This was more than polite small talk: shortly afterwards, Cockcroft and Penney heard that their establishments would be visited after Christmas by the Prime Minister.

Sure enough, at 3 p.m. on 30 December, the day after his secret tour of Aldermaston, Churchill arrived in a Humber Pullman at Harwell's security fence.[28] Wrapped in a greatcoat with huge furry lapels, he could have passed – were he not so well known – as one of the better-fed members of the Soviet politburo. After Cockcroft climbed into the back of the car, they were driven into a hangar to see the establishment's most powerful nuclear reactor. Wearing one of his capacious hats and carrying his walking stick, Churchill climbed out of the car to join a posse of officials, including the wan and doddery Lindemann.

'Take me to the neutron!' Churchill demanded.[29] It fell to Cockcroft to point to the invisible beams emerging from holes in the reactor and heading towards a target, surrounded by a tangle of detectors. After explaining what was going on, Cockcroft sat the Prime Minister down at the control desk and invited him to terminate the nuclear chain reaction raging inside the reactor by twiddling some knobs and pushing a button. Churchill had concluded his first, and last, nuclear experiment.

Cockcroft steered his visitor to some exhibits and models illustrating the benefits of Harwell's work for industry and cancer research, but Churchill seemed only intermittently interested. Entering the chemistry laboratories, he perked up in the ante-room, where he seemed surprised to have to submit to the standard safety procedures. Smoking was forbidden, two laboratory assistants told him, relieving him of his hat, overcoat and stick, before fitting him with a matching set of white accessories – a freshly laundered lab coat, rubber overshoes and a linen hat, soon discarded.[30] Churchill entered the laboratory looking like an extra from the television series *The Quatermass Experiment*.

After the tour of the laboratory's 'Plutonium Bank', containing nothing to see beyond a few specks of fissile material, it was time for Churchill to leave. By this stage of the visit, Cockcroft wrote soon afterwards, the Prime Minister 'was taking quite an intelligent interest', and was showing his anxieties about the new technology. After one last safety check, when technicians scanned his body with a Geiger counter, Churchill commented: 'I'm glad I was born when I was.'

The afternoon had been one of the highlights of Cockcroft's life, but for Churchill it was only a diversion from the business of staying in office. His Cabinet wanted him out.

APRIL 1954 TO APRIL 1955

Churchill's nuclear swansong

> 'What a wonderful thing it would have been if [Sir Winston] had called on the Members of the House of Commons yesterday to get on their knees in prayer to call humbly on God for His guidance in this dark and terrible hour.'
>
> BILLY GRAHAM, New York, 2 March 1955[1]

Eight days before Churchill visited Harwell, Anthony Eden, Harold Macmillan and five other Cabinet ministers confronted him in Downing Street, insisting on a firm date for his resignation. All of them had now had more than enough of his leadership: the vacillations, the H-bomb obsession, the ramblings in the Cabinet meetings he bothered to attend. When Churchill snapped that it was clear they wanted him out, none of them demurred. He concluded menacingly that he would reflect on their words and let them know his decision, having been given plenty to ponder over Christmas.

Whatever Churchill's shortcomings as Prime Minister, he had put his party in a strong position to win the next election, and he was not going to be bullied out of office. As promised, he had led a politically moderate administration. He had ended rationing, built more houses and put the economy back on an even keel, although not energised it, and he had not attempted to roll back the welfare state introduced by the Attlee government. But the Tories now needed a more energetic leader, in tune with the times – John Osborne was looking back in anger, and the skiffle groups were warming up teenagers for the arrival of Elvis Presley. Newspaper photos and newsreels juxtaposed images of the white-haired, shuffling Prime Minister with those of the glamorous young monarch Queen Elizabeth, six weeks older than Marilyn Monroe.

By the spring of 1954, it was clear that, for all his reputation as a great statesman, Churchill was no longer able to get things done on an international scale. The Soviets did not even bother to respond to his speeches calling for a summit, and his initiatives were largely ignored by the Eisenhower administration. Churchill's obsession with 'the special relationship' with America, his relative indifference to European alliances and his fixation on the power blocs that dominated the Second World War combined to make him appear outmoded.

His domestic authority was also on the wane, as his colleagues saw in the Commons debate on the H-bomb on 5 April 1954. Beforehand, Michael Foot and other Labour MPs taunted him for his failure to confront the Americans' apparently blasé attitude to the H-bomb's arrival. Furious, Churchill resolved to 'put Attlee on his back'. The opposition leader opened the debate with a dignified speech, seeking 'no party advantage', and complimenting Churchill on his summit initiatives to lessen the threat of thermonuclear war.[2] Blind to this goodwill, Churchill turned on his Labour opponents, denouncing what he believed to be the previous government's abandonment of the Quebec Agreement. Repeatedly interrupted, he struggled to make himself heard above opposition cries of 'Disgraceful!', 'Shocking!', 'Resign!' Behind him, rows of Tory MPs sat glum and silent.[3] With Attlee quivering with anger, Churchill fought on to the end of his prepared text, the pitch of his voice ascending to a squeak.

Away from this and other Punch and Judy spats in the Commons, the Prime Minister kept up with developments in nuclear politics, including Oppenheimer's security hearing. After Churchill requested a briefing on the American physicist, the

Prof supplied a balanced assessment of the case on 13 April, the day after the proceedings began in Washington.[3] Oppenheimer appeared to have 'vaguely left-wing sympathies' and 'sort of feeling of guilt about having made the original atom bombs', the Prof wrote, concluding that it was 'very unlikely that he should ever have betrayed any secrets.' Churchill read the letter and forwarded it to Anthony Eden, asking him to return it. A month later, the Prime Minister was once again shaken by another insight into destructive potential of H-bombs, this time in an article that explained how it could easily be converted into a 'suicide device'.[4] The piece in the *Manchester Guardian* reported comments by G. P. Thomson on 'an imaginative attempt to do the worst' by jacketing the Bomb with a cobalt-based chemical. If such a device were detonated, the entire upper atmosphere of the Earth would be poisoned. The idea had been conceived in the United States by Leó Szilárd during a live television broadcast, when he said the modified weapon was so dreadful that no nation would dare use it.[5] Asked by Churchill to comment on the idea, Lindemann confirmed that it was quite feasible – with a cobalt-enhanced H-bomb 'It might well be possible to poison the entire world,' he wrote.[6]

Churchill had another shot at convincing Eisenhower of the need for a summit in late June, when they met in Washington. The Bomb was at the top of the draft agenda, alongside tensions in the Middle East and Indo-China, but the Prime Minister persuaded the American leader to add 'Possibility of high-level talks with the Soviets'. Even after Churchill made a heartfelt proclamation of his fears of thermonuclear war,[7] Eisenhower refused to give ground. A summit was premature, he said, and would give Malenkov a chance 'to hit the free world in the face'.

Churchill backed off, but was unbowed and as determined as ever to talk with the Soviet leadership. Sailing home on the *Queen Elizabeth*, he was in a pensive mood, dining in the Verandah Grill, whiling away hours playing bezique with Anthony Eden and their aides, and absorbed in his first reading of Harold Nicolson's *Public Faces*. The book was 'remarkable', Churchill thought, as it was 'all about the atomic bomb'[8] but had been written in 1932. He seemed to have forgotten that he had also written about the Bomb before Nicolson had begun the novel.

Churchill and Eden were not getting on well. Eden had hoped in vain to secure a firm promise for the date of his succession, while Churchill's mind was on higher things. Deciding to be bold and ignore the views of Eisenhower, Eden and the rest of the Cabinet, Churchill sent a telegraph to the Soviet Foreign Minister Vyacheslav Molotov, proposing to meet him and his colleagues. The result, a few days later, was that Churchill lost control of his Cabinet.[9]

On 7 July, the day after he returned to London, he received a mildly encouraging reply from Molotov and informed his Cabinet that he had sent the message.[10] His colleagues were dismayed. They seethed at this *fait accompli*, and were even angrier a few minutes later to hear of yet another one: Churchill disclosed that he had approved the decision to build the H-bomb in England and that work on it was under way. In his diary, Harold Macmillan described the scene: Leader of the Commons Harry Crookshank 'made a most vigorous protest at such a decision being communicated to the Cabinet in such a cavalier way', before rising from his seat and walking out, followed in dribs and drabs by his incensed colleagues.

Never before had Churchill made such a mess of Cabinet business. The irony was that he had said a few months earlier that he wanted the entire Cabinet to be involved in the H-bomb decision. But these good intentions fell by the wayside during his consultations with the Chiefs of Staff, senior Cabinet colleagues, William Penney, John Cockcroft and the bevy of experts secretly brought together by Cabinet Secretary Norman Brook. In a powerful peroration at one meeting, Churchill had explained why it was essential to acquire the H-bomb. With it, Britain could preserve its global influence, make defence cuts and avoid giving the impression of disarmament and doing anything to 'weaken our power to influence United States policy'.

Had he not been so ham-fisted, Churchill would have got his policy through the Cabinet without difficulty. But now his colleagues were going to make him suffer. On the day after the Cabinet walked out, they met again, this time for an equable discussion about the wisdom of building the weapon. They returned twenty-four hours later to consider the Prime Minister's approach to the Kremlin. Seeing nothing to apologise for, Churchill forced them to read the text of his draft telegram like a class of obeisant schoolboys. The proceedings soon degenerated into the most dramatic Cabinet meeting Macmillan had ever attended, the acrimony abating only after Churchill adjourned it. Unknown to anyone outside Whitehall, the British government was on the point of collapse.

Stability returned only at the end of the month. On 26 July, Churchill withdrew his proposal to travel to Moscow and his H-bomb policy was nodded through. Macmillan wrote in his diary at the end of the month: 'All of us, who really have loved as well as admired him, are being slowly driven into something like hatred . . .'[11]

Knowing that Churchill was looking for a suitably grand departure, Eisenhower sent him a long and thoughtful letter suggesting how he might stage a 'fitting climax' to his career.[12] The President set aside the possibility of a détente summit, but suggested that Churchill counter the Kremlin's tendency to pre-empt the right 'to speak for the small nations of the world' by making a big speech to renounce colonialism. The Prime Minister politely declined, admitting 'I am a laggard' on colonial policy, having been brought up 'to feel proud of much that we had done'. Denying unconvincingly that he was casting about for a way to make 'a dramatic exit', Churchill reiterated his détente agenda: 'It will seem astonishing to future generations – such as they may be – that with all that is at stake no attempt was made by personal parley between the Heads of Government to create a union of consenting minds on broad and simple issues.' Eisenhower ignored this when he replied.

Despite all the signs that his cause was lost, Churchill persevered until the last vestige of hope for his initiative was crushed. The Soviet news agency announced in early February 1955 that Malenkov had been demoted.[13] In truth, he had been forced out, partly for his warnings of a global nuclear holocaust. The Soviet government's Central Committee now regarded that view as unacceptable, as it encouraged 'the emergence of a feeling of hopelessness . . . [that benefited only] . . . the imperialist advocates of a new world war'.[14] The new leadership in the Kremlin had no time for Churchill – they well remembered that he had wanted the Soviets killed at birth, and believed that his 'iron curtain' speech had begun the Cold War.[15] The summit project, the longest and least successful of Churchill's career, was now over.

Between the summer of 1954 and the end of the year, Churchill's government was treading water. Asked by his doctor if he was doing much work, Churchill replied with a grin: 'I do nothing . . . I am pretty skilful now at avoiding things.'[16] The one topic that still held his interest was the H-bomb. A chilling intelligence report he read in December said that the only warning of a thermonuclear attack that Britain could count on was the first twinkling dot on radar screens showing the approach of a Soviet bomber.[17] It was time to look four-square at how Britain might survive such an onslaught – the megadeaths, the fall-out, the catastrophic damage to the economy. Churchill quickly approved the preparation of a full report, commenting, 'Please keep me informed of the details of every step.'[18]

In mid-February 1955, the government published a White Paper to explain why the UK should acquire the H-bomb. Churchill was expected to present the policy in Parliament's defence debate on 1 March – the perfect opportunity for his swansong. He had not yet disclosed his plan to leave to the despondent Eden, preferring to keep the Cabinet on tenterhooks. Every day, his colleagues scrutinised his words and gestures for signs that he might have decided to stand down – this was anything but a dignified departure. Churchill often talked over matters alone with Rab Butler, who later recalled their huge dinners 'followed by libations of brandy so ample that I felt it prudent on more than one occasion to tip the liquid into the side of my shoe'.[19] Their subjects were always the same: Churchill's retirement, the prospect of a summit with the Soviets and his new preoccupation with space travel: 'He was very irritated', Butler recalled, 'by the idea of going to the moon, which he regarded as a waste of time and money.'

Many politicians and commentators wanted an open debate on Britain's acquisition of the H-bomb. A longstanding agreement with the BBC ensured that any subject to be discussed in Parliament would not be aired on radio and television within the two weeks before the debate was scheduled to take place.[20] BBC executives and several opposition MPs wanted the agreement relaxed, but Churchill refused to budge, telling the Commons that 'the bringing on of exciting debates in these vast, new robot organisations of television and BBC broadcasting' might have 'very deleterious effects on our general interests'.[21] In his view, a bunch of chattering nabobs appointed by the left-leaning BBC should never be allowed to pre-empt parliamentary debate.[22]

*

On the morning of 1 March, a few hours before he was due to deliver his speech, he was in his Downing Street bedroom, so excited and so busy polishing his lines that he broke his habit of reading the morning newspapers over breakfast. They lay unopened in a pile on his bed-table, as his doctor was surprised to see when he paid a visit. 'I've taken a hell of a lot of trouble over this speech,' Churchill declared, 'twenty hours over its preparation and eight hours checking the facts.'[23] Flicking through the seventy-odd pages of typescript, he read out some of his favourite turns of phrase, before making a few final revisions. His secretary Jane Portal delivered the final script minutes before he was driven to the Commons. MPs were hoping that the Old Man still had it in him to brighten up a chilly, overcast day with one last memorable matinée.

Well before the debate began at 3.45, the Chamber was full of chattering MPs, dozens of them sitting on the floor, squatting on the stairs, craning their necks round the Speaker's chair for a better view. Among those looking down from the galleries were several of Churchill's closest friends and family, including, unusually, his wife.[24]

His audience's eyes locked on him from the moment he entered the Chamber. He picked his way along the front bench, appearing to totter before finally flopping into his place.[25] When the Speaker called on him to speak, Members on all sides of the House roared him to the dispatch box. By the time he had ended his first sentence, on 'obliterating weapons of the nuclear age', the Chamber knew it was in for some vintage Churchill.

After acknowledging decades of advice from Lindemann, he risked losing his audience by quoting a long passage from 'Fifty Years Hence', which had looked forward to the release of nuclear energy, seven years before the discovery of fission. 'I hope the House will not reprove me for vanity or conceit,' he commented before reading the extract. In the climate of the Cold War, 'which we detest but have to endure', he argued that the 'only sane policy' was for Britain to have the H-bomb, preserving its influence on the United States. It was essential, he argued, for the Soviets to know that if they attacked the West, they would suffer immediate retribution, but that the door was always open for friendly talks.

He took pains to avoid alienating his political opponents, stressing instead that, in pursuing the next stage of Britain's programme of nuclear deterrence, he had 'tried to live up to [the leader of the opposition's] standard'. The House was silent. This was an event too momentous to be spoiled by heckles and points of order – the Members were as attentive to his every word as they had been in the summer of 1940. On this form, Churchill's delivery had a grand sweep, a soaring baritone that glided and then plunged into a near-whisper, his hands grasping the dispatch box one moment, gesturing expressively the next.

In a thrilling *coup de théâtre*, he set out the stakes of the debate, pointing out that it 'does not matter so much to old people, they are going soon anyway'. His voice trembling, he switched his focus to children growing up in the nuclear age, adding a comment he had made nine years before to Bernard Shaw about God's reaction to it all:[26]

> I find it poignant to look at youth in all its activity and ardour and, most of all, to watch little children playing their merry games, and wonder what would lie before them if God wearied of mankind.

He had long sought, he pointed out superfluously, for a summit where the superpowers could discuss these matters 'plainly and bluntly'. Then it may be that 'we shall by a process of sublime irony, have reached a stage in this story where safety will be the sturdy child of terror, and survival the twin brother of annihilation'. With that phrase, probably polished for hours, he had reached a fitting crescendo.

For the moment, he pointed out, the H-bomb was now an unavoidable reality, so what should Britain do? 'The best defence would of course be bona fide disarmament,' he said, adding that 'sentiment must not cloud our vision'. The only sane policy at that time was 'defence through deterrence', he argued, with each side of the Cold War holding a gun to the other's head. He was careful to acknowledge that this reasoning by no means guaranteed the country's safety: 'The deterrent does not cover the case of lunatics or dictators in the mood of Hitler when he found himself in his final dugout.'

After forty-five minutes, his voice was still strong. As he approached his conclusion, he set out his credo – that 'the growing sense of unity and brotherhood between the United Kingdom and the United States and throughout the English-speaking world' should be preserved at all costs. In the climax, he strove to be optimistic about the Cold War: 'All deterrents will improve and gain authority during the next ten years', when 'the deterrent may well reach its acme and reap its final reward'. If his audience was disappointed by those words, his parting ones were irresistible reminders of his finest hour: 'Never flinch, never weary, never despair.'

He sat down to reverberating cheers. The first opposition MP to reply was Manny Shinwell, a former minister and for several years a respectful adversary of Churchill's: 'We shall all agree that the Right Honourable Gentleman has made an impressive speech – one which will undoubtedly make its impact on many millions of people in our country and throughout the world.' For the last time in the Commons, Churchill's words had worked their magic: the opposition's vote of censure was overwhelmingly rejected.

After Shinwell's speech, Churchill returned to his room in

the Commons, panting and eager to know how his performance had gone down with journalists.[27] The Tory Party's Chief Press Officer Christopher O'Brien walked in and assured him excitedly that all was well: 'If you never made another speech, that was a very fine swansong.' Churchill's face fell before he commented gloomily, 'I may not make many more speeches in the House.' He was right. In the next few weeks, he spoke a few times in comparatively minor debates, but he was winding down, his heart no longer in it. His final premiership had been successful if much less distinguished than the first, and certainly quieter. When he left the Cabinet Office for the last time, the sheaf of ACTION THIS DAY labels he had been given in October 1951 was still there. None of them had been used. He stood down on 6 April and never spoke in the Commons again.

EPILOGUES

1954 ONWARDS

1: Churchill's nuclear scientists

> 'Some of us, who were called to take part in the war projects, often thought of Rutherford and modestly strove to act in the way which we imagined he himself would have taken.'
>
> NIELS BOHR, 1958[1]

Churchill's swansong speech went down well with the cleverest inmate and chess champion of Wakefield Prison: Klaus Fuchs.[2] Talking with an assistant governor of the jail, Fuchs drew attention to the Prime Minister's most striking passage, pointing to the 'sublime irony' that safety and survival may well follow, now that the superpowers had the horrible new weapon. 'I suppose the process of sublime irony won't extend to my being released early,' Fuchs commented. He was right.

Fuchs's former peers in the scientific community were rather less sanguine about the future of the world in the nuclear age. Two weeks before, in February 1955, Einstein enthusiastically supported a manifesto drafted by Bertrand Russell in the UK. The philosopher implored world leaders to acknowledge publicly that 'their purpose cannot be furthered by a world war, and we urge them, consequently, to find peaceful means for the settlement of all matters of dispute between them'.[3] In early April, Einstein signed a modified version of the document, a few days before he died. The Russell initiative focused the attention of many Manhattan Project scientists on their role in helping to build the first Bomb. They included most of the nuclear physicists who had worked for Churchill's government during the war. A decade on, there was still no consensus among them on the wisdom of their collective role.

After the war, many of Rutherford's former 'boys' often got together to reminisce. For several of them, the romance of their golden research years at the Cavendish had been succeeded by a workaday grind at the new nexus of nuclear physics and geopolitics. Among these physicists, Cockcroft and Oliphant met especially often. Although they had very different personalities, they got on well and talked for hours about the future. It was often suggested that, had Rutherford lived another decade, 'the course of history might have been rather different, perhaps rather better', as Oppenheimer remarked.[4]

Oliphant had been appointed a research director at the new Australian National University in Canberra. A Pickwickian figure – with twinkling eyes, gold spectacles and a paunch – he tried to build the most powerful sub-atomic particle accelerator in the world, but managed only to deliver what became known locally as 'the white Oliphant'. By then, he was a self-proclaimed 'belligerent pacifist' and an energetic campaigner against nuclear weapons.[5] When he read about a Commons debate on the Quebec Agreement, he was shocked to hear Churchill deny that it enabled the United States to have a virtual monopoly on nuclear energy. Oliphant wrote: 'This was the first time I had ever heard a PM tell a deliberate lie and it so shocked me that I could never regain my wartime regard for him.'[6]

After the war, Blackett drifted away from his Cavendish friends. In the late 1950s, he moved to Imperial College London and made several valuable contributions to the nuclear debate, following many years of silence.[7] Looking back on his controversial book *Military and Political Consequences of Atomic Energy*, he accepted that he had made some minor factual mistakes, but still believed he had been right: 'I had committed the unforgivable sin of being a premature military realist.'[8] There was now a 'strategic atomic stalemate' between the Soviet Union and the US, he accepted, and the main danger was not so much nuclear war as a great waste of national resources.

Having been persona non grata to Attlee's government for several years, Blackett re-established his influence on the Labour Party's thinking on science and technology. This was mainly through his friendship with Harold Wilson, who became Party leader in 1963 and Prime Minister a year later.[9] Blackett turned down an offer of a ministerial appointment, preferring to exercise his influence as President of the Royal Society. Having declined official honours for decades, he began to accept them. In 1974, after Blackett accepted a place in the House of Lords, the elderly A. V. Hill considered commenting to him, 'How are the mighty fallen',[10] but thought better of it.

Although James Chadwick was only six years older than Cockcroft and Blackett, they saw him as a remote figure, probably because for many years they had to answer to him as Rutherford's deputy. Two years after Chadwick returned to Liverpool from the United States in the summer of 1946, sallow and spent, he accepted the Mastership of his former college in Cambridge. He was joined by a former MAUD colleague in 1952, when G. P. Thomson became the Master of Corpus Christi College. Both men saw their moves to Cambridge as a way of moving into calmer pastures. For Thomson, the move worked out well – his combination of pugnacity and charm made him a popular figure. He also became an accomplished populariser in the press and on the radio, a formal though lucid speaker, with a talent for making striking analogies. In one of his first talks on the H-bomb, he told his Third Programme audience that unstable heavy atoms were like 'overgrown empires which are ripe for dissolution', an analogy Churchill had made independently almost a quarter of a century before, when he first learned about the nucleus.[11]

Chadwick did not fare so well – Gonville and Caius College never took him to its heart, nor did its Fellows take him to theirs.[12] Regarded by his colleagues as backward-looking and authoritarian, he was often ill and overcome by lassitude. During the nine unhappy years of his mastership he sought no pleasure in nostalgic visits to the Cavendish, which he visited only once for a routine meeting. In December 1958, he and his wife left Gonville and Caius, retired and moved to the deep peace of rural north Wales. There, he cultivated his hobby of gardening, began to edit a collected edition of Rutherford's papers – a task he would never complete – and reflected on his involvement in Tube Alloys, whose director Wallace Akers had died in 1954. Although Chadwick never regretted his role in making the Bomb, or the nuclear attacks on Japan, he worried about the next generation of weapons:[13]

> The H-bomb can hardly be classed as a weapon at all. Its effect is out of all proportion to its military effect. Most people would agree that it would be morally wrong to use the H-bomb – the only possible exception could be to use it in retaliation, even then only to the extent necessary to make the enemy stop its use.

Chadwick wanted to put his years on Britain's nuclear project behind him. He agreed, however, to comment on the first full official account of its history when the UK Atomic Energy Authority's historian Margaret Gowing sent him her draft manuscript in 1960.[14] They later became friends. In Chadwick's notes were surprisingly forthright comments on the meeting between Churchill and Bohr in May 1944. 'I still think that

[Bohr] was right,' he told her.[15] The problem was that Churchill did not understand Bohr's argument, Chadwick wrote. And even if the Prime Minister *had* understood, 'The US military machine would have stopped it and Roosevelt would not have been able to go against them.' Chadwick's former colleague Robert Oppenheimer also continued to lament the outcome of Bohr's meetings with Churchill and Roosevelt – it showed, Oppenheimer believed, 'how very wise men dealing with very great men can be very wrong'.[16]

Among Gowing's interviewees were Peierls – who became a close friend – and Frisch. After a brief spell at Harwell, in 1947 Frisch accepted a professorship at the University of Cambridge, where his research never had a second wind and eventually lost momentum (on one occasion, Genia Peierls bawled him out for allowing his work to drift).[17] He was in demand as a science populariser, always ready to talk about any subject in nuclear science except his time on the Manhattan Project, which he seemed to want to put firmly behind him. By contrast, Rudi Peierls was always willing to talk about the development of the Bomb. He was unwavering in his belief that, because of the Nazi threat, physicists in Britain had no choice but to work on the weapon. But he regretted that he and his colleagues 'did not insist on more dialogue with military and political leaders'.[18] Whether such conversations would have made much difference was another matter – he doubted it. Although he never quite recovered from the shock of Fuchs's betrayal, it was typical of Peierls's generosity that shortly before the former spy was released from prison in June 1959, Peierls wrote to him offering to help find him a job in England.[19] Fuchs did not reply, nor did he answer any other correspondence from his former colleagues.

In 1963, Peierls moved to Oxford University to head its theoretical physics department, but did not quite repeat his success as a research leader in Birmingham. He was committed to the international movement campaigning for a freeze on nuclear armaments: on Saturday mornings at local shopping centres, whatever the weather, he manned his makeshift stall in a suit and tie, ready to explain his views to all comers, whether or not they agreed with him. Few of the people who chatted with him seemed to be aware that he and his wife had been accused several times in books and in the press of being spies. The charges, traumatic for the Peierls family, all proved groundless.[20]

Margaret Gowing's scholarly book highlighted the roles of the Frisch–Peierls memorandum, the MAUD committee and the Tube Alloys project in the story of the Bomb. Yet they were hardly mentioned in General Groves's best-selling account of the Manhattan Project *Now It Can Be Told*, published two years before. He did, however, set out his views on the extent of the British contribution to the venture. For him, the most important role was played by Churchill, 'probably the best friend that the Manhattan Project ever had [as well as its] most effective and enthusiastic supporter'. Groves singled out Chadwick for special praise, but was rather less generous about the work of Churchill's other scientists:[21]

> On the whole, the contribution of the British was helpful but not vital. Their work at Los Alamos was of high quality but their numbers were too small to enable them to play a major role.

Groves's views did not seem to upset the elderly Chadwick, who regarded the General as 'a very great American friend'.[22] When it was announced in January 1970 that Groves, Vannevar Bush and James Conant were to receive the Atomic Powers Award from President Nixon in the White House, Chadwick wrote to congratulate all three of them, praising their 'faith, judgement and courage' in developing the Bomb, apparently forgetting all they had done to make life difficult for Akers and his colleagues during the war.[23]

Margaret Gowing continued her project in the mid-1960s with her colleague Lorna Arnold, looking into the origins of the British government's post-war nuclear projects to build weapons and generate energy. Their two-volume account revealed the full extent of Cockcroft, Penney and Hinton's achievement in making Britain a nuclear power. Reviewers commended the breadth of the work as well as its insightfulness – 'This history is not only official: it is authoritative,' the military historian Michael Howard concluded.[24]

Hinton, despite acclaim from his peers all over the world, was an anonymous figure outside his own industry, in keeping with Britain's traditional indifference to its engineers. He became the first chairman of the new Central Electricity Generating Board in 1956, and a year later had to deal with a fire in one of the reactors at Windscale.[25] It was the world's first serious nuclear accident and would almost certainly have triggered panic, had the disaster not been eclipsed in the press by the launch of the *Sputnik* satellite. Commentators all over the Western press dreaded that the Soviets, now demonstrably a leader in advanced technology, might soon be able to mount attacks from above the Earth's atmosphere. A few months before, in May 1957, Britain detonated its first H-bomb at Christmas Island in the Pacific, having developed it with remarkably few resources and to the agreed deadline. This was another feather in Penney's cap. Britain had demonstrated its nuclear competence and was rewarded with the renewal of its close nuclear partnership with the United States, eleven years after the relationship had been sundered by the McMahon Act.

The escalating East–West nuclear tensions moved J. B. Priestley, author of the post-nuclear-apocalypse novel *The*

Doomsday Men nineteen years earlier, to argue in the *New Statesman* that Britain should stop making nuclear weapons and declare that it would never use them.[26] Within months, the article led to the formation of the Campaign for Nuclear Disarmament. CND then organised a nationally publicised Easter march, led by Bertrand Russell, from Trafalgar Square to Penney's Aldermaston. Although a clear majority of voters in Britain wanted their country to have a nuclear deterrent, an articulate opposition had found its voice.

Penney was uncomfortable with his unwanted move into the spotlight. Behind the scenes, he worked on negotiations to ban nuclear tests, but was disappointed when they ended in 1963 with an agreement to outlaw only atmospheric tests, leaving open the option of underground detonations. He retired from the nuclear industry in 1967 and accepted the post of Rector of Imperial College. After several unnerving brushes with the press, he developed an aversion to journalists, destroyed all his correspondence and spoke only rarely about his role as 'the British Oppenheimer'. He did, however, bare his feelings in 1985 during an aggressive cross-examination at an Australian Royal Commission, which wiped the rictus smile from his face and provoked a rare outburst of anger:[27]

> I thought we were going to have a nuclear war. The only hope I saw was that there should be a balance between East and West. That is why I did this job, not to make money. I did not make any money. What I really wanted to do was to be a professor.

In private, Penney was sometimes scathing about the way governments had handled nuclear weapons. When Lorna Arnold asked him why the superpowers had stockpiled more of them than they could possibly use, he replied, 'Because they were mad, mad, MAD!'[28]

To the surprise of some of his colleagues, Penney joined Pugwash, an organisation that brought together leading scientists to discuss how to minimise the threat of nuclear war, with the aim of providing apolitical advice to governments. Formed after the launch of the Einstein–Russell manifesto in July 1955, Pugwash was named after the town in Nova Scotia where its first meeting was held in January 1957. Its attendees included Mark Oliphant and Leó Szilárd, who had switched fields to biology after the war and campaigned vociferously against the spread of nuclear weapons. One disappointment for the meeting's organisers was the absence of the indisposed Niels Bohr, who neither joined Pugwash nor attended any of its meetings.[29] After August 1945, he had campaigned for all countries to share scientific information openly, urging leaders to refrain from stockpiling nuclear weapons. But he had little tangible success.

Pugwash succeeded where the Great Dane failed, mainly because of the energy and diligence of its co-founder Jo Rotblat, who had a passion equal to Bohr's and more political flair. Courteous but peppery, retiring yet assertive, Rotblat proved to be the most effective figure in Pugwash, which was continually hampered by poor funding, shambolic administration and sniping comments in the press. After he left the Manhattan Project and returned to Britain, he had changed his scientific speciality to medical physics and worked at St Bartholomew's Hospital in London. He spent most of his spare time on Pugwash, administering, chivvying his colleagues and putting a spring in its step.

Among Rotblat's Pugwash colleagues in Britain were Cockcroft, Blackett, Thomson, Frisch and Peierls, though he was just as close to scientists in other countries. Gradually, the influence of Pugwash increased and it played an influential role in the Strategic Arms Limitation Treaties of the early 1990s.

The Nobel Peace Prize was awarded jointly in 1995 to Rotblat and Pugwash 'for their efforts to diminish the part played by nuclear arms in international politics and, in the longer run, to eliminate such arms'. Rotblat did more than any other nuclear scientist to refute Churchill's principle that unelected scientists should keep out of politics.

When it came to Churchill's wartime handling of science and technology policy, nothing upset his academic scientists more than his closeness to Lindemann and the sidelining of Tizard. Although bruised, Tizard said nothing about this in public. In private, he was not above telling stories of Churchillian double-dealing. He told a colleague that Churchill once took him by the arm and said: 'You think, Henry, that I rely on [Lindemann] for my scientific advice but I don't. [Lindemann] does my calculations for me but it is on you that I rely.'[30]

It was not until April 1959, six months before he died, that Tizard went on the record with comments on Churchill's achievement, and about his influence on science and technology:[31]

> . . . in my experience [he] has had neither a great influence on science and engineering, nor indeed has he displayed any real interest in science . . . As for his interest in applied science, I think I can truthfully say that when I was quite intimately concerned with his doings in this respect . . . [he] was always pressing for the wrong developments against the advice of most scientists concerned. This does not mean that he had no influence on applied science and engineering; the very fact that he was enthusiastic about everything that in his opinion could help to win the war, was of great value.

Tizard took care to put his views in perspective: 'I think he is such a great man that it is a pity to exaggerate his doings in every direction,' concluding with a quote adapted from lines written early in the eighteenth century by the poet Matthew Prior: 'Be to his virtues ever kind and to his faults a little blind.'

At the time Tizard spoke those words, however, Churchill had not finished with science and technology. He was making his final contribution to them, supported by the Prof.

6 APRIL 1955 ONWARDS

2: Churchill and his Prof

> 'If I were God Almighty, and humanity blew itself to bits, as it most certainly could, I don't think I'd start again in case they got me too next time.'
>
> WINSTON CHURCHILL, 13 September 1957[1]

Six days after Churchill left Downing Street, he began his final science-related initiative while he was on vacation in Sicily, at the grand Villa Politi in Syracuse. Accompanied by his wife Clemmie, Jock Colville and Lindemann, Churchill planned to spend three weeks reading and painting, but it rained almost non-stop and the entire party returned home a week early. The washout gave Lindemann plenty of opportunities to press his favourite case on Churchill – the need to do something radical to increase the number of high-quality engineering graduates in Britain, that is, to train more 'technologists of the highest grade, the officer class, the people who invent and introduce new processes and products'.[2] Although Churchill was not in the best of moods, he was won over and agreed to put his name to a new institution that would help solve the problem, ideally along the lines of the Massachusetts Institute of Technology.

After Churchill returned home, he comfortably retained his seat in the General Election, which saw the Conservatives remain in power. He withdrew almost entirely from public life and entrusted his diary and the management of his affairs to the former Foreign Office diplomat Anthony Montague Browne, who became the Man Friday of his winter years. Shielded from requests for his time that arrived daily in his office, Churchill was able to complete the four-volume *History of the English-*

Speaking Peoples, which he had set aside in the early months of the war. The account ended, rather oddly, in 1900 – he had no wish 'to write about the woe and ruin of the terrible twentieth century'.[3] Nor had he any wish to write about his second term as Prime Minister. Having produced some ten million words, he finally laid down his pen.[4]

To most intents and purposes retired, he spent several months each year on the French Riviera, staying with Lord Beaverbrook and Emery Reves in their luxurious villas. Churchill spent most of his time reading and playing six-pack Rubicon bezique but was still interested in extending his palette of cultural tastes – he took a course in modern art that introduced him to Manet, Monet and Cézanne, though not, apparently, to Picasso.[5] Listening to Reves's collection of gramophone records, he gradually acquired a taste for Brahms, Mozart and Beethoven, sometimes venturing into the chillier waters of Sibelius.[6]

On these visits to the Riviera, he was seldom accompanied by Clemmie, who was unwell with neuritis and rarely felt at home in the company of millionaire voluptuaries.[7] For her, Churchill's most unattractive trait was a craving for luxury so intense that he was willing to accept hospitality from almost anyone able to offer it. Clemmie disliked Beaverbrook and had even less time for Reves's fiancée Wendy Russell, with whom Churchill was 'absolutely obsessed', according to Noël Coward: 'He followed her about the room with his brimming eyes . . . staggering like a vast baby of two who is just learning to walk.'[8] Whenever Churchill ventured outside the villas – usually to visit nearby casinos and Michelin-starred restaurants – he was rarely left alone, well-wishers lining up to touch the hem. More often than not, he resented the importunities of passing celebrities, as when a stranger bounced up to him, vigorously shook him by the hand and exclaimed: 'I've wanted to do that for twenty years.' Churchill bellowed, 'Who the hell was that?' It was Frank Sinatra.[9]

One of Churchill's rare British visitors on the Riviera was Lindemann, who amused the locals when he marched along the sunny promenades dressed in a business suit, overcoat and bowler hat.[10] With the Prof now in his twilight years, the British establishment knew it was time to give him his valedictory honours. The Royal Society awarded him its Hughes Medal for the science he had done decades before, but he refused to collect what he called this 'leaving present' from an organisation that had never shown him any more than polite respect.[11] Much more gratifying to him was his appointment as a Viscount on the recommendation of Anthony Eden, an honour that meant he then ranked 'above all those damned science barons'. Oxford's back-room scribblers ensured that a few lines were soon circulating round the college common rooms:[12]

And now a greater honour yet:
He gets a leg up in Debrett.
For he becomes a nobler lord
Than Ernest Baron Rutherford.

Lindemann took up his new, more elevated position in the House of Lords, with his sponsor Lord Waverley (formerly Sir John Anderson) at his side. It was one of the last times they met.

After Lindemann left the government and returned to Oxford University, he retired and successfully lined up his old friend Francis Simon as his successor as Chair of Experimental Philosophy. However, less than a month after taking the post, Simon died of coronary heart disease, leaving Lindemann stricken and the Clarendon Laboratory bereft. This was one of the worst times of the Prof's life. Grieving for Simon, and worried about the future of his physics department, he also witnessed Britain's humiliating withdrawal from its Suez operation after an ill-considered invasion, while the United States kept its distance from the farce. This ignominy dispelled once and for all the illusions of Eden, the last British Prime Minister to believe that Britain was still a first-rank power. He soon left office and was succeeded by Harold Macmillan.

Lindemann was not down in the mouth for long. In May, he was on his most caustic form on the letters page of *The Times*, a few days after Bill Penney and his team tested Britain's first hydrogen bomb in the Pacific. Contemptuously dismissing 'the expected outcry in left-wing journals with the discreditable personal attacks on myself', he ridiculed critics of the tests, including the Pope.[13] He could still stun a roomful of dons with one of his barbs or with an outrageous turn of phrase. The most important event of the age will be seen to be 'the abdication of the white man', the Prof once announced portentously. He appeared to believe this was true in his favourite sport, too, as he demonstrated in the college's television room during the Wimbledon Championships, on the hot and sticky afternoon of 2 July 1957.[14] When he saw the great African-American tennis player Althea Gibson trouncing the British Christine Truman, he heckled Truman unmercifully for letting the side down. Feeling unwell, Lindemann went for a walk in Christ Church Meadow and then made a few final adjustments to his will, before retiring to bed. The next morning his valet found him dead of a heart attack.

'He is gone and I am left to linger on,' Churchill murmured when he heard the news.[15] At the funeral in Christ Church Cathedral, the mourners stood in unison when Churchill and his wife entered the building.[16] As the Prof's friend Roy Harrod later recalled, after the service Churchill walked in the procession up the cemetery path and beyond, advancing 'over the difficult tufts of grass, with unfaltering but ageing steps' towards the awaiting coffin. Several friends had suggested that he return to Chartwell after the service, but he insisted repeatedly: 'I *must* go to the grave.'

Sir John Anderson died six months later. Normally the epitome of discretion, he was sometimes surprisingly blunt towards the end of his life, telling Lord Moran in August 1956: 'Left to himself, Winston's judgment was a menace.'[17] A year later, Anderson told the visiting Robert Oppenheimer that 'he had never been reconciled to the fact that Bohr's counsel had not been followed' by Churchill and Roosevelt.[18]

After Lindemann's death, Churchill often pointed out to his friends the Prof's wisdom and insights on nuclear matters. There was an instance of this late in the summer after Lindemann's death, when Churchill read Nevil Shute's recent novel *On the Beach*, set after a nuclear apocalypse in which most of the bombs were encased with a cobalt compound to maximise their destructiveness. Lindemann had first told Churchill of this loathsome idea, which would later feature in the movie *Dr Strangelove*, as the basis of the 'doomsday machine'.[19] Churchill now believed that 'the Earth will soon be destroyed by a cobalt bomb'. More in hope than expectation, he sent a copy of Shute's book to Khrushchev but not to Eisenhower, who was, in Churchill's view, now too 'muddle-headed' to benefit from it.[20]

Churchill was as disconcerted as anyone in the West by the Soviets' launch of *Sputnik* in October 1957. As he wrote to his wife, it was proof of Lindemann's thesis 'of the forwardness of Soviet Science, compared to the American'.[21] 'We have fallen hopelessly behind in technical education,' he wrote, adding his own spin to the Prof's explanation: 'The necessary breeding ground has failed. We must struggle on; & looking to the Union with America.'

Although he believed that the world was destined to end in thermonuclear war, he was adamant that a strategy of deterrence was the best way to prevent a catastrophe. He kept an eye on the British H-bomb project and invited Sir Edwin Plowden – head of the UK Atomic Energy Authority – to dinner soon after Britain first detonated one of the bombs. By that time, Churchill rarely went on official visits, but he accepted Plowden's invitation to Aldermaston 'to show you something of what has resulted from your decision, just three years ago, to enter the "thermonuclear megaton" club'.[22] During the visit, on 3 December 1957, Churchill was shown rough-cut footage of the Monte Bello explosion and the first British H-bomb test. There was no soundtrack but, as Anthony Montague Browne later remembered, the images were powerful enough: 'The effect was numbing.'[23] On the journey back to London Churchill was silent, until he blurted out: 'What did you think of that?' After a perfunctory comment from Montague Browne, Churchill said he 'always feared that mass pressure in the United States might force them to use their H-bombs while the Russians still had not got any', adding, 'It's always been a tendency of the masses to drop their Hs.' The joke was, according to Montague Browne, one of the 'increasingly rare sparks in a fire that had already burnt grey'.

One of the few sustained tasks Churchill undertook towards the end of his life was to keep an eye on the progress of the project to improve the teaching of technical subjects in the UK. Despite an indefatigable fund-raising campaign by Jock Colville, the venture appeared to have run into the sands in early 1957, but it was rescued a few months later when it was yoked together with another plan, to set up an institute of postgraduate technology in Birmingham. The idea of producing a British MIT had proved too ambitious, so the project was altered to the setting up of a new Cambridge college, to be named after Churchill.[24] He was disappointed that the initiative was not going to be based at

Lindemann's university, but the Prof had thought the decision wise: 'Cambridge is the only place it could be; there is no engineering at Oxford.'[25]

Churchill chaired the meetings of the college project's Trustees, who mainly comprised academics and industrial leaders, including executives from ICI, Vickers and the Shell Oil Company. Although his role was low-key, he made several useful contributions to the planning and could always be relied upon to supply ample amounts of 'suitable lubricants for the discussion'.[26] Eventually, his voluminous correspondence was accommodated in the College in a purpose-built archives centre, which now also stores many other records, including the papers of several scientists who worked for Churchill when he was Prime Minister.The Trustees eventually agreed on the college's undergraduate entry profile: each year, seventy per cent of the new students would study science, engineering or mathematics, the remainder studying humanities. Churchill personally approved the appointment of the college's first Master, Sir John Cockcroft, who arranged for him to visit the college site on 17 October 1959.[27] After planting a tree, Churchill gave what turned out to be his penultimate public speech, praising both pure and applied research, and showing that he had changed his mind about space travel:

> Let no one believe that the lunar rockets, of which we read in the press, are merely ingenious bids for prestige . . . As with many vehicles of pure research, their immediate uses may not be apparent. But I do not doubt that they will ultimately reap a rich harvest for those who have the imagination and power to develop them, and to probe ever more deeply into the universe in which we live.

He underlined Lindemann's role in the inception of Churchill College, which could reasonably have been named after the Prof: 'His inspiration remains and he and his memory should be held bright by the scientists of tomorrow.' But this was wishful thinking: following a decent interval after Lindemann's death, his friends and enemies were about to fight an acrid and ill-tempered public battle over his reputation.[28]

Two biographies, by Roy Harrod and the Earl of Birkenhead, painted similar pictures of a man whose flaws were a price well worth paying for his brilliance, integrity and loyalty. The haters' case was made soon afterwards by C. P. Snow in *Science and Government*, most of it a knockabout account of the Tizard–Lindemann clash in the late 1930s. Stung by criticisms of his account, Snow returned to the fray a few months later and published a brief postscript that restated his case more pointedly.[29] After underlining the terrible quality of Lindemann's judgement compared with that of other wartime scientists, notably Vannevar Bush, Snow concluded: 'If you are going to have a scientist in a position of isolated power, the only scientist among non-scientists, it is dangerous, when he has bad judgement.'

One can quibble that Lindemann was not the only scientist in Churchill's company during the war, but the essence of Snow's point is correct. In science, authority comes from its communities, not from individuals, no matter how brilliant they are. Churchill made a serious error in putting so much weight on the opinion of one scientist, whose weaknesses were well known to his peers. The final sentence of the postscript was wise: 'Whatever we do, it must not happen again.'

Churchill made no public comment on this. By the end of the 1950s, the allure of the French Riviera had palled and he preferred cruising on the motor yacht *Christina O*, owned by Greek tycoon Aristotle Onassis. After a Caribbean cruise on the yacht in the spring of 1961, Churchill paid his final visit to the shores of the United States. He did not set foot on land and was unable to make the journey to Washington to meet the new President, John Kennedy.[30] Disappointed, Kennedy invited him to the White House two years later to receive an honorary citizenship of the United States, but Churchill was too weak to make the journey and watched the ceremony on television at home in London. He still paid occasional visits to the Commons, rolling into the chamber in his wheelchair, saying nothing. At the 1964 General Election, at the height of Beatlemania, he did not contest his seat, ending a parliamentary career that had begun before the Wright brothers' pioneering flight and had ended in an age when supersonic aircraft could deliver nuclear weapons.

A month after Churchill left the Commons, Margaret Gowing published the first official history of the British nuclear project during the war, laying bare for the first time the role he had played in nurturing the Bomb. Officials in the Cabinet Office had checked the penultimate draft 'to watch Sir Winston's Churchill's interests'.[31] In her description of his meeting with Bohr, Gowing gave the impression that Churchill was unaware of the implications of the Bomb in May 1944. Worried officials, unable to find anyone who could remember the meeting, sent Gowing's account to Montague Browne. He bristled at 'an implied criticism' of Churchill and asked for his side of the story to be presented more clearly:[32] 'After all, Bohr knew about as much of foreign affairs as the Prime Minister knew of nuclear physics, and perhaps lacked the latter's technical advisers.' After requests from the Cabinet Office, Gowing made several amendments.[33] The result was an account of the meeting that removed the sense of a missed opportunity and any implication that Churchill was short-sighted about the Bomb.[34] After Gowing had finalised her account of the wartime Churchill–Bohr meeting, the Cabinet Office received a letter from Churchill – almost certainly written by Montague Browne – thanking them 'for the care with which they watch his interests in these matters'.[35]

After suffering another stroke less than a year later, Churchill died in his London home on Sunday 24 January 1965. His coffin, draped in a Union flag, lay in Westminster Hall for three days, during which a third of a million mourners filed past it. Beyond requesting 'lots of military bands',[36] he had not specified the detailed arrangements for the funeral, though he would have been moved by its spectacle and by the outpouring of national grief. He was given a British state funeral in the imperial tradition, the last to be granted but the first one ever to be televised. The ceremony at St Paul's Cathedral, on the Saturday after his death, was attended by numerous heads of state – though not President Johnson – and hundreds of dignitaries, including the hunched Lord (Clement) Attlee, now able to walk only with support. There were nine military bands. A few hours later, following a brief private service, Churchill was laid to rest alongside his parents in the churchyard at Bladon, within sight of his birthplace, Blenheim Palace.

Most of his obituaries made little or nothing of his involvement in the history of the Bomb. One, however, in the *Daily Telegraph*, listed among his five main contributions to the running of the war effort 'the interweaving of science into the fabric of Government, on a scale never accomplished before'.[37] Yet even this article made no mention of the longevity of Churchill's association with nuclear weapons, as a writer and a politician. No one in the press seemed to know that the seeds of his opinions on nuclear warfare had already been sown in 1925, when he first foresaw that the weapons were coming, in the article 'Shall We All Commit Suicide?'[38]

Most revealing of all about his attitude to progress in science and technology was his early correspondence with H. G. Wells, a few years after they first met. Churchill reproved the famous author for believing naively that human beings would be able to take this progress in their stride:[39]

We shall not change so quickly as you think . . . man will [long] remain an animal with a slight balance on the side of his nobler instincts, perhaps as far below the works which his brain creates as above the species whose sufferings and struggles have created him.

Churchill even worried that science might be the death of mankind:

Till now [*Homo sapiens*] has been the most progressive feature of the world; he may end, if scientific development fulfils its promises, by being the greatest anachronism.

Churchill wrote those fearful words in October 1906, when he was an ambitious Undersecretary of State for the Colonies and was pushing to join the Cabinet. Rutherford – then at McGill University and at the peak of his career as a researcher – was working at his laboratory bench, exploring the energy that appeared to be locked up deep inside the atom.

When Wells foresaw 'atomic bombs' less than a decade later in *The World Set Free*, he believed that their destructiveness would make human beings abjure war, after they realised its pointlessness – a conclusion that Churchill probably regarded as unrealistic. By 1931, when Churchill published 'Fifty Years Hence', he feared that contemporary leaders would not be equal to the challenge of handling the weapons that scientists were about to put in their hands:

Great nations are no longer led by their ablest men, or by those who know most about their immediate affairs, or even by those who have a coherent doctrine. Democratic governments drift along the line of least resistance, taking short views, paying their way with sops and doles, and smoothing their path with pleasant-sounding platitudes.

During most of the Second World War, however, Britain had been led by its ablest politician, as Churchill himself would have been the first to agree. Yet even he – aware at the beginning of the conflict that the nuclear age was in prospect – struggled to deal effectively with the coming of the Bomb. While the articles Churchill wrote in the 1930s warning that nuclear energy might soon be harnessed are testimony to his sagacity as a writer, his handling of the technology when it arrived was not one of his great achievements as a politician.

In the 1950s, he occasionally expressed a Wellsian optimism about the potential of nuclear weapons to liberate the human race, but his words were unconvincing. The threat of thermonuclear war eventually made him revert to the pessimism that lay below the resolute hopefulness that had done much to make him a renowned wartime Prime Minister. By the second half of the twentieth century, Churchill believed, scientists had finally given international leaders weapons that were more powerful than they could handle. Science was finally becoming the master of its creator, and humanity would pay the price.

Acknowledgements

> 'Whilst writing, a book is an adventure. To begin with, it is a toy, then an amusement, then it becomes a mistress, and then it becomes a master, and then it becomes a tyrant and, in the last stage, just as you are about to be reconciled to your servitude, you kill the monster and fling him to the public.'
>
> WINSTON CHURCHILL, 2 November 1949[1]

Before a book is flung to the public, custom dictates that its author thanks everyone who made the writing more of an adventure than a struggle. I am glad to follow this admirable tradition here, for *Churchill's Bomb* has been an especially fulfilling enterprise.

Much of the research on the book was done at Churchill College, Cambridge, especially in its excellent Archives Centre. I should like to thank the Winston Churchill Memorial Trust for generously funding two stays at the college as an Archives By-Fellow, and it is a pleasure to express my gratitude to Jamie Balfour, the Trust's Director-General, for facilitating the arrangements. I could not have been treated with more friendly consideration than I received at the Archives Centre, so it is a special pleasure to thank its director, Allen Packwood, and his unfailingly helpful colleagues: Natalie Adams, Francesca Alves, Philip Cosgrove, Andrew Riley, Sarah Lewery, Sophie Bridges, Katharine Thomson, Julie Sanderson, Laure Bukh, Madelin Terrazas and Bridget Warrington, not forgetting Lynsey Darby and Caroline Herbert, who have now moved on. Churchill College's Master, Sir David Wallace, and its Fellows have made me exceptionally welcome during my numerous stays, which were essential to the development of this project.

Most of *Churchill's Bomb* was written at the Institute for Advanced Study in Princeton during four richly rewarding summers, when I was visiting its former director Peter Goddard. Over the past decade, he has been generous beyond measure in supporting my writing projects, which would otherwise not have been possible. I am duly grateful to him and to his successor as the Institute's director, Robbert Dijkgraaf, who also made me extremely welcome. From the Institute's Faculty, I have learned much about matters relating to scientific aspects of the book during conversations with Nima Arkani-Hamed, Freeman Dyson, Peter Sarnak, Nathan Seiberg and Matias Zaldarriaga. The services in the libraries and at the Shelby White and Leon Levy Archives Center at the Institute are peerless, and I am delighted to thank all the many colleagues there who have showered me with kindness and given me no end of help: Christine Di Bella, Karen Downing, Momota Ganguli, Gabriella Hoskin, Erica Mosner, Marcia Tucker, Kirstie Venanzi and Judy Wilson-Smith. Many colleagues and friends made my stays at the Institute uniquely pleasurable: Lily Harish-Chandra, Kate Belyi, Beth Brainard, Linda Cooper, Karen Cuozzo, Christine Ferrara, Catie Fleming, Michael Gehret, Helen Goddard, Jennifer Hansen, Pamela Hughes, Kevin Kelly, Camille Merger, Louise Morse (*mère et fille*), Susan Olson, Amy Ramsey, Paul Richardson, Kelly Devine Thomas, Nadine Thompson, Jill Titus, Michele Turansick, Sarah Zantua-Torres, Sharon Tozzi-Goff and former colleague Margaret Sullivan.

During stays at Trinity College, Cambridge, I made extensive use of the papers held in its Wren Library. I should like to thank the college's former Master Lord (Martin) Rees for enabling these sojourns, and Jonathan Smith for facilitating my use of these archives.

Of all the colleagues and friends who have given me information as well as expert help and advice, I am especially indebted to: Jodie Anderson, Christopher Andrew, Lorna Arnold, Joanna Batterham (*née* Chadwick), Jeremy Bernstein, Adrian Berry, Sir Michael Berry, Giovanna Bloor (*née* Blackett), Sandra Ionno Butcher, Brian Cathcart, Judith Chadwick, Chris Cockcroft, Ralph Desmarais, David Edgerton, Joyce Farmelo, Pedro Ferreira, Robert Fox, Mark Goldie, Charles Griffiths, Gaby Gross (*née* Peierls), Jonathan Haslam, Ian Hart, Lord (Peter) Hennessy, David Holloway, Jo Hookway (*née* Peierls), Ruth Horry, Steve Jebson, Gron Tudor Jones, Katharina Kraus, Rita Kravets, Richard Langworth, Dan Larson, Sabine Lee, Roy MacLeod, Alice Martin, Peter Morris, James Muller, Gros Næs, the late Sir Michael Palliser, Orhan Pamuk, Christopher Penney, Martin Penney, the late Sir Michael Quinlan, David Reynolds, Simon Schaffer, Michael Sherborne, George Steiner, Zara Steiner, Martin Theaker, Sir John Thomson, Humphrey Tizard, Jane Tizard, Lady Tizard, Alan Walton, the late Sir Maurice Wilkes and Lady Williams of Elvel.

For their careful reading of the entire manuscript, and for numerous comments and corrections, I am grateful to Andrew Brown, Brian Cathcart, Maddy Corcoran, Freeman Dyson, Paul Courtenay, Allen Packwood, Ben Sumner and David Sumner. My friend David Johnson read every draft of the text with great attention to detail and supplied me with a stream of comment and advice, all of it learned, wise and constructive. I owe him an enormous debt.

During the research for *Churchill's Bomb*, I worked in several other archives, whose staff I would like to thank: the staff at the University of Birmingham archives; the staff at the Bodleian Library at Oxford University; Finn Aaserud and Felicity Pors at the Niels Bohr Archive in Copenhagen; Laura Gardner and Rhonda Grantham at the Institute of Mechanical Engineers; the staff in the Manuscripts Room at the Library of Congress, Washington DC; Clare Kavanagh, Elizabeth Martin and Tessa Richards at Nuffield College (Lindemann archive), Oxford University; all the staff at the UK National Archives, Kew; Tatiana Balakhovskaya at the Kapitza archive in Moscow; Alan Carr, historian at the Los Alamos National Laboratory; staff at the Harold Nicolson archive in Balliol College, Oxford; Virginia Lewick and David Woolner at the Franklin D. Roosevelt Library and Museum; Peter Collins and Joanna Hopkins at the Royal Society; Lynda Corey Claassen and Matthew Peters at the Mandeville Special Collections Library (Szilárd archive) at the University of California, San Diego; Tammy Kelly at the Harry S. Truman Library and Museum; Dennis Sears at the H. G. Wells archive at the University of Illinois.

Throughout the preparation of the book, I have benefited from the unstinting encouragement and advice of my publishers. At Faber in London, I especially want to thank Neil Belton, who shepherded the project from its conception and gave me wise guidance at every stage. Kate Burton, Kate Ward and copy-editor Neil Titman were a pleasure to work with. At Basic Books in New York, Lara Heimert gave me an invaluable critique that much improved the quality of the narrative.

Finally, I should like to underline that every error of fact and judgement in the book is solely my responsibility. And I want to thank everyone I have acknowledged here for making *Churchill's Bomb* such an agreeable adventure – the book never became a marauding monster that I could not wait to fling at you, but an agreeable companion I am rather sad to see go its own way.

Princeton
February 2013

References

Please note that all references to websites in the notes were verified correct on 27 January 2013. All the Churchill documents – whose references begin with CHUR, CHAR and CHPC – are in the Churchill archive at Churchill College, Cambridge.

List of archival sources

Several of the archives listed below are in the Churchill Archives Centre (CAC) in Churchill College, Cambridge.

AAC Academic Assistance Council archive, Bodleian Library, University of Oxford, UK

AEA Albert Einstein Archives, Hebrew University of Jerusalem, Israel

AEC Atomic Energy Commission records, www.archives.gov

AHQP Archives for the History of Quantum Physics

AIP American Institute of Physics, interviews by Charles Weiner

AVHL A. V. Hill archive, CAC

BHM Birmingham University Archives, UK

BHMPHYS Archives of the Birmingham University physics department, UK

BLACKETT Patrick Blackett, Royal Society, London, UK

BMFRS Biographical Memoirs of the Fellows of the Royal Society

BRUN Papers of Frederick Brundrett archive, CAC

CHAD Papers of James Chadwick, Churchill College, University of Cambridge, UK

CHBIO Biography of Winston Churchill, by Sir Martin Gilbert, except for the first two volumes, written by Churchill's son Randolph

CHDOCS Documents in support of the biography of Winston Churchill, edited by Sir Martin Gilbert, apart from the first five volumes, which were written by Churchill's son Randolph

CHESSAYS *Collected Essays of Winston Churchill*, Bristol, Library of Imperial History

CHPC Press cuttings of Winston Churchill, CAC

CHSPCH *Winston S. Churchill: His Complete Speeches, 1897–1963*, ed. Robert Rhodes James, London, Chelsea House Publishers

CKFT Cockcroft papers, CAC

CKFTFAMILY Cockcroft family papers

CLVL Jock Colville Papers, CAC

CONANT Papers of James Conant, Harvard University, Cambridge, MA, USA

CSCT Papers of Clementine Churchill, CAC

FDRLIB Papers of Franklin D. Roosevelt, Hyde Park, New York, USA

FEAT Papers of Norman Feather, CAC

FRISCH Otto Frisch papers, Trinity College, University of Cambridge, UK

FRUS *Foreign Relations of the United States*, Department of State Publications

GPT G. P. Thomson's papers, Trinity College, University of Cambridge, UK

HGW H. G. Wells archive, University of Illinois, USA

HH Harry Hopkins's papers, Georgetown University, Washington DC, USA

HINTON Papers of Sir Christopher Hinton, Institute of Mechanical Engineers, London, UK

HLFX Lord Halifax's papers, CAC

HNKY Lord Hankey's papers, stored at CAC

HNSRD Speeches in the House of Commons and House of Lords, available online (search Hansard + "quotation")

HTT Henry Tizard archive, Imperial War Museum, London, UK

IAS Archives of the Institute for Advanced Study, Princeton, USA

LIBCON Library of Congress, Washington DC, USA

LIND Archive of Frederick Lindemann, a.k.a. Lord Cherwell,

Nuffield College, University of Oxford, UK
LOSALAMOS Archive at Los Alamos Laboratory, New Mexico, USA
MCRA Papers of Colonel Stuart Macrae, CAC
MART Papers of John Martin, CAC
NBA Niels Bohr archive, Niels Bohr Institute, Copenhagen, Denmark
NIC Harold Nicolson archive, Balliol College, University of Oxford, UK
OPPY Oppenheimer archive, Library of Congress, Washington DC, USA
PEIERLS Rudolf Peierls archive, Bodleian Library, University of Oxford, UK
PLDN Papers of Lord Plowden, CAC
RFD Rutherford archive, University Library, University of Cambridge, UK
ROSK Papers of Stephen Roskill, CAC
ROWE Papers of A. P. Rowe, Imperial War Museum, London, UK
ROWECH Papers of A. P. Rowe, CAC
RVJO Archive of R. V. Jones, CAC
SPSL Archive of the Society for the Protection of Science and Learning, formerly the Academic Assistance Council, Bodleian Library, Oxford University, UK
STIMSON Diaries of Henry Stimson, Yale University, USA
WELLS Papers of H. G. Wells, University of Illinois, USA

List of references

The documents stored in the UK National Archives in Kew are labelled NA in this list.

Aaserud, F. (1986) *Niels Bohr, Collected Works, Vol. 9*, Amsterdam, North Holland

Aaserud, F. (2005) *Niels Bohr, Collected Works, Vol. 11*, Amsterdam, North Holland

Addison, P. (1992) *Churchill on the Home Front*, London, Pimlico

Alanbrooke (2002) *War Diaries 1939–45* (eds Danchev, A., and Todman, D.) London, Phoenix

Alkon, P. (2006) 'Imagining Science: Churchill and Science Fiction', in *Winston Churchill's Imagination*, Lewisburg, Bucknell University Press

Anderson, H. (1984) 'The First Chain Reaction', in Sachs, R. (ed.) (1984)

Andrew, C. (1988) 'Churchill and Intelligence', *Intelligence and National Security*, Vol. 3, part 3, pp. 181–93

Andrew, C. (2009) *The Defence of the Realm*, London, Penguin Books

Anon (ed.) (1954) *Rutherford by Those Who Knew Him*, London, Taylor & Francis

Arms, N. (1966) *A Prophet in Two Countries*, Oxford, Pergamon Press

Arnold, L. (2001) *Britain and the H-bomb*, Chippenham, Palgrave

Arnold, L. (2003) 'The History of Nuclear Weapons: the Frisch–Peierls Memorandum', *Cold War History*, Vol. 3, No. 3, April, pp. 111–26

Arnold, L. (2012) *My Short Century*, Palo Alto, CA, Cumnor Hill Books (most easily available online)

Badash, L., Hodes, E., and Tiddens, A. (1986) 'Nuclear Fission: Reaction to the Discovery in 1939', *Proceedings of the American Philosophical Society*, Vol. 130, No. 2, pp. 196–231

Bainbridge, K. T. (1974) 'Orchestrating the Test', in *All in Our Time*, Chicago, The Bulletin of the Atomic Scientists

Baylis, J., and Garnett, J. (eds) (1991) *Makers of Nuclear Strategy*, London, Pinter

Berlin, I. (1980) *Personal Impressions*, London, Hogarth Press

Bernal, J. D. (1939) *The Social Function of Science*, London, George Routledge & Sons Ltd

Bernstein, B. J. (1975) 'Roosevelt, Truman and the Atomic Bomb, 1941–45: A Reinterpretation', *Political Science Quarterly*, Vol. 90, No. 1, pp. 23–69

Bernstein, B. J. (1976) 'The Uneasy Alliance: Roosevelt, Churchill and the Atomic Bomb', *The Western Political Quarterly*, Vol. 29, No. 2, pp. 202–30

Bernstein, B. J. (1987) 'Churchill's Secret Biological Weapons', *Bulletin of the Atomic Scientists*, January/February 1987, pp. 46–50

Bernstein, J. (2011) 'A Memorandum That Changed the World', *American Journal of Physics*, Vol. 79 (5), pp. 440–6

Best, G. (2001) *Churchill: A Study in Greatness*, Oxford, Oxford University Press

Beveridge, W. (1959) *A Defence of Free Learning*, London, Oxford University Press

Bialer, U. (1980) *The Shadow of the Bomber*, London, Royal Historical Society

Bird, K., and Sherwin, M. J. (2005) *American Prometheus*, New York, Alfred A. Knopf

Birkenhead (1930) *The World in 2030 AD*, London, Hodder and Stoughton

Birkenhead (1961) *The Prof in Two Worlds*, London, Collins

Blackburn, R. (1959) *I Am an Alcoholic*, London, Allan Wingate

Blackett, P. M. S. (1946) *The Atom and the Charter*, London, Fabian Society and the Association of Scientific Workers

Blackett, P. M. S. (1948) *Military and Political Consequences of Atomic Energy*, London, Turnstile Press

Blackett, P. M. S. (1972) 'Rutherford', *Notes and Records of the Royal Society of London*, Vol. 27, No. 1, pp. 57–9

Blake, R., and Louis, W. R. (eds) (1996) *Churchill*, Oxford, Clarendon Press

Bohr, A. (1964) 'The War Years and the Prospects Raised by the Atomic Weapons', in Rozental (ed.) (1964: 191–226)

Bohr, N. (1961) 'Reminiscences of the Founder of Nuclear Science and of Some Developments Based on his Work', *Proceedings of the Physical Society*, Vol. 78, pp. 1083–115

Bowen, E. G. (1998) *Radar Days*, London, Taylor & Francis

Boyle, A. (1955) *No Passing Glory*, London, Collins

Boyle, P. G. (ed.) (1990) *The Churchill–Eisenhower Correspondence 1953–1955*, Chapel Hill, University of North Carolina Press

Brandon, H. (1988) *Special Relationships*, London, Macmillan

Brendon, P. (1984) *Churchill: An Authentic Hero*, London, Methuen

Brown, A. (1997) *The Neutron and the Bomb*, Oxford, Oxford University Press

Brown, A. (2012) *Keeper of the Nuclear Conscience*, Oxford, Oxford University Press

Bush, V. (1972) *Pieces of the Action*, London, Cassell

Butler, R. A. (1971) *The Art of the Possible*, London, Hamish Hamilton

Campbell, J. (1999) *Rutherford: Scientist Supreme*, Christchurch NZ, AAS Publications

Cannadine, D. (ed.) (1989) *Blood, Toils, Tears and Sweat: Winston Churchill's Famous Speeches*, London, Weidenfeld & Nicolson

Cannadine, D. (1994) *Aspects of Aristocracy*, London, Yale University Press

Cantelon, P. L., Hewlett, R. G., and Williams, R. C. (eds) (1984) *The American Atom*, Philadelphia, University of Pennsylvania Press

Cathcart, B. (1994) *Test of Greatness*, London, John Murray

Cathcart, B. (2004) *The Fly in the Cathedral*, London, Viking Penguin

Catterall, P. (ed.) (2003) *The Macmillan Diaries, 1950–7*, London, Pan Books

Churchill, W. S. (1929) *The Aftermath*, New York, Charles Scribner's Sons

Churchill, W. S. (1930) *My Early Life*: republished in 1996 by Touchstone Books, New York

Churchill, W. S. (1948) *The Second World War*, Vol. 1, London, Cassell & Co.

Churchill, W. S. (1949) *The Second World War*, Vol. 2, London, Cassell & Co.

Churchill, W. S. (1951) *The Second World War*, Vol. 4, London, Cassell & Co.

Churchill, W. S. (1954) *The Second World War*, Vol. 6, London, Cassell & Co.

Clark, R. (1961) *The Birth of the Bomb*, London, Phoenix House

Clark, R. (1965) *Tizard*, London, Methuen & Co.

Clark, R. (1975) *The Life of Bertrand Russell*, London, Jonathan Cape

Clarke, I. F. (1992) *Voices Prophesying War*, Oxford, Oxford University Press

Clarke, P. (2012) *Mr Churchill's Profession*, London, Bloomsbury

Cockburn, S., and Ellyard, S. (1981) *Oliphant*, Adelaide, Axiom

Cohen, R. S., and Stachel, J. J. (eds) (1979) *Selected Papers of Léon Rosenfeld*, Boston, D. Reidel Publishing Co.

Colville, J. (1985) *The Fringes of Power*, London, Weidenfeld & Nicolson

Conant, J. B. (1970) *My Several Lives*, New York, Harper & Row

Coote, C. (1971) *The Other Club*, London, Sidgwick & Jackson

Crowther, J. G. (1965) *Statesmen of Science*, London, Cresset Press, London

Crozier, W. P. (1973) *Off the Record: Political Interviews 1933–1943* (ed. A. J. P. Taylor), London, Hutchinson

Da Costa Andrade, E. N. (1964) *Rutherford and the Nature of the Atom*, New York, Anchor Books

Davis, W., and Potter, R. D. (1939) 'Atomic Energy Released', *Science News Letter*, 11 February 1939, pp. 86–7, 93

Dilks, D. (ed.) (2010) *Diaries of Alexander Cadogan*, London, Faber

Dirac, P. A. M. (1964) 'The Versatility of Niels Bohr', in Rozental (ed.) (1964: 306–9)

Dugdale, B. E. C. (1937) *Arthur James Balfour,* Vol. 2, New York, G. P. Putnam's Ltd

Eade, C. (ed.) (1953) *Churchill by His Contemporaries*, London, The Reprint Society

Eden, A. (1965) *The Reckoning*, Cambridge, MA, Houghton Mifflin

Edgerton, D. (2006) *Warfare State*, Cambridge, UK, Cambridge University Press

Edgerton, D. (2011) *Britain's War Machine*, London, Allen Lane

Eliot, T. S. (1940) 'Journalists of Today and Yesterday', *The New English Weekly*, 8 February 1940, pp. 237–8

Eve, A. S. (1939) *Rutherford*, Cambridge, Macmillan

Feis, H. (1960) *Between War and Peace: The Potsdam Conference*, Princeton, Princeton University Press

Feis, H. (1966) *The Atomic Bomb and the End of World War II*, Princeton, Princeton University Press

Fermi, L. (1987) *Atoms in the Family*, Washington, Tomash Publishers

Ferrell, R. H. (1998) *The Dying President: Franklin D. Roosevelt 1944–45*, Columbia, University of Missouri Press

Fishman, J. (1963) *My Darling Clementine*, New York, David McKay Co., Inc.

Fort, A. (2004) *Prof: The Life of Frederick Lindemann*, London, Pimlico

Fox, R., and Gooday, G. (eds) (2005) *Physics in Oxford 1839–1939*, Oxford, Oxford University Press

Franklin, H. B. (2008) *War Stars*, Amherst, University of Massachusetts Press

French, A. P., and Kennedy, P. J. (eds) (1985) *Niels Bohr: A Centenary Volume*, Cambridge, MA, Harvard University Press

Frisch, O. R. (1979) *What Little I Remember*, Cambridge, Cambridge University Press

Gannon, J. (2002) *Stealing Secrets, Telling Lies*, New York, Brassey's

Gellately, R. (2013) *Stalin's Curse*, Oxford, Oxford University Press

Gilbert, M. (2005) *Churchill and America*, London, Free Press

Goodman, M. (2007) *Spying on the Nuclear Bear*, Stanford, Stanford University Press

Gowing, M. (1964) *Britain and Atomic Energy 1939–1945*, London, Macmillan & Co. Ltd

Gowing, M. (1974a) *Independence and Deterrence Vol. 1*, London, Macmillan

Gowing, M. (1974b) *Independence and Deterrence Vol. 2*, London, Macmillan

Gromyko, A. (1989) *Memories*, London, Hutchinson

Groves, L. R. (1962) *Now It Can Be Told*, New York, Da Capo Press

Hardy, H. (ed.) (2004) *Isaiah Berlin: Letters 1928–1946*, Cambridge, Cambridge University Press

Harris, K. (1984) *Attlee*, London, Weidenfeld & Nicolson

Harrod, R. F. (1959) *The Prof*, London, Macmillan & Co. Ltd

Hartcup, G., and Allibone, T. E. (1984) *Cockcroft and the Atom*, Bristol, Adam Hilger

Harvie-Watt, G. S. (1980) *Most of My Life*, London, Springwood Books

Haslam, J. (2011) *Russia's Cold War*, London, Yale University Press

Hastings, M. (2007) *Nemesis*, London, HarperCollins

Hastings, M. (2009) *Finest Years: Churchill as Warlord 1940–1945*, London, HarperPress

Hendry, J. (ed.) (1984) *Cambridge Physics in the Thirties*, Bristol, Adam Hilger

Hennessy, P. (1989) *Whitehall*, New York, The Free Press

Hennessy, P. (1993) *Never Again, Britain 1945–51*, New York, Pantheon

Hennessy, P. (1996) *Muddling Through*, London, Indigo

Hennessy, P. (2000) *The Prime Minister*, London, Allen Lane

Hennessy, P. (2006) *Having It So Good*, London, Allen Lane

Hennessy, P. (2007) *Cabinets and the Bomb*, Oxford, Oxford University Press

Hennessy, P. (2010) *The Secret State*, London, Penguin Books (second edition)

Hermann, A., Krige, J., Mersits, U., and Pestre, D. (1987) *History of CERN, Vol. 1*, Amsterdam, North Holland

Hershberg, J. G. (1993) *James B. Conant*, Stanford CA, Stanford University Press

Hewlett, R. G., and Anderson, Jnr, O. E. (1962) *The New World, 1939–46*, University Park, Pennsylvania, Pennsylvania State University Press

Hill, A. V. (1962) *The Ethical Dilemma of Science and Other Writings*, London, Scientific Book Guild

Hinsley, F. H., Thomas, E. E., Ransom, C. F. G., and Knight, R. C. (1981) *British Intelligence in the Second World War, Vol. 2*, London, HMSO

Hobhouse, H. (1971) *Lost London*, New York, Weathervane Books

Holloway, D. (1994) *Stalin and the Bomb*, New Haven, Yale University Press

Holt, J. R. (1988) Text of History of Science & Technology Seminar, 8 February 1988, private communication

Holt, J. R. (1999) *Chadwick: Reminiscences by John Holt*, private communication from Holt

Hopkins, B. S. (1939) *Chapters in the Chemistry of the Less Familiar Elements*, Chapter 18, Uranium, Champaign, IL, Stipes Publishing Co.

Howard, M. (1974) 'The Explosive Secret', *Sunday Times*, 8 December, p. 38

Hughes, J. (2002) *The Manhattan Project*, London, Icon Books

Ironside (1962) *The Ironside Diaries 1937–1940* (eds Macleod, R., and Kelly, D.), London, Constable

Jenkins, R. (2001) *Churchill*, New York, Farrar, Straus and Giroux

Jones, R. V. (1998) *Most Secret War*, London, Wordsworth Publishing (first published 1978)

Kennedy, C. (1986) *ICI: The Company That Changed Our Lives*, London, Hutchinson

Kennedy, P. (2013) *Engineers of Victory*, London, Allen Lane

Kevles, D. J. (1995) *The Physicists*, Cambridge, MA, Harvard University Press

Kimball, W. (ed.) (1984) *Churchill & Roosevelt: The Complete Correspondence*, Vol. 1, Princeton, Princeton University Press

Kimball, W. (1991) *The Juggler*, Princeton, Princeton University Press

Kirby, M. W., and Rosenhead, J. (2011) 'Patrick Blackett', in Assad, A. A., and Gass, S. I., *Profiles in Operations Research*, London, Springer

Kojevnikov, A. B. (2004), *Stalin's Great Science,* London, Imperial College Press

LaFollette, M. C. (2008) *Science on the Air*, Chicago, Chicago University Press

Lamont, L. (1965) *Day of Trinity*, New York, Atheneum

Lanouette, W. (1992) *Genius in the Shadows*, New York, Charles Scribner's Sons

Larres, K. (2002) *Churchill's Cold War*, New Haven, Yale University Press

Laurence, D. H., and Peters, M. (eds) (1996) *Unpublished Shaw*, University Park, Pennsylvania State University Press

Laurence, W. L. (1947) *Dawn Over Zero*, New York, Alfred Knopf

Leasor, J. (ed.) (1959) *The Clock with Four Hands*, New York, Reynal & Co.

Lee, S. (ed.) (2007) *Sir Rudolf Peierls, Selected Private and Scientific Correspondence*, Vol. 1, London, World Scientific

Lee, S. (ed.) (2009) *Sir Rudolf Peierls, Selected Private and Scientific Correspondence*, Vol. 2, London, World Scientific

Lees-Milne, J. (1994) *A Mingled Measure*, Wilby, Michael Russell

Lindemann, F. A. (1920) *Mind*, Vol. 29, No. 116, pp. 415–45

Lindemann, F. A. (1932) *The Physical Significance of the Quantum Theory*, Oxford, Oxford University Press

Lindemann, F. A. (1933) 'The Place of Mathematics in the Interpretation of the Universe', *Philosophy*, Vol. 8, No. 29, January edition

Lota, V. I. (2010) *The Test of War: Russia's Military Intelligence on the Eve of, and during, the Great Patriotic War, 1941–1945*, Moscow, Kuchkovo polie

McCullough, D. (1992) *Truman*, New York, Simon and Schuster

McGucken, W. (1978) 'The Royal Society and the Genesis of the Scientific Advisory Committee to Britain's War Cabinet', *Notes and Records of the Royal Society of London*, Vol. 33, No. 1, pp. 87–115

McMahon, R. J. (2003) *The Cold War*, Oxford, Oxford University Press

Macrae, S. (1971) *Winston Churchill's Toyshop*, Stroud, Amberley

MacDougall, D. (1987) *Don and Mandarin*, London, John Murray

Macmillan, H. (1966) *Winds of Change*, London, Macmillan

Macmillan, H. (1971) *Riding the Storm*, London, Macmillan

Maddock, S. (2010) *Nuclear Apartheid*, Chapel Hill, University of North Carolina Press

Manchester, W. (1988) *The Last Lion, Winston Spencer Churchill, 1932–40*, London, Little, Brown & Co.

Meacham, J. (2003) *Franklin and Winston*, London, Granta Books

Mendelssohn, K. (1960) 'The Coming of the Refugee Scientists', *New Scientist*, 26 May 1960, pp. 1343–4

Monk, R. (2012) *Inside the Centre*, London, Jonathan Cape
Montague Browne, A. (1995) *Long Sunset*, London, Cassell
Moran (1966) *Winston Churchill: The Struggle for Survival 1940–65*, London, Constable
Moss, N. (1987) *Klaus Fuchs*, London, Grafton Books
Mosse, W. E., and Mohr, J. C. B. (eds) (1991) *Second Chance: Two Centuries of German-speaking Jews in the UK*, Tübingen, Paul Siebeck
Mott, N. F. (1984) 'Theory and Experiment at the Cavendish circa 1932', in Hendry (1984: 125–32)
Muller, J. (ed.) (2009) edition of WSC's *Thoughts and Adventures*, Wilmington, Delaware, ISI books
Muller, J. (ed.) (2012) edition of WSC's *Great Contemporaries*, Wilmington, Delaware, ISI books
Noakes, J. (2004) 'Leaders of the People? The Nazi Party and German Society', *Journal of Contemporary History*, Vol. 39, No. 2, pp. 189–21
Nye, M. J. (2004) *Blackett: Physics, War, and Politics in the Twentieth Century*, Cambridge, MA, Harvard University Press
Oliphant, M. (1972a) *Rutherford: Recollections of the Cambridge Days*, Amsterdam, Elsevier
Oliphant, M. (1972b) 'Some Personal Recollections of Rutherford, the Man', in *Notes and Records of the Royal Society*, Vol. 27, No. 1, August 1972, pp. 7–23
Oliphant, M. (1982) 'The Beginning: Chadwick and the Neutron', *Bulletin of the Atomic Scientists*, December, pp. 14–18
Oppenheimer, J. R. (1964a) 'Ernest Rutherford', in *New York Review of Books*, 14 May edition
Oppenheimer, J. R. (1964b) 'Niels Bohr and Atomic Weapons', in *New York Review of Books*, 17 December edition
Orwell, S., and Angus, I. (eds) (1970) *The Collected Essays, Journalism and Letters of George Orwell*, Vol. 3
Overy, R. (2006) *Why the Allies Won*, London, Pimlico Books
Pais, A. (1991) *Niels Bohr's Times: In Physics, Philosophy and Polity*, Oxford, Clarendon Press
Parrinder, P. (1997) *H. G. Wells: The Critical Heritage*, London, Routledge

Parry, A. (ed.) (1968) *Peter Kapitza on Life and Science*, London, Macmillan

Pawle, G. (1963) *The War and Colonel Warden*, London, George C. Harap

Payn, G., and Morley, S. (eds) (1982) *The Noël Coward Diaries*, London, Da Capo Press

Peierls, R. (1985) *Bird of Passage*, Princeton, Princeton University Press

Pickersgill, J. W., and Forster, D. F. (eds) (1970) *The Mackenzie King Record, Vol. 4: 1947–48*, Toronto, University of Toronto Press

Pottker, S., (2004) *Sara and Eleanor*, New York, St Martin's Press

Powers, T. (1993) *Heisenberg's War*, London, Jonathan Cape

Quinault, R. (2002) 'Winston Churchill and Gibbon', in McKitterick, Rosamond, and Quinault, Roland (eds), *Edward Gibbon and Empire*, Cambridge, Cambridge University Press, pp. 317–32

Ratcliff, J. D. (1942) 'War Brains', *Collier's*, 17 January edition, pp. 28, 40

Ratcliffe, J. A. (1975) 'Physics in a University Laboratory before and after WWII', *Proceedings of the Royal Society, A* 342, pp. 457–64

Rau, U. S. (2001) 'The National/Imperial Subject in T. B. Macaulay's Historiography', *Nineteenth-Century Contexts: An Interdisciplinary Journal*, 23:1, pp. 89–119.

Raymond, J. (ed.) (1960) *The Baldwin Age*, London, Eyre & Spottiswoode

Reader, W. J. (1975) *Imperial Chemical Industries: A History, Vol. 2*, London, Oxford University Press

Reynolds, D. (1994) 'The Atlantic "Flop"', in Brinkley, D., and Facey-Crowther, D. R. (eds) (1994) *The Atlantic Charter*, London, Macmillan, pp. 129–50

Reynolds, D. (2004) *In Command of History*, London, Penguin

Rhodes, R. (1986) *The Making of the Atomic Bomb*, New York, Simon & Schuster

Rhodes James, R. (1970) *Churchill: A Study in Failure 1900–1939*, London, Weidenfeld & Nicolson

Rhodes James, R. (1986) *Anthony Eden*, London, Weidenfeld & Nicolson

Roberts, A. (1991) *The Holy Fox*, London, Phoenix Giant

Roosevelt, E. (ed.) (1950) *FDR: His Personal Letters 1928–45*, New York, Duell, Sloan and Pearce

Roosevelt, F. D. (1941) *The Public Papers and Addresses of Franklin D. Roosevelt*, 1938 volume, New York, Macmillan

Rose, N. (1994) *Churchill: An Unruly Life*, London, Simon & Schuster

Rotblat, J. (1985) 'Leaving the Bomb Project', *Bulletin of the Atomic Scientists*, August edition, pp. 16–19

Roskill, S. (1974) *Hankey, Man of Secrets*, Vol. 3, London, Collins

Rowlands, P. (2001) *120 Years of Excellence: the University of Liverpool Physics Department 1881 to 2001*, Liverpool, U–P L Communications/PD Publications

Rozental, S. (ed.) (1964) *Niels Bohr: His Life and Work*, London, Interscience

Russell, D. S. (2005) *Winston Churchill: Soldier*, London, Brassey's

Sachs, R. G. (ed.) (1984) *The Nuclear Chain Reaction: Forty Years Later*, Chicago, The University of Chicago

Schilpp, P. A. (ed.) (1997) *Albert Einstein: Philosopher-Scientist*, La Salle, IL, Open Court Publishing

Sherborne, M. (2010) *H. G. Wells*, London, Peter Owen Publishers

Sherwood, R. (2008) *Roosevelt and Hopkins*, New York, Enigma Books

Shuckburgh, E. (1986) *Descent to Suez*, London, Weidenfeld & Nicolson

Sime, R. Lewin (1996) *Lise Meitner*, Berkeley, University of California Press

Skemp, J. B. (1978) 'Mastership of Sir James Chadwick', in *Biographical History of Gonville and Caius College*, Vol. 7, pp. 485–502

Smith, A. M. (1946) *Thank You, Mr President*, London, Harper and Brothers

Smith, D. C. (1986) *H. G. Wells: Desperately Mortal*, New Haven, Yale University Press

Smith, D. C. (1989) 'Winston Churchill and H. G. Wells: Edwardians in the Twentieth Century', in *Cahiers, Victoriens & Edouardiens*, No. 30, October 1989, Montpellier

Smith, D. C. (ed.) (1998a) *Correspondence of H. G. Wells, Vol. 2, 1904–1914*, London, Pickering & Chatto

Smith, D. C. (ed.) (1998b) *Correspondence of H. G. Wells, Vol. 4, 1935–1946*, London, Pickering & Chatto

Smyth, H. D. (1976) 'The Smyth Report', *Princeton University Library Chronicle*, Vol. 37, No. 3, pp. 173–89

Snow, C. P. (ed.) (1938) *Background to Modern Science*, Cambridge, Cambridge University Press

Snow, C. P. (1960a) *Science and Government*, London, Oxford University Press

Snow, C. P. (1960b) *Rutherford and the Cavendish*, in Raymond, J., *The Baldwin Age*, London, Eyre & Spottiswoode

Snow, C. P. (1962) *A Postscript to Science and Government*, London, Oxford University Press

Soames, M. (1979) *Clementine Churchill*, London, Doubleday

Soames, M. (ed.) (1998) *Speaking for Themselves: The Personal Letters of Winston and Clementine Churchill*, London, Black Swan

Soames, M. (2011) *A Daughter's Tale*, London, Doubleday

Soddy, F. (1909) *The Interpretation of Radium*, London, John Murray

Szasz, F. M. (1984) *The Day the Sun Rose Twice*, Albuquerque, University of New Mexico Press

Szasz, F. M. (1992) *British Scientists and the Manhattan Project*, New York, St Martin's Press

Szanton, A. (1992) *The Recollections of Eugene P. Wigner*, New York, Plenum Press

Teller, E. (1962) *The Legacy of Hiroshima*, New York, Doubleday

Thorpe, C. (2006) *Oppenheimer: The Tragic Intellect*, Chicago, University of Chicago Press

Toye, R. (2008) 'H. G. Wells and Winston Churchill: A Reassessment', in McLean, Steven (ed.) *H. G. Wells: Interdisciplinary Essays*, Cambridge Scholars Publishing, Newcastle, pp. 147–61

Truman, H. S. (1955) *Memoirs, Vol. 1*, New York, Doubleday & Co., Inc.

Truman, M. (ed.) (1981) *Letters from Father*, New York, Arbor House

Villa, B. (1977) 'The Atomic Bomb and the Normandy Invasion', *Perspectives in American History*, Vol. 11, pp. 463–502

Walsh, J. J. (1998) 'Postgraduate Technological Education in Britain: Events Leading to the Establishment of Churchill College, Cambridge, 1950–58', *Minerva*, Vol. 36, pp. 147–77

Ward, G. (ed.) (1995) *Closest Companion*, Boston, Houghton Mifflin

Weart, S. (1976) 'Scientists with a Secret', *Physics Today*, February, pp. 23–30

Weart, S. (1979) *Scientists in Power*, Cambridge, MA, Harvard University Press

Weart, S. (1988) *Nuclear Fear*, Cambridge, MA, Harvard University Press

Weart, S., and Szilard, G. W. (eds) (1978) *Leó Szilárd: His Version of the Facts*, Cambridge, MA, MIT Press

Welles, S. (1950) *Seven Decisions That Shaped History*, New York, Harper & Brothers

Wells, G. P. (ed.) (1984) *H. G. Wells in Love*, London, Faber

Wells, H. G. (1901) *Anticipations of the Reaction of Mechanical and Scientific Progress upon Human Life and Thought*, New York and London, Harper & Brothers

Wells, H. G. (1914) *The World Set Free*, New York, E. P. Dutton & Co.

Wells, H. G. (1934) *Experiment in Autobiography, Vol. 1*, London, Faber

Wells, H. G. (1940) 'Churchill, Man of Destiny', *Collier's*, 2 November 1940, pp. 17, 18 and 57

Wheeler-Bennett, J. (1958) *King George VI: His Life and Reign*, London, Macmillan

Wheeler-Bennett, J. (1962) *John Anderson*, London, Macmillan & Co. Ltd

Wheeler-Bennett, J. (ed.) (1969) *Action this Day*, London, St Martin's Press

Wilson, D. (1983) *Rutherford*, Cambridge, MA, MIT Press

Young, J. W. (1996) *Winston Churchill's Last Campaign*, Oxford, Clarendon Press

Zachary, G. P. (1992) 'Vannevar Bush Backs the Bomb', *Bulletin of the Atomic Scientists*, December edition, pp. 24–31

Zachary, G. P. (1999) *Endless Frontier*, Cambridge, MA, MIT Press

Zhukov, G. K. (1971) *The Memoirs of Marshal Zhukov*, New York, Delacorte Press

Zimmerman, D. (1996) *Top Secret Exchange*, Stroud, Sutton Publishing

Zimmerman, D. (2001) *Britain's Shield*, Stroud, Sutton Publishing

Zimmerman, D. (2006) 'The Society for the Protection of Science and Learning and the Politicization of British Science in the 1930s', *Minerva*, No. 34, pp. 25–45

Notes

In these references, 'WSC' is an abbreviation of Winston Churchill, and 'f.' is short for folio (plural 'ff').

Epigraphs

1 Churchill, W. S., 'Mankind Is Confronted by One Supreme Task', *News of the World*, 14 November 1937.
2 'Scientists in Birmingham', *Sunday Mercury*, Birmingham, UK, 21 April 1940, p. 3.

Prologue

1 The final text of the speech is in CHUR 5/57A.
2 Moran (1966: 530).
3 Moran (1966: 634).
4 Jenkins (2001: 874–84).
5 WSC, 'Fifty Years Hence', *Strand Magazine*, December 1931. Reproduced in http://teachingamericanhistory.org/library/index.asp?document=1914.
6 Letter from WSC to Swinton, 19 November 1936, CHAR 25/7, ff. 63–4.
7 *Daily Telegraph*, 'Splitting the Atom', 16 March 1933.
8 Moran (1966: 578).
9 FRUS 1952–4, Vol. 6, Memorandum of meeting between WSC and Eisenhower, 25 June 1954, pp. 1085–6. The report 'Devastation and the Hydrogen Bomb', *Manchester Guardian*, 18 February 1954, p. 1.
10 Interview with Sir Maurice Wilkes, 24 March 2009.
11 Edgerton (2011: 89); Kennedy (2013: 56–7, 269–71).
12 Reynolds (2004: 400).
13 Pickersgill and Forster (eds) (1970: 112–13).
14 See, for example, CHSPCH, Vol. 8, p. 7943.
15 WSC BMFRS (1966), p. 99.
16 Description of WSC's physique from Moran (1966: 621–22), Lees-Milne (1994: 49).
17 Interview with Lady Williams, 19 October 2010.

Wells and his liberating 'atomic bombs'

1 Moran (1966: 328).
2 WSC's 'H. G. Wells', *Sunday Pictorial*, 23 August 1931, reprinted in Muller (ed.) (2012: 372–8) see p. 377.
3 Alkon (2006: 167).
4 CHBIO, Vol. 1, WSC to his mother, 6 April 1897, pp. 316–19.
5 Wells took first-class honours in zoology and second-class honours in geology. I thank Michael Sherborne for this information.
6 Quoted in Sherborne (2010: 167).
7 Parrinder (1997: 330).
8 WSC to Commons, 13 May 1901, HNSRD; Toye (2008: 151).
9 Quotations in this passage are from Wells (1901: 201, 207, 213, 222).
10 WSC to Wells, 17 November 1901, HGW archive, C.238.30.
11 Wells to WSC, 19 November 1901, CHAR 1/29, ff. 54–5.
12 Wells to WSC, 21 November 1901, CHAR 1/29, ff. 56–7.
13 Postcard from Wells to WSC, 5 March 1902, CHAR 1/33, ff. 64–5.
14 Smith (1989) is a rewarding study of the relationship between WSC and Wells.
15 CHBIO, Vol 1, p. 353.
16 Letter from Churchill to Wells, 9 October 1906, HGW, C.238.2.
17 Soames (1979: 238).
18 *Daily News*, 21 April 1908, pp. 5–6.
19 Smith (1986: 83).
20 Wells (ed.) (1984: 87); Smith (1986: 371).
21 Weart (1988: 17–35).
22 Soddy (1909: 224).
23 Soddy (1909: 232, 244).
24 Soddy (1909: 4).
25 Smith (1986: 84).
26 Text of *Penguin Island*: http://www.gutenberg.org/files/1930/1930-h/1930-h.htm#2H_4_0064.
27 Wells (ed.) (1984: 89).
28 Wells (ed.) (1984: 93).
29 Wells (1914: 50–1).
30 In Section 3 of Chapter Two, Wells refers to a 'long, coffin-shaped box which contained in its compartments the three atomic bombs': http://www.gutenberg.org/files/1059/1059-h/1059-h.htm.
31 *New York Times*, 29 March 1914, p. BR 141.
32 Text of Wells's 'The World Set Free': http://www.gutenberg org files/1059/1059-h/1059-h.htm.
33 Advertisement in *New York Tribune*, 6 August 1914, p. 2.
34 Churchill (1930: 44).
35 Andrew (1988: 182).

36 Committee of Imperial Defence, 25 February 1909, CAB 2/2, NA; Sherborne (2010: 185–6).
37 On WSC's leading role promoting the tank – 'Mr Lloyd George on the "Tanks"', *Daily Chronicle*, 9 September 1916, p. 5; WSC to Wells, 1 October 1916, WELLS c.238.7a; Smith (1989: 99–100); Muller (ed.) (2012: 377).
38 Smith (1989: 100).
39 Sherborne (2010: 241); Kennedy (2013: 82–3).
40 See Wells's letters to *The Times*, 11 June and 22 June 1915.
41 Rose (1994: 146).
42 Quoted in Larres (2002: 40).
43 Wells wrote: 'Apart from individual atrocities, it did on the whole kill for a reason and to an end.' Quoted in Sherborne (2010: 259).
44 Toye (2008: 150).
45 Smith (1989: 102).
46 Smith (1989: 102).
47 Alkon (2006: 169–71).
48 Dugdale (1937: 337).
49 WSC to Lindemann, 3 April 1924, LIND, K62/1.

Churchill glimpses a nuclear future

1 Muller (ed.) (2009: 294)
2 Lindemann to his father, 19 August 1921, LIND A93/f8.
3 Fort (2004: 1–10).
4 Lindemann to Fowey Montmorency, 30 October 1936, LIND, A.30/f1. Lindemann's characteristics: Fort (2004: 91–2); Birkenhead (1961: 24–6); Harrod (1959: 10, 29–32, 53, 89).
5 Lindemann to WSC, 29 June 1922, LIND A93/f6.
6 J. A. Little to Lord Carr, 19 December 1985, RVJO B.390.
7 WSC to Lindemann, 21 April and 10 May 1924, LIND, K62/2–3. Text of 'Daedalus': http://cscs.umich.edu/~crshalizi/Daedalus.html.
8 Muller (ed.) (2009: 259–66).
9 Bolsheviks – WSC annotation to memo to D.C.I.G. S., 9 April 1919, WO 32/5749, NA. Mesopotamian tribes – WSC departmental minute, 12 May 1919, CHAR 16/16A.
10 CHBIO, Vol. 5, p. 50–2. For a comment on the article in the US, see *American Review of Reviews*, Vol. 70, July–December 1924, pp. 537–8.
11 Russell (2005: 25).
12 Churchill (1930: 25–7).
13 Rau (2001: 92, 93, 94, 97).
14 Churchill (1930: 59).

15 Quinault (2002: 317–18).
16 Churchill (1930: 109); WSC to his mother, 14 January 1897, CHAR 28/23/10–11.
17 Churchill (1930: 112).
18 WSC to his mother, 31 March 1897, CHAR 28/23/29; Churchill (1930: 211).
19 Churchill (1930: 112); WSC to his mother, 6 April 1897, CHAR 28/23/31–3A.
20 WSC to his mother, 14 January 1897, CHAR 28/23/10–11.
21 Churchill (1930: 117).
22 This is the introduction to a lecture given by Lindemann on 16 March 1933: LIND E5/1.
23 Wheeler-Bennett (ed.) (1969: 28).
24 Churchill to Lindemann, 4 April 1926. There are two sources of this document: LIND K62 4/5 (on Chartwell notepaper but without the memo) and CHAR 1/188, ff. 14–25 (full document).
25 The busts are still on WSC's desk at Chartwell.
26 Rose (1994: 192–4). For a full description of Chartwell see CHDOCS, Vol. 13, pp. 972–6.
27 Churchill to Lindemann, 4 April 1926, CHAR 1/188, f. 14.
28 Schilpp (ed.) (1997: 47).
29 WSC, *Nash's Magazine*, 83, no. 435, August 1929, reprinted in Muller (2012: 46–60).
30 Hastings (2009: 223).
31 Gilbert (2005: 9); CHBIO, Vol. 1, p. 542.
32 Fishman (1963: 89).
33 Gilbert (2005: 98).
34 'Cruiser and Parity', 20 July 1927, CHAR 22/182.
35 WSC, 'The American Mind and Ours', *Strand Magazine*, August 1931, pp. 140–50. See p. 150.
36 Jenkins (2001: 427).
37 'Mr Churchill's 57th Birthday', *East Anglian Times*, 1 December 1931.
38 See the press article by John Bull on 6 May 1933: 'He annoys the House because he only comes on State occasions, makes his speech and then disappears until there is again an opportunity to shine.' CHPC 13.
39 Wenden, D. J., 'Churchill, Radio and Cinema', in Blake and Louis (eds) (1996: 215–39), see pp. 217, 219.
40 WSC, speech in the House of Commons, 12 November 1936.
41 Clarke (2012), chapters 4 and 5. See also Muller (ed.) (2012).
42 WSC, 'A Very Poor Form of Revolutionary', *The Times*, 26 January 1934, p. 9.
43 Coote, C., 'Churchill the Journalist', in Eade (ed.) (1953: 116).

44 Churchill (1929: 11–12).
45 *Strand Magazine* to WSC, CHAR 8/292, ff. 1–2.
46 WSC to Lindemann, 8 February 1931, LIND K64/9–10.
47 Coote (1971: 3–16).
48 Birkenhead (1930).
49 Lindemann to WSC, 18 February 1931, and accompanying notes, CHAR 8/301, ff. 2–13.
50 Text of WSC's 'Fifty Years Hence': http://teachingamericanhistory.org/library/index.asp?document=1914; Muller (ed.) (2009: 283–95).
51 http://theotherpages.org/poems/tenny02.html.
52 The article had been published slightly earlier in Canada, on 15 November 1931, in *Maclean's*, Vol. 7, pp. 66–7.
53 Muller (ed.) (2009: 2).
54 Birkenhead (1961: 162); Harrod (1959: 30); Manchester (1988: 16); Fort (2004: 234).
55 Baldwin to the Commons, 10 November 1932, HNSRD.
56 WSC to Lord Riddell, 18 October 1932, CHAR 8/311.
57 The dining customs of the Churchills are described in Rose (1994: 192–4).
58 Churchill (1930: 127).
59 CHBIO, Vol. 5, pp. 442–3; Lindemann (1932).
60 Birkenhead (1961: 280).

Rutherford: nuclear sceptic

1 Rutherford, E. (1915) 'The Constitution of Matter and the Evolution of the Element', in *Popular Scientific Monthly*, August 1915, Vol. 86, New York, The Science Press, pp. 105–42. For the quotation, see pp. 127–8.
2 Rutherford quotation: Rhodes James (1970: 242). See also Crowther (1965: 353), Snow (1960a: 22). A second-hand report of Einstein's assessment of Lindemann as a physicist has more than a ring of truth: 'Lindemann [is] essentially an amateur; he [has] ideas, which he never [works] out properly; but he [has] a thorough comprehension of physics. If something new [comes] up, he [can] rapidly assess its significance for physics as a whole, and there are very few people who [can] do that', Harrod (1959: 48). See also Birkenhead (1961: 159).
3 Lindemann to Rutherford, 24 January 1932, LIND C62/12.
4 See the view Rutherford expressed at the Royal Academy banquet on 30 April 1932 in the *Observer*, 1 May 1932, p. 20.
5 Mott (1984: 131).

6 Edwin Kemble to Garrett Birkhoff, 3 March 1933, AHQP.
7 Nandor Balazs, private communication, 18 August 1989.
8 Lindemann (1933); review of the book by the Cambridge physicist Nevill Mott: *Nature*, 1932, No. 3279, Vol. 130, pp. 330–1; Nandor Balazs, private communication, 18 August 1989.
9 'Lord Rutherford, Physicist, Is Dead', *New York Times*, 20 October 1937, pp. 1, 18. See p. 18.
10 Raymond (ed.) (1960: 236–8); Obituary of Lord Rutherford by J. J. Thomson, *Cambridge Review*, 5 November 1937, pp. 64–5; see p. 64. Oliphant (1972b: 9–10).
11 Campbell (1999: 445).
12 Oliphant (1972a: 139).
13 Max Born to Chadwick, 11 August 1954, CHAD IV 13/1.
14 Report by Rutherford and Cyprian Bridge on the visit, 19 May to 9 July 1917 to the British Board of Invention and Research, 18 July 1917, ADM 293/10, p. 2, NA.
15 Bernal (1939: 9). See also the autobiographical notes of G. P. Thomson (GPT A5): 'Rutherford classified knowledge into physics (which included straightforward mathematics), chemistry, and stamp collecting!'
16 'Atomic Energy', text of talk delivered by G. P. Thomson in 1945, GPT F154.
17 Lindemann to Rutherford, 9 May 1919, L.105, RFD.
18 Fort (2004: 155).
19 Raymond (ed.) (1967: 246–7).
20 Parry (ed.) (1968: 121).
21 Harrrod (1959: 49, 50); Fox and Gooday (eds) (2005: 272).
22 Oliphant (1972b: 20–1).
23 Campbell (1999: 356).
24 Anon. (1954: 28).
25 Chadwick to Feather, 22 October 1959, FEAT 23/6.
26 *Guardian*, 27 February 1932, p. 10.
27 *The Times*, 29 February 1932, p. 9.
28 *Spectator*, 14 June 1930, p. 979.
29 Lindemann told Rutherford of a possible loophole in the argument, but it subsequently proved to be irrelevant. Lindemann to Rutherford, 9 June 1919, quoted in Wilson (1983: 435); Rutherford to Lindemann, 10 June 1919, LIND D 218/6.
30 Cathcart (2004: 242–3).
31 Cathcart (2004: 252).
32 Arms (1966: 71).
33 *Proceedings of the Royal Society of London*, Series A, Vol. 136, No. 830, 1 June 1932, pp. 735–62.

34 *Observer*, 1 May 1932, pp. 17 and 20. See also CHBIO, Vol. 5, pp. 428–31.
35 Parry (ed.) (1968: 114).
36 For Rutherford's view on the British Empire see Anon. (1954: 33–4).
37 Parry (ed.) (1968: 114).
38 'Political Painters', CHSPCH, Vol. 5, pp. 5153–4. CHBIO, Vol. 5, pp. 428–9.
39 Eve (1939: 353–4).
40 Cathcart (2004: 244–50).
41 *Daily Mirror*, 3 May 1932.
42 This point was, however, correctly stressed in the article in *The Times*, 2 May 1932 (p. 11) and in the review by Waldemar Kaempffert in the *New York Times*, 8 May 1932, p. xxi.
43 Oliphant (1972a: 141). The same remark features in Ritchie Calder's report on his conversation with Rutherford in 'The Truth About the Atom', *Daily Herald*, 27 June 1932, p. 8. Rutherford's comments are well reviewed in Badash et al. (1986), where the authors note Lord Hankey's recollection of Rutherford's commenting privately in 1930 that one day nuclear physics might have relevance to warfare. As the authors say, 'It is difficult to evaluate whether Rutherford was engaging in idle chatter or revealing an intuitive feeling, but Hankey believed the latter' (p. 204).
44 Wilson (1983: 572).
45 *Wings Over Europe* opened at the Martin Beck Theater on 10 December 1928 and later transferred to the Alvin Theater in New York. The production ran for ninety performances.
46 The *New Statesman and Nation*, 7 May 1932, pp. 584–5.
47 The dinner was on 17 December 1932. Da Costa Adrade (1964: 48, 162).
48 The paper announcing Blackett and Occhialini's discovery of cosmic-ray showers was received by the Royal Society on 7 February 1933.
49 See for example: Anon. (1954), Parry (ed.) (1968: 75–99). Einstein quotation: Blackett (1972: 58).
50 Wilson (1983: 565).

The Prof advises 'a scientist who missed his vocation'

1 Letter from Isaiah Berlin to Stephen Spender, 20 June 1936, Hardy (ed.) (2004: 175).
2 James Tuck, 'Lord Cherwell and His Part in World War II', 9 March 1961, in RVJO B395.
3 Brendon (1984: 144).
4 *Country Life*, 6 June 1931, pp. 736–8.

5 Reader (1975: 82). The Research Council was dissolved towards the end of 1939.

6 Harrod (1959: 93, 94–5). Birkenhead (1961: 114, 116). For another description of Lindemann's laugh: Harrod (1959: 84).

7 Birkenhead (1961: 116), Harrod (1959: 84).

8 Draft text of talk, LIND E5/1–5.

9 Morrell, J., 'The Lindemann Era', in Fox and Gooday (eds) (2005: 233–66), see p. 257.

10 Several members of the audience will probably have seen the report on the Blackett–Occhialini discovery in 'New Light on the Atom' by John Cockcroft, *Spectator*, 24 February 1933, pp. 245–6.

11 Cameron, N., 'The Owl and the Pussycats: Science, Politics and the Late War', in *The Idler*, Vol. 2, No. 9, September 1986, pp. 31–40. See p. 32.

12 *Daily Telegraph*, 'Splitting the Atom', 16 March 1933. See also the report in *The Times* on the same day (p. 6). Both articles are in WSC, CHPC 13.

13 Letter from Violet Bonham Carter to R. V. Jones, 25 July 1966, RVJO B.238.

14 Harrod (1959: 104).

15 Fox and Gooday (eds) (2005: 236).

16 BMFRS of Lindemann, November 1958, pp. 47–8.

17 Fort (2004: 57–66).

18 Crowther (1965: 352), Harrod (1959: 45).

19 Lindemann (1920: 437–45).

20 See, for example, Crowther (1965: 353), Snow (1960a: 20).

21 See comments by G. P. Thomson, GPT, G56, February 1961.

22 Harrod (1959: 91).

23 *The Times*, 15 March 1933, p. 15.

24 Noakes (2004: 189).

25 Coote (1971: 81).

26 WSC to Commons, 13 April 1933, HNSRD.

27 WSC, *Daily Mail*, 14 October 1933.

28 'Winston Churchill – The Tragic Truth', *John Bull*, 6 May 1933, CHPC 13.

29 Strauss, H., 'Jewish Emigration in the Nazi Period: Some Aspects of Acculturation' in Mosse and Mohr (eds) (1991), pp. 81–95, see p. 83. Strauss notes that between 278,000 and 300,000 emigrated.

30 Birkenhead (1961: 23–5); Fort (2004: 91); Harrod (1959: 107–11).

31 Birkenhead (1961: 121–2).

32 Letter from Lindemann to Einstein, 4 May 1933, LIND D57. Lindemann says he spent 'four or five days' in Berlin, and that he saw 'a great many' of Einstein's colleagues.

33 Beveridge (1959: 4–5).
34 For the case of Rutherford and his colleagues at the AAC, see the letter from its General Secretary to Szilárd, 19 December 1935: 'If German scientists with senior qualifications compete for these junior posts it will be regarded by British graduates and other members of staff as competition . . . an abuse of the displaced scholar's circumstance and would be an unfair act against younger British competitors.' Fritz London file, SPSL.
35 Leaflet is in SPSL 21/1/113. See also the accompanying letter from A. V. Hill to G. P. Thomson, 14 February 1939, SPSL 21/1/114.
36 Zimmerman (2006: 36); Fort (2004: 117–18).
37 Einstein to Lindemann, 1 May 1933, LIND D57/6.
38 Einstein to Lindemann, 7 May 1933, LIND D57/12.
39 The meeting took place on 22 July 1933, as indicated in the letter from Einstein's main host Locker-Lampton to Lindemann, 20 July 1933, LIND D57/22. Martin Gilbert's *Churchill and the Jews* specifies that Lindemann was present.
40 Einstein to his wife Elsa, July 1933, AEA 143–50 (undated).
41 Mendelssohn (1960: 1343); Harrod (1959: 31–2).
42 See the introduction to the catalogue of SPSL. Note that the refugees came not only from Germany but also from Austria, Italy, Spain and a few from the USSR.
43 Einstein set sail from Southampton on the following Saturday evening: *The Times*, 9 October 1933, p. 16.
44 Einstein to Lindemann, 17 December 1933. LIND D57/27.
45 *New York Times*, 29 December 1934, pp. 1 and 7.

Szilárd's nuclear epiphany

1 Szanton (1992: 94).
2 Lanouette (1992: 59–60). See also the remark on Szilárd by Enrico Fermi in the November 1955 edition of *Physics Today*, pp. 12–16.
3 Szanton (1992: 93).
4 Szilárd had arrived in London in mid-April 1933: see the letter from Szilárd to Sir William Beveridge, 22 April 1933, SPSL 167/1–2. For a description of Szilárd's accommodation, see Hobhouse (1971: 199, 207).
5 Szilárd (1969: 97–8). Beveridge's history of the Academic Assistance Council does not mention Szilárd (Beveridge (1959)).
6 Szanton (1992: 93–4); Lanouette (1992: 121–7).
7 Manchester (1988: 65).
8 *The Times*, 12 September 1933, p. 7. The word covered by the ellipsis is 'the', referring to atoms that are split.

9 Rutherford's coolness on the prospect of harnessing nuclear energy was reported on the front page of the *New York Times*, 12 September 1933 (report continued on p. 18). See also the *Los Angeles Times*, 12 September 1933, p. 3.
10 *Manchester Guardian*, 12 September 1933, p. 12, and 13 September, p. 7. See also 'Sayings of the Week' in the *Observer*, 17 September 1933, p. 15 (see also p. 16).
11 Weart and Szilárd (eds) (1978: 17).
12 Weart and Szilárd (eds) (1978: 16).
13 Weart and Szilárd (eds) (1978: 17); Lanouette (1992: 134).
14 Lanouette (1992: 136).
15 Lanouette (1992: 137–8, 140–2). Szilárd soon afterwards consulted Rutherford's colleague Mark Oliphant: Weart and Szilárd (eds) (1978: 47).
16 Letter from Rutherford to Walter Adams of the AAC Council, 30 May 1934, AAC, Szilárd papers.
17 Rhodes (1986: 200–2, 209–13).
18 Ratcliffe (1975: 462).
19 Draft letter from Chadwick to Bainbridge, dated February 1974, CHAD 13/1.
20 Szilárd chose to frame the chain reaction not in terms of neutrons, as he had previously done, but using helium nuclei. Rutherford saw immediately that such a reaction would be impossible.
21 Weart and Szilárd (eds) (1978: 46).
22 Badash et al. (1986: 209), Bainbridge (1974: 203), correspondence in CHAD IV 13/2.
23 Lanouette (1992: 143).
24 Weart and Szilárd (eds) (1978: 18n).
25 Lanouette (1992: 145–7).
26 Thomson to Walter Adam at the AAC, 26 July 1933, GPT J113.
27 Weart and Szilárd (eds) (1978: 20).
28 Morrell, J., 'The Lindemann Era', in Fox and Gooday (eds) (2005: 233–66), see p. 262.
29 Arms (1966: 68). Other members of the group included Kurt Mendelssohn and Nicholas Kurti. See the letter from the Oxford University authorities concerning one of the group's grant applications to the Rockefeller Foundation, 5 August 1933, LIND B15/3.
30 Arms (1966: 82).
31 Hoch, P. K. (1991) 'Some Contributions to Physics by German-Jewish Emigres in Britain and Elsewhere', in Mosse et al. (eds) (1991) pp. 229–54, see p. 236.
32 Szilárd to Lindemann, 3 June 1935: Weart and Szilárd (1978: 41–2).
33 Lanouette (1992: 170–1).

34 Fort (2004: 148–58).
35 Lindemann to Groves, 12 July 1945, http://www.dannen.com/decision/lrg-fal.html.

Churchill fears war – and that nuclear energy will soon be harnessed

1 WSC speech to Commons concerning Science in War, 21 March 1934: HNSRD.
2 WSC to Commons, 7 February 1934, HNSRD.
3 WSC to Commons, 30 July 1934, HNSRD.
4 'Angels of Peace', *Manchester Guardian*, 7 February 1934, p. 12.
5 WSC to Commons, 28 November 1934, HNSRD.
6 Brendon (1984: 220).
7 WSC, 'The Effect of Air Transport on Civilization', *News of the World*, 8 May 1938 (CHPC 17). CHESSAYS, Vol. 4, pp. 427–34.
8 WSC to Commons, 7 June 1935, HNSRD.
9 Lindemann to WSC, 22 January 1935, CHAR 2/243/10–11.
10 Birkenhead (1961: 178–181).
11 Snow (1960a: 6).
12 Radar was developed simultaneously in several other countries, including the US and Germany. The committee first met on 28 January 1935, see Zimmerman (2001: 55).
13 Edgerton (2011: 107–8); Reynolds (97–8).
14 Rowe to Roskill, 7 July 1968, ROSK 7/131.
15 Correspondence relating to this meeting is in CAB 21/426.
16 WSC to Swinton, 19 November 1936, CHAR 25/7/63–4.
17 Quoted in CHBIO, Vol. 5, p. 716.
18 Birkenhead (1961: 150–5).
19 Rose (1994: 230–3).
20 Clarke (2012: 135).
21 Letter from Reves to WSC, 24 October 1939, CHUR 2/386, f. 35, 36. See also Jenkins (2001: 523).
22 Kipling to WSC, 26 October 1934, CHAR 8/487, ff. 72–3.
23 The phrase 'largest circulation in the world' was printed on the newspaper's letterhead. See, for example, the letters from officials of the newspaper to WSC stored in CHAR 8/551.
24 Carr to WSC, 30 October 1937, CHUR 8/551.
25 'Vision of the Future Through the Eyes of Science', *News of the World*, 31 October 1937, p. 12.
26 Compare Muller (ed.) (2009) p. 289 with CHESSAYS, Vol. 4, pp. 410–14, p. 414.
27 Chadwick to Feather, 13 August 1971, FEAT 13/2.

28 *Nature*, 30 October 1937, pp. 754–5.
29 WSC, 'Life in a World Controlled by the Scientists', *News of the World*, 7 November 1937 (CHAR 8/567).
30 Clementine to WSC, 9 August 1937, CHAR 1/322/10.
31 Wells to Clementine Churchill, 16 August 1937: Smith (ed.) (1998b: 166). Wells visited Chartwell on 15 August 1937 and signed the visitors' book.
32 WSC to H. G. Wells, 4 July 1937, WELLS C238-11.
33 'A Federation for Peace Is the Hope of the World', *News of the World*, 21 November 1937. Republished in CHESSAYS, Vol. 4, pp. 422–6.
34 *Manchester Guardian*, 15 March 1938, p. 11.
35 'The Union of the English-Speaking Peoples', *News of the World*, 15 May 1936, CHESSAYS, Vol. 4, pp. 435–42, see p. 435; 'Europe's Plea to Roosevelt', *Evening Standard*, 10 December 1937, CHPC 16.
36 'Mr Churchill's Plea to Germany', *Daily Telegraph* 19 November 1936, CHPC 15. The words in ellipsis are 'We must try to do the work ourselves. But'.
37 'Whither Churchill?', *Daily Sketch*, 17 March 1938 (CHPC 17); 'Churchill', *Time and Tide*, 28 November 1936, pp. 1567–8, see p. 1567 (CHPC 15).
38 WSC to Sir Kingsley Wood, 9 June 1938, CAB 21/630, NA. The consequences are discussed in Zimmerman (2001: 139–41).
39 Memo from Tizard to Secretary of State for Air, 22 June 1938, CAB 64/5, NA.
40 Sir Frederick Brudrett on Sir Henry Tizard, 22 December 1959: 'This, indeed, was the only matter on which I ever found Tizard entirely unreasonable, but [Lindemann] was worse.' BRUN 1/5.
41 Birkenhead (1961: 195) Brundrett uses similar words in his testimony on 22 December 1959: 'in this matter, both these eminent gentlemen behaved like a couple of spoiled children', BRUN 1/5.
42 Cabinet Minutes, 12 September 1938, CAB 23/95, NA. See CHDOCS, Vol. 13, p. 1156.
43 Bialer (1980: 158).
44 Macmillan (1966: 522).
45 'The Promotion of Peace', *Nature*, 8 October 1938, p. 629.
46 WSC's speeches to Parliament, 5 and 6 October 1938.
47 WSC to Lindemann, 27 October 1937, LIND K 67/4.
48 WSC, 'What Other Secrets Does the Inventor Hold?', *News of the World*, 23 October 1938.
49 Review in the *Observer*, 24 July 1938.
50 'J. B. Priestley's American Novel', *New York Times*, 31 July 1938, p. 72.

Bohr thinks the Bomb is 'inconceivable'

1 French and Kennedy (eds) (1985: 185).
2 Bohr (1961: 1115); Bohr to Rutherford, 9 January 1924, RFD.
3 Snow (ed.) (1938) pp. 49–76. See pp. 71–2.
4 Dirac (1964: 306–9).
5 'Prohibition of *Nature* in Germany', *Nature*, 22 January 1938, p. 151.
6 'Reich Will Last 1000 Years', *Manchester Guardian*, 15 December 1938, p. 11.
7 *Daily Telegraph*, 11 November 1938.
8 Wheeler, J., 'Of Historical Note', *Institute for Advanced Study Newsletter*, Spring 2010.
9 Kevles (1995: 282).
10 This was a meeting of the physicists' 'journal club'. See Cohen and Stachel (eds) (1979: 343).
11 Badash, Hodes and Tiddens (1986: 211).
12 Laurence (1947: vii).
13 Rhodes (1986: 274–5).
14 'Revolution in Physics', *New York Times*, 3 February 1939, p. 11.
15 'The Presidency: Wives', *Time* magazine, 13 February 1939.
16 LaFollette (2008: 271, note 3).
17 LaFollette (2008: 186–7).
18 Davis and Potter (1939). This article appears to have been based on the article 'Is World on Brink of Releasing Atomic Power?', *Science Service*, 30 January 1939.
19 'Blown to bits': Davis and Potter (1939: 86).
20 Anderson (1984: 27).
21 Weart and Szilárd (eds) (1978: 53).
22 Lanouette (1992: 189).
23 Weart and Szilárd (eds) (1978: 54).
24 Pais (1991: 456).
25 Letter from Bohr to Abraham Flexner, 31 May 1939: IAS, Director's Office Member Series, Box 12a, 1936–47.
26 'Armies on March', *New York Times*, 15 March 1939, pp. 1 and 14.
27 Wheeler, J., 'A Few Memories of Bohr and Heisenberg' (remarks made on 27 March 2000, at a symposium on the play *Copenhagen*) – private communication from Peter Goddard. See also Wheeler, J., 'Of Historical Note', *Institute for Advanced Study Newsletter*, Spring 2010.
28 At a public meeting in late April 1939, however, reports suggest that Bohr did consider that a bomb was possible in principle, if not in practice: 'Vision Earth Rocked by Isotope Blast', *New York Times*, 30 April 1939, p. 35; 'Physicists Here Debate Whether Experiment Will Blow Up 2 Miles of the Landscape', *Washington Post*, 29 April 1939, p. 30.
29 Badash, Hodes and Tiddens (1986: 215).

30 Weart (1976: 29–30).
31 Bohr was in the UK with his son Aage from *c.*23–28 June 1939 – see correspondence in the NBA.
32 See, for example, 'If War Came', *Manchester Guardian*, 26 June 1939, p. 12.
33 'Foreign Affairs', *The Times*, 27 June 1939, p. 19.
34 'Incomparable Promise or Awful Threat?', *Scientific American*, July 1939, p. 2; Badash, Hodes and Tiddens (1986: 217).
35 Sime (1996: 277). The meetings began in April 1939.
36 Bohr, N., 'Recent Investigations of the Transmutations of Atomic Nuclei', 6 December 1939. Translation in Aaserud (1986: 443–66).
37 'English War Cabinet with Churchill and Eden?', *Politiken*, 2 September 1939, p. 2.

Churchill – nuclear weapons will not be ready for the war

1 Churchill to Sir Kingsley Wood, 13 August 1939, AIR 19/26, NA (copy in CHAR 25/17).
2 *Discovery*, September 1939, Editorial, pp. 443–4.
3 It is tempting to suggest that Bohr hazarded some early version of these views when he visited Cambridge in June 1939 and spoke about nuclear fission at the Kapitza Club. If so, Snow would probably have heard about them directly, or through the grapevine.
4 CHBIO, Vol. 5, pp. 1101–2. Manchester (1988: 499).
5 Manchester (1988: 519).
6 Edgerton (2011: 36).
7 Jenkins (2001: 552).
8 Colville (1985: 3).
9 Manchester (1988: 606).
10 Edgerton (2011: 5, 7, 29, 32, 37).
11 See, for example, 'That "Secret" Weapon', *Daily Express*, 21 September 1939, p. 2; 'Hitler's New Weapon' – letter to *The Times* from Leopold Loewenstein-Wertheim, 22 September 1939, p. 9.
12 Gowing (1964: 38); crop-eating-locusts suggestion is in the *Daily Mirror*, 9 November 1939, p. 4; death-ray suggestion – Chatfield to Hankey, 27 November 1939, CAB 21/1262, NA.
13 'At Random', *Observer*, 5 February 1939, p. 13.
14 'Scientists Make an Amazing Discovery', *Sunday Express*, 30 April 1939, p. 17.
15 Fort (2004: 234).
16 'Note on uranium' by Tizard, *c.*26 April 1939, CAB 21/1262, NA.
17 Lindemann refers to 'the recently discovered chain processes', whereas scientists had ascertained only that the processes were possible.

18 CHDOCS, Vol. 13, p. 1587.
19 WSC to Sir Kingsley Wood, 13 August 1939, AIR 19/26, NA. See also the copy in CHAR 25/17. Harrod (1959: 174).
20 Note by D. R. Pye on 17 August 1939, AIR 19/26, NA. See also the notes in CAB 21/1262, NA.
21 Jenkins (2001: 553, 559).
22 Manchester (1988: 552–3).
23 ADM1/10459, NA.
24 Harrod (1959: 179).
25 Harrod (1959: 179), Fort (2004: 233).
26 Zimmerman (2001: 177).
27 Fort (2004: 179–180) Comments on Lindemann's ideas on aerial mines in the memorandum, 2 October 1939, CHAR 25/17.
28 Tizard to Lindemann, 10 September 1939, LIND D243/32.
29 Lindemann to Tizard (draft), 11 September 1933, LIND D243/33.
30 LIND E21/8, undated document.
31 Manchester (1988: 554).
32 Zuckerman, 'Scientific Advice During and Since World War II', *Proceedings of the Royal Society of London*, Vol. 342, No. 1631, pp. 465–80, see p. 478.
33 Jenkins (2001: 561–2).
34 WSC to Wells, 11 September 1939, WELLS c238.20; WSC invited Wells to join the Club five years before: WSC to Wells, 1 March 1934, CHAR 2/233/2–4.
35 Eliot (1940: 237). Smith (1998b: 235–55) See, in particular, Wells's letter to *The Times*, 23 October 1939.
36 Manchester (1988: 606).

Chadwick doubts that the Bomb is viable

1 Aileen Chadwick to Feather, 30 August 1974, FEAT 13/2.
2 Letter from Chadwick to Bohr, 9 April 1961, NBA.
3 Brown (1997: 153).
4 Chadwick interview, 17 April 1969, AIP: http://www.aip.org/history/ohilist/3974_3.html.
5 Peierls to Andrew Brown, 2 December 1991: Lee (ed.) (2009: 1013).
6 Holt (1988).
7 Brown (2012: 22).
8 Chadwick interview, 16 April 1969, AIP: http://www.aip.org/history/ohilist/3974_2.html.
9 Brown (1997: 39–44).
10 'The Voice of Science', *Nature*, 9 September 1939.
11 Chadwick to Feather, 15 October 1939, FEAT 23/6.

12 Chadwick to Appleton, 31 October 1939, CAB 104/186, NA.
13 Holt (1999).
14 Brown (1997: 174).
15 Brown (1997: 175).
16 Chadwick to Appleton, 5 December 1939, CAB 21/1262, NA.
17 Lord Hankey to Admiral of the Fleet Lord Chatfield, 12 December 1939: 'I gather that we may sleep fairly comfortably in our beds.' CAB 21/1262, NA.
18 G. P. Thomson, text of 'Atomic Energy' lecture, 1945, GPT F154.
19 Elliot, W. to Bridges, E., *c.*27 December 1939, CAB 21/1262, NA.
20 Clark (1965: 184–6). Tizard to G. P. Thomson, 2 October 1957, and G. P. Thomson to Tizard, 5 October 1957, GPT D25.
21 Hopkins (1939: 2).
22 Rowlands (2001: 19–20); Brown (1997: 181–7)
23 Chadwick to Cockcroft, 8 January 1940, CKFT 20/5. A few weeks later, on 24 February, Chadwick made a similar comment in a letter to his former colleague Norman Feather: 'I am very much afraid the [centre of gravity] will move to the USA.' FEAT 23/6.
24 AIP interview with Weisskopf, 21 September 1966, pp. 12–14.

FDR receives a nuclear warning

1 Bernard Baruch to WSC, *c.*30 March 1954, CHUR 2/210A/349.
2 FDR to Lord Tweedsmuir (formerly the author John Buchan), 5 October 1939, quoted in Roosevelt, E. (ed.) (1950: 934).
3 Kimball (1991: 7).
4 FDR, 3 September 1939: http://millercenter.org/president/speeches/detail/3315.
5 'Only Five Per Cent Favour Sending US Army to Fight Nazis', *Boston Globe*, 7 October 1939, p. 3.
6 Kimball (ed.), Vol. 1 (1984: 24–5).
7 FDRLIB: WSC dates his dedication 8 October 1933.
8 FDR: *Day by Day – The Pare Lorentz Chronology*, 11 October 1939, FDRLIB. See also 'The Day in Washington', *New York Times*, 12 October 1939, p. 16.
9 Entry for Sachs in *Dictionary of American Biography*, New York, Charles Scribner's Sons, 1994.
10 Weart and Szilard (eds) (1978: 15–17).
11 'Einstein Chats About Sea', *New York Times*, 26 January 1934, p. 15.
12 Lanouette (1992: 198–204).
13 Rhodes (1986: 313–14).

14 FDR to Einstein, 19 October 1939: http://www.mphpa.org/classic/COLLECTIONS/MP-PFIL/Pages/MPP-PFIL-020.htm.
15 Ferrell (1998: 168).
16 Pottker (2004: 156).
17 During FDR's first year at Harvard (1900–1), he took Geology 4: Elementary Geology and Geology 5: Elementary Field and Laboratory Geology. In the following year, he took Geology 14: General Palaeontology. I thank David Woolner and his colleagues at the FDR library for this information.
18 FDR talk, 5 December 1938: Roosevelt (1941: 615).
19 WSC article 'Europe's Plea to Roosevelt', *Evening Standard*, 10 December 1937.
20 Kevles (1995: 252–66), see p. 258 for FDR's quotation.
21 FDR's second inaugural address: http://teachingamericanhistory.org/library/index.asp?document=90.
22 Hewlett and Anderson (1962: 20–1).
23 Zachary (1992: 24–5).
24 Zachary (1992: 26).
25 Zachary (1999: 108–14).
26 Ratcliff (1942: 28).
27 Zachary (1999: 189–90).

Frisch and Peierls discover how to make the Bomb

1 From the first part of the Frisch–Peierls memorandum, AB 1/210, NA.
2 Bernstein (2011: 440); Arnold (2003: 112).
3 'The Scientist Works for Industry', *Birmingham Gazette*, 6 July 1938, BHMPHYS.
4 '£60,000 to Split Atom', *Daily Express*, May 1939; 'Physics Department's Needs', *Birmingham Post*, June 1938, BHMPHYS.
5 'They Will Split the Atom to Save Lives', *The Referee*, May 1939, BHMPHYS.
6 Frisch to Oliphant, 27 May 1939, FRISCH A48/11.
7 Chadwick to Feather, 15 October 1939, FEAT 23/6.
8 Frisch to Margaret Hope, 10 October 1939, FRISCH F61/9.
9 Frisch to the Aliens' Department of the Home Office, 27 January 1940, FRISCH A116/10.
10 Ministry of Labour to Frisch, 28 November 1939, FRISCH A53/11.
11 Frisch (1979: 76–9).
12 Frisch to Margaret Hope, 10 October 1939 and 5 February 1940, respectively FRISCH F61/9 and F62/3.
13 Documents in FRISCH A86/2 and A116/5.
14 Peierls (1985: 130–5).

15 This was recalled by Sir Charles Darwin in an intelligence report on Peierls after the war: KV2/1661, f. 285a, NA.
16 Peierls to the Home Office, 31 August 1939, and letter from Peierls to Appleton, 13 September 1939: Lee (ed.) (2007: 678, 682).
17 Peierls (1985: 63).
18 Frisch to Margaret Hope, 10 October 1939, FRISCH F61/9.
19 'Great Britain in War-time, Birmingham and District', *The Times*, 10 February 1940, p. 3.
20 Frisch to T. Bjerge, 29 February 1940, FRISCH B96/2.
21 Peierls (1985: 145).
22 Frisch to T. Bjerge, 29 February 1940, FRISCH B96/2.
23 UK Home Office to Frisch, 13 February 1940, FRISCH A116/23.
24 Lee (ed.) (2007: 701).
25 This account of the inception of the Frisch–Peierls memorandum is taken mainly from Peierls (1985: 154–5) and from Frisch (1979: 125–6). For information on the approximate dates on which the memo was written, see letter from Peierls to Mr Murphy, 2 May 1995, D110, and letter from Peierls to T. E. Allibone, 5 January 1983, D15, both in PEIERLS.
26 For Peierls's assessment of his contribution see his letter to Margaret Gowing, 30 October 1961, Lee (ed.) (2009: 647–52).
27 Peierls (1985: 154–5).
28 Peierls (1985: 154–5).
29 Clark (1965: 218).
30 Pais (1991: 476). See, for example, the report 'The Nazi Invasion' in the *Manchester Guardian*, 10 April 1940, p. 7.
31 'Mr Lloyd George's Onslaught on the Prime Minister', *Manchester Guardian*, 9 May 1940, p. 7.
32 Autobiographical notes of G. P. Thomson, GPT, A7, p. 18.
33 Weart (1979: 136–7); Gowing (1964: 49–50) and G. P. Thomson's draft memoir, GPT A7.
34 Thomson to Chadwick, 16 April 1940, CHAD I 19/7.
35 Peierls (1985: 155).
36 Interview with Peierls by Charles Weiner, AIP, 12 August 1969, p. 96.
37 Letter from Peierls to the chairman of the 'U-bomb' committee, 22 April 1940: see Lee (ed.) (2007: 702–3). On Peierls's being granted British nationality: letter from Peierls to the Academic Assistance Council SPSL 335/9, 28 March 1940.
38 A copy of G. P. Thomson's letter, written on 3 May 1940, is in KV2/2421, NA.
39 Frisch to Genia Peierls, 30 April 1978, FRISCH F94/9.
40 Peierls (1985: 157).
41 Cockburn and Ellyard (1981: 86–7, 163).

42 Hartcup and Allibone (1984: 95).
43 Meacham (2003: 51).

Churchill has more pressing problems

1 Mallalieu, J. P. W., 'Churchill Commentary', *The Tribune*, 8 September 1950.
2 Jenkins (2001: 606–10).
3 Quoted in Cannadine (1994: 130). See also Ironside (1962: 301–2).
4 Diary of Harold Ickes, XXXI, 12 May 1940, LIBCON.
5 Muller (ed.) (2012: 358–68), see p. 367.
6 Jenkins (2001: 632).
7 Hinsley, F. H., 'Churchill and the Use of Special Intelligence', in Blake and Louis (eds) (1996: 407–26), see p. 411.
8 Quoted in Reynolds (2004: 166).
9 Reynolds (2004: 179).
10 Churchill's radio speeches during this era are available at http://archive.org/details/Winston_Churchill.
11 Wenden, D. J. 'Churchill, Radio and Cinema' in Blake and Louis (eds) (1996: 215–239), see p. 216.
12 Jenkins (2001: 591).
13 Birkenhead (1961: 210).
14 Fort (2004: 227). Desmond Morton to WSC, 30 May 1940, PREM 7/2, NA; see also the draft of this note, with Lindemann's annotations, 29 May 1940, LIND G 526/9.
15 Birkenhead (1961: 215); Best (2001: 200).
16 Macdougall (1987: 26, 36).
17 Harrod (1959: 180–237).
18 Macdougall (1987: 26) and Birkenhead (1961: 216–17).
19 'A New Source of Power', *The Times*, 7 May 1940, p. 5.
20 'Vast Power Source in Atomic Energy Opened by Science', *New York Times*, 5 May 1940, p. 1; Pimm to Lindemann, 7 May 1940, LIND D230/1.
21 Clark (1961: 105).
22 Macrae (1971: 144).
23 Edgerton (2011: 260). See a similar story in Fort (2004: 227).
24 Jenkins (2001: 597).
25 Jenkins (2001: 610). Draft attached to note from Morton to WSC, 30 May 1940 PREM 7/2, NA.
26 Meacham (2003: 51).
27 Colville (1985: 109).
28 J. H. Peck to J. J. Balfour, 11 May 1940, FO 371/24255, NA; Zimmerman (1996: 61–70).

29 Lindemann's private secretary to A. V. Hill, 21 June 1940, LIND G 319/19.
30 Memo dated 8 May 1940, ADM 116/4302, NA.
31 WSC to Archibald Sinclair, 30 June 1940, responding to Sinclair to WSC, 25 June 1940, both FO 371/24255, NA.
32 Reynolds (2004: 172).
33 Crozier (1973: 178). See also the comments of Churchill's colleague Sir Archibald Sinclair (Crozier: 1973: 172).
34 Soames (ed.) (1998: 454).
35 Comments in John Martin's diary, May and June 1940, MART 1.
36 Colville (1985: 98–125).
37 Clark (1965: 227).
38 Jones (1998: 92–105), WSC to R. V. Jones, 14 December 1946, RVJO B.216.
39 Churchill (1949: 339–40).
40 Zimmerman (1996: 67), see reference 48.
41 G. P. Thomson to Sir Harold Hartley, concerning C. P. Snow's 'Science in Government', 29 May 1961, GPT G112.
42 R. V. Jones (from Aberdeen) to Mr Kelly, 16 May 1950, RVJO B.219.

Thomson and his MAUD committee debate policy on the Bomb

1 MAUD minutes are in AB 1/8, NA.
2 Cockcroft to his wife, 18 April 1940, CKFTFAMILY.
3 Cockcroft to his wife, 13 January 1941, CKFTFAMILY.
4 BMFRS of Thomson, September 1975, pp. 531, 549–50.
5 Gowing (1964: 80–9).
6 G. P. Thomson officially requested permission to consult with the 'enemy aliens' Frisch and Peierls on 3 May 1940 but then had to wait for it to be granted: KV 2/1658.
7 Interview with Peierls by Charles Weiner, AIP, 12 August 1969, p. 94.
8 Cockcroft to his mother, 24 May and 16 June 1940, CKFTFAMILY.
9 'How I Became a Member of Parliament in 1940', in *Memories and Reflections*, Vol. 2, unpublished and undated (in the Roskill Library of Churchill College, Cambridge).
10 Zimmerman (1996: 52–3); Hill to Roskill, 16 January 1973, ROSK 7/13. 'Letter from A. V. Hill', undated, HTT 706.
11 Memo by A. V. Hill: 'Uranium – "235"', 16 May 1940, AB1/9, NA. See also HTT 241.
12 It is not clear if Hill met Szilárd; if he did, there is no evidence of Szilárd's views in Hill's correspondence.

13 Autobiographical notes of G. P. Thomson, GPT A7, pp. 25–6.
14 Sime (1996: 284–5).
15 Fowler to Tizard, 24 May 1940, HTT 241.
16 Peierls to Frisch, 31 July 1940, FRISCH B125/1.
17 Gowing (1964: 47, 54–5).
18 Gowing (1964: 49–51).
19 Autobiographical notes of G. P. Thomson, GPT A7, p. 17.
20 Letter from Max Born to Lindemann, 9 July 1940, LIND D24/7.
21 Home Office file F.962 on Frisch, KV2/2421, NA.
22 Hill (1962: 231–2).
23 Clark (1965: 244–5).
24 Macrae (1971: 9–10).
25 *Nature*, 11 January 1941, pp. 35–6, see p. 35.
26 Jenkins (2001: 630).
27 Zimmerman (1996: 83–4).
28 WSC had asked Ismay a few days before, 'Are we going to throw our secrets into the American lap, and see what they give us in exchange? If so, I am against it.' WSC to Ismay, 17 July 1940, PREM 3/475/1 NA.
29 Zimmerman (1996: 88–94).
30 Kevles (1995: 302–3).
31 Hartcup and Allibone (1984: 98).
32 Clark (1965: 298).
33 Bowen (1998: 155).
34 See, for example, 'Blast Rocks London, Toll Feared Great', *Washington Post*, 6 September 1940, p. 1.
35 Clark (1965: 269).
36 Fowler to Hill, 11 September 1940, and Hill to Fowler, 4 October 1940, AVHL I 3/19.
37 Chadwick to Lindemann, 20 June 1940, LIND D 230/8. Gowing (1964: 47).
38 Gowing (1964: 47); Peierls to T. E. Allibone, 5 January 1983, PEIERLS D.15.

In his finest hour, Churchill begs America for help

1 Wells (1940: 57, 17).
2 Colville (1985: 315); CHBIO, Vol. 6, p. 1038. Tennyson's poem: http://theotherpages.org/poems/tenny02.html.
3 Description taken from 'Nazi Bombers Stage Longest Raid of the War', *Washington Post*, 6 September 1940, p. 1.
4 Wells (1901: 213).
5 Hall, D. J. (2000) 'Bulldog Churchill', *Finest Hour*, No. 106, pp. 18–20.

6 Cannadine (ed.) (1989: 4).
7 Edgerton (2011: 68–9).
8 A second blitz was rendered virtually impossible by 1944 by the advent of radar-controlled guns firing shells with proximity fuses.
9 Jenkins (2001: 630–1).
10 Jenkins (2001: 641).
11 Wells (1934: 99–100).
12 Smith (1986: 470–1).
13 Sherborne (2010: 342–3). 'The Illusion of Personality', *Nature*, 1 April 1944, pp. 395–7. Example of correspondence with WSC: WSC to Wells, 17 September 1940, c.238.19 WELLS.
14 For descriptions of Hankey, see the press cuttings in HNKY 2/5. See *Daily Express*, 31 May 1938. Roskill (1974: 656).
15 WSC to Neville Chamberlain, 27 September 1940, CAB 21/829, NA.
16 McGucken (1978: 112).
17 During the Tizard mission, from 9 September to 30 November 1940, there appear to have been only three discussions about nuclear weapons; see 'Activities of British Technical Mission', HTT 706.
18 Tizard to WSC, 19 October 1940, PREM 3/475/1, NA.
19 'WM (40), Cabinet minutes, 293rd conclusions', 21 November 1940, AVIA 22/2286, NA.
20 Charles Lindemann to Frederick Lindemann, 16 December 1940, LIND D146/9.
21 War Cabinet minutes, 2 December 1940, CAB 65/10, NA, p. 152.
22 'President's Call for a Full Response on Defense', *New York Times*, 30 December 1940, p. 6.
23 'Hopkins Given London Task by Roosevelt', *New York Times*, 4 January 1941, p. 3.
24 Report by W. Ridsdale, 25 January 1941, FO 371/26179, NA.
25 Note by Alex Cadogan, 29 January 1941, FO 371/26179, NA.
26 Hershberg (1993: 142).
27 'Conant: Man of the Week', *Akron Beacon Journal*, 5 November 1941, see Box 155 CONANT.
28 Conant's first national radio broadcast was on 29 May 1940, CONANT, Box 26. 'Conant Urges Aid to Allies at Once', *New York Times*, 30 May 1940, p. 24.
29 The lunch was on 6 March 1941. Conant (1970: 253–5); Hershberg (1993: 145).
30 Conant (1970: 262).
31 Wenden, D. J., 'Churchill, Radio and Cinema', in Blake and Louis (eds) (1996: 215–39), see p. 233. See also Dilks (ed.) (2010: 402).
32 Conant (1970: 252–6, 266). 'Conant Urges That the US Take Place Beside England in War', *Atlanta Constitution*, 5 May 1941, p. 1.

33 Churchill's nomination papers: http://www2.royalsociety.org/DServe/dserve.exe?dsqIni=Dserve.ini&dsqApp=Archive&dsqCmd=ImageView.tcl&dsqDb=Catalog&dsqImage=EC_1941_21.jpg; *Nature*, 21 June 1941, p. 772; *Nature*, 7 June 1941, p. 704.
34 See, for example, letter from A. V. Hill to Sir Henry Dale, 17 September 1941, HTT 58.
35 Clark (1965: 271).
36 Colville (1985: 344–5).
37 Clark (1965: 293).
38 Reynolds (1994: 249).
39 Halifax to WSC, *c.*15 June 1941, HLFX, A.4.7.
40 Churchill and Roosevelt had met once before, in 1918, though this escaped Churchill's memory.
41 Colville (1985: 368).
42 Muller (2009: 262).

Chadwick believes Britain should build its own Bomb

1 Interview with Chadwick, 20 April 1969, AIP: http://www.aip.org/history/ohilist/3974_4.html.
2 Interview with Chadwick's daughters, 10 May 2012. See also Brown (1997: 204).
3 Peierls to G. P. Thomson, 14 May 1941, AB 1/580, NA; Peierls to Andrew Boyle, 13 November 1978, PEIERLS D.53.
4 Fuchs officially joined the MAUD team on 28 May 1941. Report on Fuchs by Chief Constable of Birmingham, 6 May 1942, KV 2/1246, NA; Gowing (1964: 53n).
5 Hastings (2009: 86).
6 Kennedy (1986: 93).
7 Slade to Lindemann, 1 December 1939, LIND D101/1; 'Anglo-American Economic Cooperation' by Melchett, D109/20; Melchett to Lindemann, 14 January 1941, LIND D105/14; Melchett to Lindemann, 3 July 1941, LIND D107/1.
8 Oliphant to Appleton, 18 June 1940, LIND G526/12.
9 Gowing (1964: 75).
10 Minutes of the MAUD committee, 19 May 1941, AB1/8, NA.
11 Minutes of the MAUD technical committee, 9 April 1941, AB1/8, NA.
12 Letter from Cockcroft to his wife, 21 March 1941, CKFTFAMILY.
13 A few days later, Conant was startled to hear more from the American Ambassador's assistant Ben Cohen about the British programme to develop a nuclear weapon. Hershberg (1993: 146).
14 Chadwick to Dickens, 25 and 30 June 1941, AB 1/222, NA.

15 Minutes of MAUD technical committee, 2 July 1941, AB 1/8, NA.
16 The MAUD report is reprinted in full in Gowing (1964: 394–436).
17 Chadwick to Feather, 16 July 1941, FEAT 23/6.
18 Chadwick to Feather, 30 August and 6 September 1941, FEAT 23/6.
19 Gowing (1964: 76–89).
20 Tizard to Hankey, 5 August 1941, CAB 104/207, NA.
21 Blackett to Pye, 5 August 1941, AB 1/238, NA; Gowing (1964: 78).
22 Darwin to Hankey, 2 August 1941, CAB 104/227, NA.
23 Hankey to Jonathan, 'Uranium', undated but evidently August 1941, CAB 126/330, NA.
24 Thomson to Chadwick, 11 August 1941, CHAD I 19/7.

Lindemann backs a British Bomb

1 Testimony by Lindemann to the Scientific Advisory Committee to the War Cabinet, 17 September 1941, CAB 90/8, p. 2, NA.
2 Gowing (1964: 96). Lindemann to WSC, 27 August 1941, CAB 126/330, NA. Lindemann copied the letter to Sir John Anderson: CAB 126/39, NA. Lindemann's recommendations differed little from the conclusion of the Defence Services Panel of the Scientific Advisory Committee, finalised on 25 September 1941, CAB 104/227, NA.
3 CHBIO, Vol. 6, pp. 1176–7.
4 Reynolds (1994: 130).
5 'President Debarks', *New York Times*, 17 August 1941, p. 1.
6 WSC to Hopkins, 28 August 1941, PREM 3/224/2, NA.
7 Lindemann to WSC, 27 August 1941, CAB 126/330, NA.
8 WSC to Ismay, 30 August 1941, PREM 3/139/8A, NA; WSC to Anderson, 23 August 1945, CHUR 2/3.
9 Gowing (1964: 106).
10 Ismay to Anderson, 4 September 1941, CAB 126/330, NA.
11 Clark (1965: 300–1).
12 Anderson to Hankey, 12 August 1941, CAB 126/330, NA.
13 Reader (1975: 173–4, 292).
14 *Dictionary of National Biography*, entry for Anderson by G. C. Peden.
15 Anderson to Hankey, 6 August and 12 August 1941, CAB 126/330, NA.
16 Anderson to Moore-Brabazon, 23 October 1941, CAB 126/330, NA.
17 Birkenhead (1961: 256).
18 Fort (2004: 233–4). Lindemann's chauffeur was George Topp.
19 I thank Peter Sarnak of the Institute of Advanced Study, Princeton, for giving me the benefit of his expert opinion on Lindemann's con-

tribution to the theory of prime numbers. Lindemann's letter to the editor of *Nature* completed on 13 September 1941 was published on 11 October 1941 (Vol. 148, p. 436).

20 A. V. Hill to Dale, 17 September 1941, HTT 58.

21 Minutes of the 12th Meeting of the War Cabinet's Scientific Advisory Committee, 17 September 1941, CAB 90/8, NA.

22 Scientific Advisory Committee's report on MAUD, 24 September 1941, CAB 90/8, NA.

23 Hankey's report also recommended that pilot plants should be built in Britain and Canada.

24 Moore-Brabazon to Hankey, 27 August 1941, CAB 104/227, NA.

25 Hankey to Moore-Brabazon, 27 August 1941, CAB 104/227, NA.

26 WSC quoted, on 2 May 1948, in Colville (1985: 587).

27 FDR to WSC, 11 October 1941, CAB 126/330, NA; PREM 3/139/8A, NA.

Oliphant bustles in America

1 Cockcroft to his wife, 18 February 1941, CKFTFAMILY.

2 Cockburn and Ellyard (1981: 29).

3 Cockburn, S., and Ellyard, S. (1981: 74).

4 Oliphant (1982: 17).

5 Oliphant to Appleton, 27 October 1941, CHAD I 19/3.

6 Oliphant to G. P. Thomson, 9 August 1941, AB 15/6077, NA.

7 Oliphant (1982: 17).

8 Rhodes (1986: 372–8).

9 Monk (2012: 296).

10 Bernstein (1976: 205).

11 Villa (1977: 478).

12 FDR to WSC, 11 October 1941, PREM 3/139/8A/574, NA.

13 Weart and Szilárd (eds) (1978: 146).

14 Gowing (1964: 109).

15 Oliphant to Cockcroft, 3 November 1941, AB 1/157, NA.

16 Oliphant to Appleton, 27 October 1941, CHAD I 19/3.

17 Oliphant to Chadwick, 27 October 1941, CHAD I 19/3.

18 Minutes of a meeting relating to Tube Alloys, 16 October 1941, CAB 126/46, NA.

19 Oliphant to Chadwick, 27 October 1941, CHAD I 19/3.

20 Chadwick to Oliphant, 10 November 1941, CHAD I 19/3.

21 Oliphant to Cockcroft, 3 November 1941, AB 1/157, NA. The American scientists were Harold Urey and George Pegram.

22 Oliphant to Chadwick, 12 November 1941, CHAD I 19/3. Some of the early briefing papers that Akers commissioned are in AB 1/256, NA.

23 Oliphant to Chadwick, 14 January 1942, CHAD I 19/3.
24 Cockburn and Ellyard (1981: 90–1, 163).

Churchill talks about the Bomb with FDR

1 Eden (1965: 491).
2 WSC to Roosevelt, December 1941, PREM 3/139/8A, NA.
3 'Tube Alloys', report by Norman Brook, 27 November 1941, PREM 3/139/8A, NA.
4 Gilbert (2005: 242–4).
5 Roberts (1991: 280).
6 Jenkins (2001: 672).
7 WSC to Lord Privy Seal, 3 January 1942, CAB 120/29, NA.
8 Gilbert (2005: 249).
9 See the essays on WSC and Roosevelt in Berlin (1980). Also Hastings (2009: 224).
10 Moran (1966: 420); Hastings (2009: 224).
11 WSC to Mr Curtin, 13 January 1942, CAB 120/29, NA.
12 He refers to 'the stir' in his note to Attlee, 11 January 1942, CAB 120/29, NA.
13 Martin to WSC, 9 January 1942, PREM 3/139/8A, f. 568, NA.
14 Norman Brook to John Martin, 10 December 1941, CAB 126/330, NA.
15 War Cabinet, 17 January 1942, CAB 65/25/8, NA.
16 Wheeler-Bennett (1958: 535). Churchill actually used the phrase 'walking out' rather than 'courting'.
17 Roosevelt to Bush, 11 March 1942, AEC 32.
18 Hankey, 'Bacteriological Weapons', 6 December 1941, PREM 3/65. Bernstein (1987: 47–8).
19 Hastings (2009: 236).
20 Churchill (1951: 81).
21 Jenkins (2001: 682–3).
22 Harvie-Watt (1980: 63).
23 Hankey to Halifax, 1 May 1941, Halifax papers CAC, ref. A4.410.4.5; Roskill to Rowe, 12 July 1968, C11, ROWE.
24 Reynolds (2004: 334).
25 Churchill (1951: 341).
26 FDR to Bush, 11 July 1942, FDRLIB, Bush file.
27 Reynolds (2004: 334).
28 Welles (1950: 215).
29 Churchill (1951: 343).
30 Hastings (2009: 297).
31 Alanbrooke (2002: 269).
32 Notes by Hankey, 29 June 1942, HNKY 13/1.

33 Anderson to WSC, 30 July 1942, PREM 3/139/8A, NA. Churchill wrote on the document 'As proposed' on 31 July.

Akers attempts a merger

1 Akers to Perrin, 21 December 1942, CHAD I 28/2.
2 Gowing (1964: 108–10).
3 BMFRS, Akers, November 1955, pp. 1–4.
4 Gowing (1964: 131).
5 Gowing (1964: 128, 141).
6 Brown (1997: 205–8).
7 Interview by GF of Harold Agnew and Al Wattenburg, 30 November 1992.
8 Gowing (1964: 139).
9 For Akers's view on the threat of the German project: memo to Anderson from his assistant, 11 June 1942, CAB 126/166, NA.
10 Gowing (1964: 144–5, 437–8).
11 Gowing (1964: 145).
12 Akers to Perrin, 21 December 1942, CHAD I 28/2.
13 Sachs (ed.) (1984: 43–53); Rhodes (1986: 433–42); interview by GF of Harold Agnew and Al Wattenburg, 30 November 1992.
14 Atomic Heritage Foundation website is very useful: http://www.atomicheritage.org/index.php?option=com_content&task=view&id=295.
15 Lanouette (1992: 237–8).
16 Bird and Sherwin (2005: 185–6).
17 Groves (1962: 125–6).
18 Akers to Halban, 1 January 1943, CHAD I 28/2.
19 Conant to Bush, 'Some thoughts concerning the correspondence . . .', 25 March 1943, FDRLIB.
20 Gowing (1964: 176).
21 Hewlett and Anderson (1962: 270).
22 Peierls (1985: 173).
23 Minutes of 13th Technical Committee of Tube Alloys, 22 January 1943, CAB 126/46, NA.
24 Gowing (1964: 217).
25 Chadwick to Fowler, 3 December 1942, CHAD IV 1/9; Peierls to Chadwick, 22 September 1942, CHAD I 19/6; Clark (1961: 184).
26 Letter to *The Times* from A. V. Hill, 1 July 1942 (p. 5).
27 Major General E. N. C. Clarke to Hill, 5 October 1942, AVHL I 2/2.
28 Kirby and Rosenhead (2011: 11–15).
29 Joubert to A. V. Hill, 4 March 1942, AVHL I 2/2.
30 'Combined Operations and a Great General Staff', undated, probably 1942, AVHL I 2/1.

31 E. J. S. Clarke to G. P. Thomson, 4 June 1958, GPT G58/3.
32 Akers to Groves, 31 December 1942, CHAD I 28/2; obituary of Akers, *Chemistry and Industry*, 20 November 1954, p. 1449.
33 Conant to Mackenzie, 2 January 1943, PREM 3/139/8A, NA.
34 Conant memo, dated 7 January 1943 by Akers, PREM 3/139/8A, f. 460–1, NA; Gowing (1964: 155–7).
35 Anderson to WSC, 23 July 1942, PREM 3/139/8A, NA.
36 Bernstein (1976: 211).
37 See, for example, Peierls to Akers, 1 May 1943, in Lee (ed.) (2007: 791–4).

Bush aims for an American monopoly

1 Bush (1972: 280–1).
2 Hershberg (1993: 180).
3 Bernstein (1976: 211).
4 Cockcroft to his mother, 11 July 1943, CKFTFAMILY.
5 Zachary (1992: 24); Zachary (1999: 61–4, 90–5).
6 Ratcliff (1942: 28, 40).
7 Bush (1972: 280).
8 Entry for Stimson, *American National Biography*, Oxford University Press (1989); Bernstein (1976: 210–11); Hewlett and Anderson (1962: 265–6).
9 Bernstein (1976: 210).
10 Hewlett and Anderson (1962: 267).
11 http://www.atomicheritage.org/index.php/atomic-history-main-menu-38/the-manhattan-project-begins-81942-121942-main-menu-209/fdr-gives-final-approval-mainmenu-212.html.
12 WSC to Hopkins, 1 April 1943, PREM 3/139/8A/523, NA.
13 Bernstein (1976: 213).
14 Conant to Bush, 'Some thoughts concerning the correspondence . . .', 25 March 1943, FDRLIB.
15 Sherwood (2008: 159).
16 Bush, V., Memorandum of conference with Mr Harry Hopkins and Lord Cherwell at the White House, May 25, 1943. FDRLIB 'Atomic Bomb file'.
17 Bernstein (1976: 214–15).
18 Hewlett and Anderson (1962: 274).
19 Hewlett and Anderson (1962: 275) give an account of the meeting based on contemporary sources. Bush's recollections, written some twenty-six years later, are in Bush (1972: 282). See also Gowing (1964: 168n).
20 STIMSON, Notes of Overseas Trip, 1943, entries for 17 July 1943,

Brief Report of Certain Features of My Overseas Trip, 4 August 1943.

21 This account of the meeting is taken from Hewlett and Anderson (1962: 275–7) and references therein, and from Bush (1972: 282–5).

Churchill's nuclear deal with FDR

1 Soames (2011: 275–6).
2 'Note on Tube Alloys' by Lindemann, undated, probably January or February 1943, PREM 3/139/8A, f. 494, NA.
3 See, for example, WSC to Hopkins, 1 April 1943, PREM 3/139/8A, NA.
4 'Tube Alloys', Lindemann to WSC, 7 April 1943, PREM 3/139/8A, ff. 511–15, NA.
5 WSC to Anderson, 15 April 1943, PREM 3/139/8A, ff. 502–3, NA.
6 Gowing (1964: 163).
7 Anderson to WSC, 26 July 1943, PREM 3/139/8A, ff. 422–4, NA.
8 FDR to WSC, 26 July 1943, PREM 3/139/8A, f. 421, NA.
9 Lindemann to WSC, 28 July 1943, PREM 3/139/8A, ff. 419–20, NA.
10 Churchill to Tizard, 30 July 1943, CHAR 20/94A, f. 63.
11 Gowing (1964: 170).
12 Oppenheimer (1964b).
13 Akers to Perrin, 19 August 1943, AB 1/376, NA.
14 Pawle (1963: 243); CHBIO, Vol. 7, p. 461.
15 Report on meeting by Cadogan, 29 August 1942, and note from Brown to Lawford, 27 October 1942, FO 1093/247; Reynolds (2004: 327).
16 Meacham (2003: 225–8); Gilbert (2005: 278–9).
17 CHBIO, Vol. 7, p. 471.
18 CHBIO, Vol. 7, p. 470.
19 Ward (ed.) (1995: 230–1).
20 Meacham (2003: 237).
21 CHBIO, Vol. 7, p. 476.
22 Reynolds (2004: 156, 380).
23 Cantelon, Hewlett and Williams (eds) (1984: 31–3).
24 Hinsley, Thomas, Ransom and Knight (1981: 128).
25 WSC to Lindemann, 27 May 1944, PREM 3/139/11A, ff. 761–2, NA.
26 'Negotiations with the Americans after the signing of the Quebec Agreement' by Wallace Akers, 13 September 1943, AB 1/129, NA, p. 1.
27 CHBIO, Vol. 7, p. 482.
28 WSC to Smuts, 5 September 1943, CHAR 20/117.

29 See the documents in MART 2 and 5, especially the report 'Mr Churchill at Harvard' by Barbara Williams.
30 Text of WSC's Harvard speech: http://www.winstonchurchill.org/learn/speeches/speeches-of-winston-churchill/118-the-price-of-greatness.
31 Lindemann to WSC, 19 October 1943, PREM 3/139/8A, f. 312, NA.

Bohr takes a political initiative

1 Interview with Joanna Batterham (née Chadwick) about Bohr's visits to Los Alamos, 9 November 2012.
2 Gowing (1964: 245–8).
3 Pais (1991: 487–8).
4 Perrin to Gorrell Barnes, 26 October 1943, CAB 126/39, NA.
5 Brown (1997: 251).
6 Anderson to Llewellin, undated memo, probably October 1943, CAB 126/39, NA.
7 French and Kennedy (eds) (1985: 183–4, 230, 244).
8 Robert Oppenheimer, quoted in 'The Philosophy of Niels Bohr' by Aage Peterson, *Bulletin of the Atomic Scientists*, September 1963, pp. 8–14, see p. 9; French and Kennedy (eds) (1985: 182).
9 Peierls (1985: 180); Aaserud (2005: 13–15); Jones, R. V., 'Meetings in Wartime and After', in French and Kennedy (eds) (1985: 278–87); Aage Bohr's journal 'Bohr's activities October 1943 to April 1945', NBA.
10 Bohr (1964: 197); testimony of Sir Charles Darwin's wife, G41, GPT.
11 *New York Times*, 9 October 1943, p. 3.
12 Akers to Munro, 31 August 1943, AB 1/376, NA.
13 Akers to Perrin, 31 August 1943, AB 1/376, NA.
14 Anderson to WSC, 15 October 1943, PREM 3/139/8A, ff. 313–15, NA.
15 'Tube Alloys Project – negotiations with the Americans after the Quebec Agreement', 24 September 1943, AB 1/129, NA; Oliphant to Chadwick, 21 November 1943, CHAD IV 3/7; Tube Alloys collaboration report, 9 October 1943, PREM 3/139/8A, , ff. 316–21, NA.
16 Chadwick's note on his meeting with Groves, 11 February 1944, CHAD IV 3/2.
17 Monk (2012: 395, 397n); Powers (1993: 244, 536, note 11).
18 Chadwick, J., 'Comments on Mrs Gowing's 2nd Volume', CHAD IV 14/11, p. 3; Oliphant to Chadwick, 21 November 1943, CHAD IV 3/7.
19 Anderson to WSC, 21 March 1944, PREM 3/139/2, NA.
20 Anderson to Lewellin, undated but probably October 1943, CAB 126/39, NA; Journal 6.02 of Bohr's activities, October 1943 to April 1945, NBA.
21 Untitled Anderson memo, 25 November 1943, CAB 126/39, NA.

22 Szasz (1984: 75); Powers (1993: 240–1).
23 Bohr (1964: 199).
24 Hughes (2002: 65).
25 Chadwick's notes from meeting with Groves, 11 February 1944, CHAD IV 3/2.
26 Gowing (1964: 249).
27 Aydelotte to Hanson, 20 December 1943; Aydelotte to Bohr, 22 December 1943: Niels Bohr papers in DO Members' Box 12a-36-47, IAS.
28 Akers to Perrin, 27 January 1944, CAB 126/332, NA.
29 Powers (1993: 246–8).
30 Pais (1991: 462).
31 Szasz (1992: 32–34); Peierls (1985: 187–203).
32 Brown (2012: 45).
33 Szasz (1992: 19).
34 Oppenheimer (1964b).
35 Teller (1962: 211).
36 Szasz (1992: 148–9); note from Manhattan authorities to General Gee, 6 August 1946, FRISCH A58A.
37 Oppenheimer (1964b).
38 Akers to Perrin, 27 January 1944, CAB 126/332, NA.
39 Powers (1993: 257).
40 Bohr's journal after he left Denmark in 1943, 6.02, pp. 3–5, NBI.
41 Aaserud (2005: 20).
42 Anderson to WSC, 21 March 1944, PREM 3/139/2, NA. This reference and all others relating to the preparations for the WSC–Bohr meeting are in Aaserud (2005: 20–3).
43 Anderson to WSC, 27 April 1944, PREM 3/139/2, NA.
44 Kapitza to Bohr, 28 October 1943, CAB 126/39, NA. Even on that date, Kaptiza believed Bohr was still in Sweden: see 'Correspondence between Kapitza and B.', 2 May 1945, CAB 126/39.
45 Bohr (1964: 205).
46 Wheeler-Bennett (1962: 297).
47 Jones (1998: 476).
48 Dale to WSC, 11 May 1944, CAB 126/39, NA.
49 Comment on Gorrell Barnes to Anderson, 19 May 1944, CAB 126/39, NA.
50 Sir Henry Dale to Lindemann, 11 May 1944, CAB 127/201, NA.
51 Aaserud (2005: 25).

The Bulldog meets the Great Dane

1 HNSRD.
2 Brendon (1984: 187).

3 Hastings (2009: 477).
4 Jenkins (2001: 732–3).
5 Dilks (ed.) (2010: 621).
6 WSC to Eden, 21 October 1942, PREM 4/100/7, ff. 265–6, NA.
7 Hastings (2009: 431–4).
8 Reynolds (2004: 380–2).
9 Moran (1966: 145–54).
10 Hastings (2009: 484–5).
11 Reynolds (2004: 417).
12 This account of the Bohr–Churchill meeting is based on 'The interview with the Prime Minister' in Aage Bohr's journal 'Bohr's activities October 1943 to April 1945', NBA; Jones (1998: 475–7), Jones, R. V., 'Meetings in Wartime and After', in French and Kennedy (eds) (1985: 278–87); Aaserud (2005: 23–7); Gowing (1964: 355).
13 Bohr (1964: 204).
14 Lindemann to WSC, 24 May 1944, PREM 3/139/11A, f. 764, NA; the text of the letter from Bohr to WSC is in Aaserud (2005: 96–8).
15 WSC to Lindemann, 20 September 1944, PREM 3/139,8A, ff.298–9, NA.
16 WSC to Lindemann, 27 May 1944, PREM 3/139/11A/761, NA. The word [later] in the quote replaces the original 'after', which appears to be a typographical error.
17 WSC to Ismay, 19 April 1945, PREM 3/139/11A, f. 817.
18 See, for example, Anderson to WSC, 25 May 1944, PREM 3/139/11A, NA.
19 CHBIO, Vol. 7, pp. 793–4.
20 CHBIO, Vol. 7, p. 808.
21 WSC to Anderson, 6 October 1943 CHAR 20/94B.
22 See the V1 and V2 timeline at http://www.flyingbombsandrockets.com/Timeline.html and Overy (2006: 293–5).
23 Leasor (ed.) (1959: 69).
24 Fort (2004: 294–5); BMFRS of Lindemann, November 1958, pp. 63–4.
25 Colonel J. D. Wyatt to Colonel Macrae, 30 May 1961, and A. L Bonsey to *The Times*, 24 April 1961, MCRA 2/21.
26 Lindemann to WSC, 25 February 1944, PREM 3/65, NA; Bernstein (1987: 46–7).
27 WSC to Ismay, 6 July 1944, PREM 3/89, NA.
28 Bernstein (1987: 46–7).
29 Cited in 'Potentialities of Weapons of Bacteriological Warfare during the Next Ten Years', 6 November 1945, DEFE 2/1252, NA.
30 Anderson to WSC, 1 June 1944, PREM 3/139/2, NA. Bernstein (1976: 223–4).

31 WSC to Anderson, 13 April 1944 and Anderson to WSC, 14 April 1944, PREM 3/139/2, NA; Gowing (1964: 298–301, 444–5).
32 Moran (1966: 179).
33 Lindemann to WSC, 25 July 1944, PREM 3/139/11A, ff. 744–5, NA.
34 Lindemann to WSC, 12 September 1944, PREM 3/139/8A, f. 309, NA.
35 Aide-mémoire signed by Roosevelt and WSC, 18 September 1944, PREM 3/139/8A, f. 310, NA.
36 WSC to Lindemann, 20 September 1944, PREM 3/139/8A, ff. 298–9, NA.
37 Lindemann's 'Notes on Conversation between President, Admiral Leahy, Bush and self', 22 September 1944, RVJO B.396.
38 Lindemann to WSC, 23 September 1944, CAB 126/39, NA.
39 Lindemann to WSC, 17 November 1944, PREM 3/139/11A, f. 836, NA; Lindemann to Oppenheimer, 21 October 1944, OPPY, Box 26, Folder 6.

Chadwick witnesses the first nuclear explosion

1 Gowing (1974a: 47).
2 Szasz (1984: 39–40); Brown (1997: 260–1).
3 Brown (1997: 254–7).
4 Brown (1997: 268).
5 Tube Alloys salaries, AB 1/267, NA.
6 Webster to Appleton, 24 February 1944, CHAD IV 3/7.
7 Chadwick to Akers, 24 June 1944, AB 1/615, NA.
8 A pithy overview of the technical challenges of the project is given in Hughes (2002: 68–83). See also Kevles (1995: 327–34).
9 Peierls (1985: 199–201).
10 Gowing (1964: 262).
11 Note from Manhattan authorities to General Gee, 6 August 1946, FRISCH A58A.
12 This account of the British mission draws on the interviews with Rudolf Peierls and several others in the late 1980s, available at LOSALAMOS. See also documents in FRISCH A60–62; Szasz (1992: 32–46).
13 MI5 security report on Peierls, 24 January 1951, KV 2/1661, NA; Peierls (1985: 197, 205).
14 Peierls to Tom Sharpe, 3 September 1993, PEIERLS D.58.
15 Szasz (1992: 27).
16 Clark (1992: 157).
17 Groves (1962: 166).
18 Some of Bohr's technical work is available from LOSALAMOS, attached to the letter from Bohr to Oppenheimer, 19 June 1944; on

Oppenheimer's view of Bohr: Thorpe (2006: 257). Quote is from the *New York Herald Tribune*, 19 November 1962.

19 Jones (1998: 477).
20 Oppenheimer (1964b).
21 Brown (2012: 45–6, 295 n9).
22 Rotblat gives the date of the dinner at the Chadwicks as March 1944: Rotblat (1985: 19).
23 Brown (2012: 54–5).
24 A few months later, the Chadwicks moved to Q Street – interview with Chadwick's daughters, 9 November 2012.
25 Chadwick to Appleton, 21 March 1945, AB 1/615, NA; Szasz (1992: 62).
26 Szasz (1992: 22, 25); Peierls (1985: 200–1).
27 Churchill to Anderson, 28 January 1945, CAB 126/30, NA.
28 Oliphant to Akers, 21 January 1945, CAB 126/30, NA.
29 Oliphant to Akers, 8 February 1945, CAB 126/59, NA.
30 Gowing (1964: 330); Oliphant to Akers, 21 January 1945, CAB 126/30, NA.
31 Chadwick to Anderson, and 'Future TA Policy and Programme', 23 March 1945, CAB 126/59, NA.
32 Chadwick to Akers, 31 May 1945, CHAD IV 11/1.
33 Oliphant to Hill, 25 June 1945, AVHL I 3/67.
34 Birkenhead (1961: 295).
35 Peierls to Kitty Oppenheimer, 18 January 1972: Lee (ed.) (2009: 741).
36 'People's Faces Show of Death News', *Washington Post*, 13 April 1945, p. 1.
37 Szasz (1992: 29).
38 Copy of message from Robert Patterson, dated 4 May 1945, in FRISCH A59.
39 Bernstein (1975: 38); http://www.dannen.com/decision/franck.html.
40 Stimson diary, 14 May 1945, STIMSON.
41 Bernstein (1975: 36–7).
42 Kevles (1995: 336).
43 'Memorandum by Sir Henry Dale', 20 June 1963, CKFT 25/25.
44 This account draws from: Szasz (1984: 79–114); Lamont (1965: 203–56); Monk (2012: 437–9); the Trinity Test, http://www.cddc.vt.edu/host/atomic/trinity/projtrinity.html; Chadwick, J. 'The Atomic Bomb', *Liverpool Daily Post*, 4 March 1946 (CHAD IV 9/3); Peierls (1985: 200–2).
45 See the account by Frisch, reproduced in Gowing (1964: 441–2).
46 Lamont (1965: 238).
47 Groves's memorandum for Stimson, 18 July 1945, reproduced in Feis (1960: 165–71), see p. 166. Chadwick donation to the Natural History Museum: CHAD I 24/2.

48 Hughes (2002: 97); McMahon (2003: 6).
49 Hastings (2007: 489); Schwartz (ed.) (1998: 562).

Churchill says yes to dropping the Bomb

1 Moran (1966: 280).
2 Gowing (1964: 362–75). For one of Anderson's briefings, see 'Notes on TA for "TERMINAL"', 29 June 1945, PREM 3/139/8A, NA.
3 Bohr journal 6.02, activities from October 1943 to April 1945, p. 19, NBA.
4 Truman (1955: 415).
5 Ferrell (1998: 3–4, 151). See also Truman to his daughter, 3 March 1948: Truman (ed.) (1981: 106–7).
6 Truman (1955: 10–11).
7 Smith (1946: 286).
8 McCullough (1992: 64–5).
9 Franklin (2008: 52–3).
10 Gromyko (1989: 99–100).
11 Jenkins (2001: 783–5).
12 Haslam (2011: 50–2).
13 Moran (1966: 347).
14 Harris (1984: 241–3).
15 Churchill College Archives Guide 'Cosmos Out of Chaos' (2009: 32–3).
16 Churchill to Truman, 12 May 1945, CHAR 20/218, f. 109. H. G. Wells coined the phrase in his 1904 novel *The Food of the Gods*, in Book III, near the beginning of Chapter Four: http://ebooks.adelaide.edu.au/w/wells/hg/food/chapter11.html.
17 See the papers in CAB 120/691, NA, in the period May to June 1945.
18 WSC to Sir Orme Sargent, 10 June 1945, CAB 123/147, NA. Of the scientists scheduled to travel to Moscow, there was certainly a case for preventing the eminent crystallographer J. D. Bernal from making the trip, as he was a follower of Stalin and therefore a security risk (e-mail from Andrew Brown, 13 December 2012).
19 WSC to Commons, 14 June 1945, HNSRD.
20 See CAB 123/147, NA, and the Prime Minister's papers PREM 3/139/7, NA. Woolton to Blackett, 18 June 1945, CAB 123/147, NA. Hill (1962: 301–2). See also BLACKETT G.91.
21 Jenkins (2001: 792).
22 McCullough (1992: 415).
23 CHBIO, Vol. 8, p. 61.
24 Hastings (2007: 506–7).

25 Gilbert (2005: 356–7).
26 The message is decoded somewhat differently, but with the same essential meaning, in Feis (1960: 164). I am using the phrase quoted in Harvey Bundy's 'Remembered Words', *Atlantic*, March 1957, Vol. 199, No. 3, pp. 56–7, see p. 57, and am assuming that this is the message that Stimson showed Churchill on the next day. Churchill recalled that the message was 'Babies satisfactorily born' (CHBIO, Vol. 8, p. 62), but it is unlikely that the plural was used.
27 Bernstein (1975: 41).
28 For a rare mention of Tube Alloys to Attlee by WSC, see CHBIO, Vol. 7, p. 482; Harris (1995: 277–8).
29 McCullough (1992: 416–20).
30 Truman (1955: 343).
31 Rhodes James (1986: 307–8). For Truman's reaction: McCullough (1992: 423).
32 On Stalin's fondness for Chopin: FRUS, Conference of Berlin (Potsdam) 1945, p. 1530.
33 This purpose emerges from Stalin's encrypted correspondence with the Soviet embassy in Tokyo: Haslam (2011: 60).
34 STIMSON: diary entry for 19 July 1945.
35 STIMSON: diary entry for 22 July 1945.
36 This quote is from Stimson's assistant Harvey Bundy, who accompanied Stimson to the meeting: 'Remembered Words', *Atlantic*, March 1957, Vol. 199, No. 3, pp. 56–7, see p. 57.
37 'Vision of the Future Through the Eyes of Science', *News of the World*, 31 October 1937.
38 Moran (1966: 280); WSC to Lindemann, 27 May 1944, PREM 3/139/11A, ff. 761–2, NA.
39 Wells, H. G., 'Churchill Must Go', *Tribune*, 15 December 1944; WSC's bouquet reached Wells on 10 May 1945: Toye (2008: 151).
40 Alanbrooke (2002: 709).
41 Truman (1955: 416). In his account, written years after the incident, he writes that Stalin showed no special interest: 'All he said was that he was glad to hear it and hoped we would "make good use of it against the Japanese".' This is contradicted by the first-hand account of Stalin's reaction given by Pavlov, written soon after the event – this is the account used in the text.
42 Pavlova V. N., 'Avtobiograficheskie zametki', *V. N. Pavlova – Perevodchika I. V. Stalina' Novaya i Noveishaya Istoriya*, No. 4, July–August 2000, p. 110.
43 Haslam (2011: 62); Holloway (1994: 117); Feis (1966: 101–2).
44 Churchill (1954: 579–80); Truman (1955: 416); Gellately (2013: 162–5).

45 Goodman (2007: 82–3); Kojevnikov (2004: 136–8).
46 Haslam (2011: 62).
47 Chikov, V. (1997). Dos'e KGB no. 13676, Nelegaly, Vol. 1, Operatsiya 'Enormous', Moscow, pp. 17–19.
48 Haslam (2001: 17).
49 Kojevnikov (2004: 140).
50 Lota (2010: 656–66), see especially p. 658.
51 Stalin made this remark on 27 July 1945: FRUS, Conference of Berlin (Potsdam) 1945, p. 1531.
52 Zhukov (1971: 674–5).
53 Haslam (2001: 61).
54 Gromyko (1989: 109).
55 CHBIO, Vol. 8, pp. 106–7.
56 Churchill's colleague William Mabane described him as 'bewildered, hurt and rather stunned': entry in Harold Nicolson's diary, 14 August 1945, NIC. Randolph Churchill's quoted words are in Harold Nicolson's diary, 5 September 1945, NIC.
57 Chiang Kai-shek, the President of China, also signed the declaration, in absentia. http://pwencycl.kgbudge.com/P/o/Potsdam_Declaration.htm. FRUS, Conference of Berlin (Potsdam) 1945, pp. 1474–6.
58 Truman (1955: 421).
59 WSC gave his agreement on 1 July 1945: documents in CAB 126/46, NA; Lindemann to WSC, 28 January 1953; 'Events leading up to the use of the atomic bomb, 1945', PREM 11/565, NA.
60 CHBIO, Vol. 8, pp. 111–12.
61 Colville (1985: 577).
62 Lindemann to WSC, 26 July 1945, PREM 3/139/9, f. 640, NA.
63 Draft of WSC announcement on the Bomb, 29 July 1945, PREM 3/139/9, NA. The text of the final speech was published in, for example, the *Manchester Guardian*, 7 August 1945, p. 5.

Blackett: nuclear heretic

1 Blackett (1948: 127).
2 BBC's 9 p.m. bulletin on 6 August 1945: http://www.bbc.co.uk/radio4/history/august1945.shtml (3:03 into sequence).
3 E-mail from Blackett's daughter Giovanna, 5 December 2011.
4 Blackett returned to Manchester in the summer of 1945: Nye (2004: 85).
5 Chadwick, J., 'Comments on Mrs Gowing's 2nd Volume', CHAD IV 14/11, p. 3.
6 Orwell, G., 'London Letter', autumn 1945, in Orwell and Angus (eds) (1970: 452).

7 Diary entry by Harold Nicolson, 7 August 1945, NIC.
8 Shaw, G. B., 'The Atom Bomb', *Sunday Express*, 12 August 1945. It was reprinted in, for example, the *Washington Post*, 19 August 1945. On 21 August, *The Times* in London published a letter from Shaw on the subject.
9 Report on the press conference, 12 August 1945, CHAD IV 3/13.
10 Smyth (1976: 199).
11 Reprinted in 'Statements Relating to the Atomic Bomb', *Reviews of Modern Physics*, Vol. 17, No. 4, October 1945, pp. 472–90.
12 Gowing (1974a: 58–9).
13 Chadwick, J., 'Comments on Mrs Gowing's 2nd Volume', CHAD IV 14/11, p. 3.
14 Attlee 'The Atomic Bomb', 28 August 1945, CAB 130/3 (in the GEN 75/1 papers), NA.
15 Attlee to Truman, 25 September 1945, reprinted in Gowing (1974a: 78–81).
16 Blackett to Polanyi, 3 November 1941, BLACKETT, J65.
17 Oliphant to Chadwick, 1 October 1945, CHAD I 25/1.
18 Speech to Parliament by Captain Blackburn, 30 October 1945, HNSRD; 'Man-to-Man Talks on the Atomic Bomb', *Manchester Guardian*, 31 October 1945, p. 6.
19 Rickett to Anderson, 29 October 1945, CAB 126/304, NA.
20 Briefing for Chancellor of the Exchequer, 25 October 1945, CAB 126/304, NA.
21 'All Topics Open for Discussion', *Manchester Guardian*, 1 November 1945, p. 5.
22 See Attlee's comments on Blackett, 'Atomic energy: an immediate policy for Great Britain', 6 November 1945, PREM 8/115, NA.
23 Gowing (1974a: 87–92).
24 Orwell, G. , 'You and the Atomic Bomb', *Tribune*, 19 October 1945: http://tmh.floonet.net/articles/abombs.html.
25 Blackett's notes for his reply to Oliphant's letter of 22 January 1946, BLACKETT D.192.
26 WSC to Commons, 7 November 1945, HNSRD.
27 Minutes of the GEN 163 Committee on 8 January 1947, CAB 130/16, NA.
28 Cathcart (1994: 88).
29 Interview with Sir Maurice Wilkes, 24 March 2009.
30 Chadwick, J., 'Comments on Mrs Gowing's 2nd Volume, CHAD IV 14/11, p. 11.
31 'Atomic Science', *Manchester Guardian*, 18 May 1946, p. 6.
32 McMahon (2003: 21–8).
33 Interview with R. Gordon Erneson, 21 June 1989, Truman Library,

http://www.trumanlibrary.org/oralhist/arneson.htm; Hennessy (1996: 101).

34 Stimson to Maitland-Wilson, 3 July 1945, PREM 3/139/9, f. 660, NA.
35 Hennessy (1996: 102).
36 Szasz (1992: 48–9).
37 Gowing (1974a: 114).
38 Gowing (1974a: 90).
39 Blackett (1946).
40 Blackett to Peierls, 22 October 1946, PEIERLS C.22.
41 Quoted in Peierls to Oppenheimer, 26 August 1946, reproduced in Lee (2009: 65–6).
42 Barnes to Attlee, 14 November 1946, PREM 8/684, NA.
43 See, for example, Mott to Blackett, 11 November 1946, BLACKETT D.175.
44 Gowing (1974a: 115); interview with Sir Maurice Wilkes, 24 March 2009.
45 Hennessy (1993: 267–9).
46 Blackett (1948: Chapter X).
47 Blackett to Tizard, 16 September 1948, BLACKETT H.37.
48 Exchange between George Jeger and the Minister of Defence, 12 May 1948, HNSRD; Cathcart (1994: 88–9); Gowing (1974a: 212–13).
49 Gowing (1974a: 241–72).
50 'Moscow Hotly Protests Treaty', *Los Angeles Times*, 1 April 1949, p. 1.
51 Information about 'Consequences' in BLACKETT A.10a.
52 Kirby and Rosenhead (2011: 4, 24).
53 Orwell to Celia Kirwan, list compiled *c.* April 1949, FO 1110/189, NA.
54 Peierls to Blackett, 15 November 1948, PEIERLS C.23.
55 Thomson, G. P. to Blackett, 16 October 1948, BLACKETT H.37; Thomson, G. P., 'Russia and the Bomb', *Spectator*, 22 October 1948, p. 532.
56 Rabi, I. I., 'Playing Down the Bomb', *Atlantic Monthly*, April 1949, pp. 21–4.
57 There appears to have been only one brief exchange of correspondence between Lindemann and Blackett: about uncontroversial funding matters, in May 1946, BLACKETT D.197.
58 Interview with Peierls, 12 August 1969, pp. 99–100 of transcript.
59 BMFRS on Blackett, November 1974, pp. 35–6.
60 Lindemann to WSC, 13 March 1949, LIND E13/1.

Churchill the Cold Warrior

1 Laurence and Peters (eds) (1996: 188).
2 CHBIO, Vol. 8, pp. 119, 125.

3 Moran (1966: 316).
4 Moran (1966: 328).
5 Moran (1966: 289).
6 WSC to Commons, 21 August 1945, HNSRD.
7 Desmond Morton to Sir David Petrie, 20 March 1941, and Petrie to Morton, 22 March, KV 2/3217 1941, NA.
8 WSC to Anderson, 23 August 1945, CHUR 2/3.
9 Anderson to WSC, 31 August 1945, CHUR 2/3.
10 WSC to Anderson, 7 September 1945, CHUR 2/3.
11 Morris to WSC, 23 March 1946 – WSC crossed out Blackett's nomination, CHUR 2/302.
12 Gowing (1974a: 32).
13 Attlee to WSC, September 1945 (undated), CHUR 2/3.
14 WSC'S Fulton speech, 5 March 1946: http://www.churchill-society-london.org.uk/Fulton.html; Reynolds (2004: 41–6); Jenkins (2001: 811–13).
15 Gilbert (2005: 373).
16 McCullough (1992: 486–90); Harris (1995: 298).
17 Interview with Stalin in *Pravda*, 14 March 1946: translation in CHBIO, Vol. 8, p. 211.
18 'Most important speech' – quoted in Reynolds (2004: 42).
19 WSC speech at University of Zurich, 19 September 1946, http://www.churchill-society-london.org.uk/astonish.html; 'An Ill-timed Speech?', 20 September 1946, *Manchester Guardian*, p. 5.
20 WSC to Attlee, 10 October 1946, CHUR 2/4.
21 Moran (1966: 315).
22 Blackburn (1959: 104–5).
23 The lunch took place on 30 July 1946. CHBIO, Vol. 8, p. 249.
24 CHBIO, Vol. 8, pp. 253–4.
25 Smith (ed.) (1998b: 531–2). Wells's note is undated, but was probably written *c.* May 1946.
26 Conversation took place on 8 August 1946: Moran (1966: 315).
27 Pickersgill and Forster (eds) (1970: 112–13).
28 WSC speech in Parliament, 23 January 1948, HNSRD.
29 WSC to Eden, 12 September 1948, CHUR 2/68A.
30 CHBIO, Vol. 8, p. 315.
31 Jenkins (2001: 808).
32 Churchill (1948): 'Theme of the Volume', printed at the beginning.
33 Churchill (1948: 301–2).
34 Reynolds (2004: 481–2).
35 Reynolds (2004: 136–44).
36 Coward to WSC, 9 December 1948, CHBIO, Vol. 8, pp. 449–50.
37 Reynolds (2004: 97–8).

38 Clementine Churchill to WSC, 5 March 1949, CHUR 1/46. See CHBIO, Vol. 8, pp. 461–2.
39 Lindemann to WSC, 13 March 1949, CHUR 2/81A. See Lindemann's letters to *The Times*, 30 November and 9 December 1949.
40 'Churchill Arrives Beaming', 24 March 1949, *New York Times*, p. 1.
41 'Churchill Hailed by Capital Crowd', 25 March 1949, *New York Times*, p. 1.
42 WSC to Truman, 29 June 1949, CHUR 2/158.
43 Truman to WSC, 2 July 1949, CHUR 2/158, f. 22.
44 WSC to Lindemann, 20 September 1949, LIND J84/7, and the reply on 1 October, J84/8.
45 The text is in CHUR 4/390B, ff. 212–16.
46 Lindemann speech in House of Lords, 5 July 1951, HNSRD.

Peierls and 'the spy of the century'

1 Genia Peierls to Klaus Fuchs, 4 February 1950, KV 2/1661, NA; for the 'spy of the century' quote, see Gannon (2002: chapter 13).
2 *New York Times*, 31 January and 1 February 1950, both p. 1; 'Work to Begin on Hydrogen Bomb', *Manchester Guardian*, 1 February 1950, p. 7.
3 Extract from the *Daily Worker*, 2 February 1950, KV 2/1661, NA, PF.109567.
4 Moss (1987: 184).
5 'Note' of J. H. Marriott, 6 February 1950, KV 2/1661, NA.
6 KV 2/1661, NA, extract PF.109.567; Peierls (1985: 223); Moss (1987: 184–7); Peierls interview, 13 August 1969, AIP, p. 153 of transcript. Peierls expressed his initial view that Fuchs 'could not possibly be guilty' to the security forces: see 'Note' of J. H. Marriott, 6 February 1950, KV 2/1661, NA.
7 Communications with Gaby Gross (née Peierls), 20 October 2011 and 23 January 2013.
8 E-mail from Gaby Gross (née Peierls), 24 January 2013.
9 Phone tap on the Peierlses, KV 2/1661, NA, report dated 6 December 1950.
10 Phone tap on the home of Herbert Skinner, 3 February 1950, KV 2/1661, NA.
11 KV 2/1661, NA, extract PF.109.567, report 6 December 1950; Metropolitan Police report, 6 February 1950.
12 Moss (1987: 239–48), see p. 244.
13 This account largely follows the account of the conversation given by Peierls in his letter to Commander Burt, written on 5 February 1950, and J. H. Marriott's report on the conversation, 6 February 1950,

KV 2/1661. See also the letter Genia Peierls wrote to Fuchs on 4 February 1950 (PEIERLS, D.52) and Moss (1987: 186–90).

14 Peierls to Commander Burt, 5 February 1950, KV 2/1661, NA.

15 Genia Peierls to Fuchs, undated but clearly written on 4 February 1950, D.52 PEIERLS. Lee (ed.) (2009: 209–11).

16 Fuchs to Genia Peierls, 6 February 1950, PEIERLS D.52. Lee (ed.) (2009: 212–13).

17 See, for example, *Daily Mirror*, *New York Times* and *Washington Post*, 4 February 1950.

18 'Wartime Atom Chief Tells Congress . . .', *Washington Post*, 5 February 1950, p. M1.

19 Quoted in Moss (1987: 210).

20 'The Lesson of the Fuchs Case', *c.* March 1950, reproduced in Lee (2009: 219–25), see p. 222.

21 'Atomic Scientists' Association', report by A. K. Longair, 23 September 1947, KV 2/1658, NA.

22 'Atomic Energy Train Exhibition', *Nature*, 13 September 1947, p. 358; see also *Nature*, 15 November 1947, pp. 668–9. Brown (2012: 88–90).

23 MI5 report by Capt. A. C. M. Bennett, 18 March 1948, KV 2/1658.

24 Interview with Peierls, 12 August 1969, AIP, see p. 103 of transcript.

25 E-mail from Dyson, 28 October 2011.

26 Peierls to Bethe, 15 February 1950, PEIERLS C.17.

27 Peierls to Bohr, 14 February 1950, NBA, supplementary Peierls archive.

28 Peierls to Derek Curtis-Bennett, 28 February 1950, KV 2/1661, NA.

29 Note by J. C. Robertson, 6 March 1950, PF. 62251, Vol. 9, KV 2/1661, NA.

30 Szasz (1992: 84); Moss (1987: 194–203).

31 'Mr Churchill Assails Foreign Policy', *Manchester Guardian*, 15 February 1950; 'First Major Statement on Foreign Policy', *Daily Telegraph*, 15 February 1950. CHSPCH, Vol. 8, pp. 7936–44.

Churchill softens his line on the Bomb

1 'Mr Churchill Assails Foreign Policy', *Manchester Guardian*, 15 February 1950.

2 CHSPCH, Vol. 8, p. 7943–4.

3 Handwritten note by Attlee, 3 March 1950, PREM 8/1279, NA.

4 Attlee to Commons, 6 March 1950, HNSRD; 'Swift Plunge into Controversy', *Manchester Guardian*, 7 March 1950, p. 7.

5 Jenkins (2001: 836–7).

6 McMahon (2003: 50); Jenkins (2001: 833).

7 WSC to Attlee, 4 August 1950, CHUR 2/28.

8 WSC to Eden, 12 December 1951, PREM 11/1682, NA; Maddock (2010: 60).
9 Lindemann to WSC, 2 October 1950, CHUR 2/36; Lindemann to WSC, 23 July and 1 August 1950, CHUR 2/28.
10 Arrangements for WSC's visit to Copenhagen and its press coverage: CHUR 2/276 and CHUR 2/278.
11 Bohr's presence was pointed out in WSC's notes for the trip: CHUR 2/276. Bohr's Open Letter: Aaserud (2005: 173–85).
12 WSC (1951) Københavns Universitets Promotionfest, Bianco Lunos Bogtrykkeri, p. 14.
13 WSC (1951) Københavns Universitets Promotionfest, Bianco Lunos Bogtrykkeri, p. 21.
14 'Mr Churchill Calls for a United Europe', *Birmingham Post*, 12 October 1950, p. 1.
15 Report in *Vendsyssel Tidende*, a north Jutland newspaper, 30 November 1950, p. 1.
16 McCullough (1992: 820–2).
17 Commons session on 14 December 1950: HNSRD.
18 WSC to Lindemann, 3 December 1953, EG 1/36, NA.
19 Parliamentary exchanges on 14 December 1950 and 30 January 1951, HNSRD.
20 Lindemann to WSC, 6 December 1950, CHUR 2/28, ff. 118–19.
21 Blackburn (1959: 96).
22 Attlee to WSC, 3 December 1950, PREM 8/1559, NA.
23 See Roger Makins's account of the meeting, 31 January 1951, and WSC's subsequent letter to Attlee, 12 February 1951, PREM 8/1559, NA.
24 Confidential Foreign Office note, 'Atomic Energy', 28 December 1950, PREM 8/1559, NA.
25 See correspondence in CHUR 2/28: WSC to Truman, 10 and 12 February 1951; Truman to WSC, 16 February 1951, delivered to WSC on 26 February via the US Embassy; Truman to WSC, 24 March 1951.
26 Murrow, E. R. 'Churchill: The Hinge of Fate', *Atlantic Monthly*, Vol. 187, pp. 70–3, see p. 71.
27 Reynolds (2004: 334).
28 Churchill (1951: 341–2). Further evidence of the implausibility of WSC's claim: Gowing (1964: 162–3).
29 Churchill (1951: 723).
30 Lindemann to WSC, 1 August 1950, CHUR 2/28.
31 Peierls (1985: 283).
32 Top Secret document 'Professor Peierls', 15 January 1951, KV 2/1661, NA.

Penney delivers the British Bomb

1 Howard (1974).
2 Hennessy (1989: 713); see correspondence in PREM 11/292, NA.
3 BMFRS of Penney, Vol. 39, February 1994, p. 288.
4 Boyle (1955: 261).
5 Penney, Samuels D. E. J. and Scorgie G. C. (1970) 'The Nuclear Explosive Yields at Hiroshima and Nagasaki', *Transactions of the Royal Society of London*, A266, pp. 357–424, see p. 419.
6 Cathcart (1994: 29–30, 82).
7 Telephone conversation with Tam Dalyell, 9 January 2005.
8 Cathcart (1994: 39–40).
9 Hennessy (1989: 712–13).
10 Goodman (2007:13).
11 Gowing (1974a: 224).
12 'Strategy' memorandum by Tizard, 4 November 1949, DEFE 9/34, p. 2, NA.
13 'Strategy' memorandum by Tizard, 4 November 1949, DEFE 9/34, NA.
14 CHSPCH, Vol. 8, p. 7943; WSC 'Press Notice', February 1951, CHUR 2/28, ff. 91–3.
15 Cathcart (1994: 129).
16 Quoted in Hennessy (2007: 59).
17 Lindemann to WSC, 17 June 1951, CHUR 2/113A.
18 Speech by Lindemann in the House of Lords, 5 July 1951, HNSRD.
19 'Development of Atomic Energy in Great Britain', *Nature*, 14 July 1951, p. 61.
20 The Bomb should be ready to test between 'July and October 1952' Portal told Sir William Elliot, 10 August 1950, AB 16/1132, NA.
21 'Note on Dr Penney', unsigned, 25 May 1951, AB 16/942, NA.
22 Cathcart (1994: 164).
23 Minutes of Chiefs of Staff Committee, 17 September 1951, DEFE 32/2, NA.
24 'Pledge to Cut Spending, End Nationalisation', *Daily Telegraph*, 29 September 1951.
25 WSC speech on 23 October 1951: CHSPCH, Vol. 8, pp. 8281–6.
26 Item 6 in 'Some Reflections on the Present Situation', undated but probably January 1951, DEFE 9/34, NA (Sir Frederick Brundrett's supportive response is dated 9 January 1951). Gowing (1974a: 229–30) refers to this passage but dates it 1949.
27 Gowing (1974a: 230).
28 The plans for the first British Bomb are in ES 1/11, NA; see also Cathcart (1994: 202–35).
29 Minutes of Chiefs of Staff Committee, 24 October 1951, DEFE 32/2,

NA; Cathcart (1994: 150).
30 Cathcart (1994: 148).
31 Gowing (1974a: 405).
32 Lindemann speech to the House of Lords, 5 July 1951, HNSRD.

Churchill – Britain's first nuclear Premier

1 FRUS, 1952–4, Vol. 5, p. 1721.
2 WSC's speech to Commons, 6 November 1951, HNSRD.
3 Colville (1985: 604).
4 Colville (1985: 658).
5 Colville (1985: 595); Shuckburgh (1986: 160–1).
6 Wheeler-Bennett (1969: 119).
7 Fort (2004: 318).
8 Taylor, A. J. P., 'Lindemann and Tizard: More by Luck than Judgment?', *Observer*, 9 April 1961, p. 21.
9 Birkenhead (1961: 277–9, 302–5).
10 Birkenhead (1961: 310).
11 Lindemann to WSC, 13 November 1951, PREM 11/292, NA.
12 WSC to Lindemann, 15 November 1951, PREM 11/292, NA. The only precedent for the views expressed here by WSC is in the note he drafted for the press in the previous February, CHUR 2/28, untitled, but beginning 'There seems to be some misunderstanding . . .'.
13 Moran (1966: 352).
14 WSC to Sir Edward Bridges, 8 December 1951, PREM 11/297, NA. WSC uses the word 'millions' rather than the correct 'million'.
15 'Atomic Energy Expenditure' briefing for the Prime Minister, 12 December 1951, PREM 11/297, NA.
16 WSC endorses the policy on the minute Lindemann to WSC, 14 November 1951, PREM 11/292, NA.
17 Hennessy (2000: 180–3, 188).
18 Coote (1971: 102–10).
19 Moran (1966: 349).
20 CAB 21/3058, NA; CHBIO, Vol. 8, p. 674.
21 'Subjects for discussion at Washington', November 1951, CAB 21/3058, NA.
22 *New York Times*, 14 December 1951, article by James Reston, p. 8. Young (1996: 72).
23 McCullough (1992: 874); Larres (2002: 168); Young (1996: 72–82).
24 Young (1996: 75); CHBIO, Vol 8, p. 675.
25 McCullough (1992: 874–5).
26 Young (1996: 80).
27 See WSC–McMahon correspondence in November 1948, CHUR

2/69B; and their correspondence in April 1949, CHUR 2/84B.
28 Moran (1966: 359).
29 British record of the WSC–Truman meeting, 18 January 1952, CAB 21/3058, NA.
30 Brandon (1988: 94).
31 Colville (1985: 604); Birkenhead (1961: 279); Moran (1966: 382).
32 WSC to the Commons, 23 October 1952, HNSRD.
33 Gowing (1974b: 37, 56).
34 'Soviet atomic capabilities: appreciation for SACEUR', 14 November 1952, DEFE 21/62, NA.
35 Colville (1985: 596).
36 Hennessy (2006: 174–5).
37 Moran (1966: 400).
38 Birkenhead (1961: 310); Lindemann to WSC, 26 September 1952, LIND J122/99–100.
39 Colville (1985: 614).
40 Lindemann's atomic energy papers for WSC: LIND J122/4–16 (see also PREM 11/561, NA).
41 Arnold (2001: 37–8).
42 Confidential Annex to D.(52) 12th Meeting, 12 December 1952, CAB 131/12, NA.
43 Lindemann to WSC, 'Atomic Energy – Future UK Programme', LIND J122/4–6.
44 The estimate is from the Ridley Committee: see the text of Hinton's talk on 30 October 1953 in AB 19/85, NA.
45 Quote is from Cockcroft's draft memoir, CKFT 25/6 p. 41.

Hinton engineers nuclear power

1 Sir Christopher Hinton's unpublished memoir, p. 61, HINTON.
2 Hennessy (2006: 329) Gowing (1974b: 20) and BMFRS of Hinton, December 1990, p. 226.
3 Lindemann to WSC, 17 September 1952, PREM 11/292, NA. Gowing (1974b: 12).
4 Gowing (1974b: 191).
5 Wells (1914: 51).
6 BMFRS of Hinton, Vol. 36, December 1990, p. 220.
7 Gowing (1974b: 22–3).
8 Gowing (1974b: 20–1).
9 Sir Christopher Hinton's unpublished memoir, p. 61, HINTON.
10 Hinton, C., 'Atomic Energy in Industry' conference, New York, 30 October 1953, text in AB 19/85, NA.
11 Hinton BIOFRS, December 1990, p. 227.

12 'Talks on Exchange of Atom Data End', *New York Times*, 17 October 1953.
13 The text of Hinton's talk on 30 October 1953: *Bulletin of the Atomic Scientists*, December 1953, pp. 366–8, 390. Conference proceedings, 'Atomic Energy in Industry', National Industrial Conference Board (1954), section 2.
14 Zachary (1999: 361–3).
15 Bird and Sherwin (2005: 472–5).
16 *Time*, 21 September 1953. Hinton's clip of the article 'The Atom', pp. 13–15: AB 19/85, NA.
17 Larres (2002: 192, 199–205).
18 Hinton to Plowden, 22 December 1953, PLDN 5 1/2.

Churchill the nuclear missionary

1 Moran (1953: 403).
2 Larres (2002: 182–4).
3 Quoted in Larres (2002: 189).
4 Larres (2002: 195–6).
5 Boyle (ed.) (1990: 31–52).
6 Eden wrote Eisenhower's 'tiresome' comment in his diary on 4 March 1953: Avon papers, AP 20/1/29, BHM. Larres (2002: 182–3).
7 Larres (2002: 199).
8 Larres (2002: 222–32).
9 WSC to Commons, 11 May 1953, HNSRD.
10 CHBIO, Vol. 8, pp. 846–7. Jenkins (2001: 860–1).
11 Telephone conversation with Lady Williams of Elvel (née Jane Portal), 2 February 2012.
12 Moran (1966: 415).
13 Moran (1966: 427, 437).
14 Moran (1966: 448, 451, 457). Colville (1985: 633) apparently misremembers the anecdote about WSC's amazement at reading news of the Soviet H-bomb in August 1953 and that this anecdote refers to WSC's reaction to Sterling Cole's speech in February 1954.
15 Butler (1971: 174).
16 Lovell, R., letter to the *British Medical Journal*, 10 June 1995, p. 310.
17 Birkenhead (1961: 311–16).
18 Birkenhead (1961: 332).
19 Allén, S. (2005) 'If You Have No Misgivings: Churchill's Nobel Prize in Literature', *European Review*, Vol. 13, No. 4, pp. 591–5.
20 Reynolds (2004: 439); Jenkins (2001: 864).
21 WSC to Commons, 3 November 1953, HNSRD.
22 Lindemann to WSC, 12 November 1953, LIND J138/66.

23 Larres (2002: 308–9).
24 Young (1996: 224–5); Larres (2002: 310).
25 British record of the Bermuda conference, 4 December 1953, FO 371/125138, NA. See also FRUS 1952–4, Vol. 5, p. 1761.
26 Notes on meeting between Lindemann and Strauss, 4 December 1953, EG 1/36, NA.
27 FRUS 1952–4, Vol. 5, pp. 1767–9.
28 Macmillan (1971: 324).
29 FRUS 1952–4, Vol. 5, 'Western European Security', p. 1739; Montague Browne (1995: 156–7).
30 CLVL 1/8, 'The Bermuda Conference', 7 and 8 December 1953.
31 FRUS 1952–4, Vol. 5, p. 1768; CLVL 1/8, 6 December 1953.
32 Eisenhower speech to UN, 8 December 1953: http://greatspeeches.wordpress.com/2008/10/02/dwight-d-eisenhower-atoms-for-peace-8-december-1953/.
33 CHBIO, Vol. 8, pp. 940–1.
34 Wheeler-Bennett (ed.) (1969: 121–2).
35 Hennessy, P. (2001) 'Churchill and the Premiership', *Transactions of the Royal Historical Society*, Vol. 11, pp. 295–306, see pp. 305–6. Interview with Lady Williams, 19 October 2010.
36 WSC to Eisenhower, 9 March 1954, quoted in Boyle (ed.) (1990: 124).
37 Butler (1971: 173).

Cockcroft becomes a confidant of the Prime Minister

1 Cockcroft to his mother, 17 December 1954, CKFTFAMILY.
2 Interview with Maurice Wilkes, 24 March 2009.
3 Chadwick to Appleton, 2 June 1945, CHAD IV 2/1.
4 Cockcroft lecture on 'Nuclear Physics since Rutherford', CKFT 4/21.
5 Cockcroft lecture to McMaster University, 15 May 1947, CKFT 25/24, p. 2.
6 Cockcroft talk to Manchester Grammar School, 15 July 1952, CKFT 4/19, p. 2.
7 Cockcroft talk, 'The influence of Lord Rutherford on the modern world', text of BBC broadcast, 12 December 1950, p. 5. Cockcroft lecture to McMaster University, 15 May 1947, CKFT 25/24, p. 2.
8 Personal conversation with George Steiner, 24 May 2011.
9 Personal conversations with Chris Cockcroft 8–10 May 2009, 9 March 2012; 'Thoughts on our father' by Cockcroft's children, undated, passed to GF 8 March 2012.
10 Hartcup and Allibone (1984: 191).
11 Cockcroft, J. D. (1959) 'The scientist and the public', *Harlequin* magazine, Christmas edition, pp. 25–6.

12 Letter from Cockcroft to his brother Eric, 26 August 1945, CKFT-FAMILY. Chadwick strongly disagrees with this, believing ICI's record in the project to be 'admirable': Chadwick to Akers, 31 May 1945, CHAD IV 11/1.
13 Cockcroft to his mother, 16 June and 6 December 1940, 12 March 1941, CKFTFAMILY; Hartcup and Allibone (1984: 166).
14 Cockcroft, draft memoir, CKFT 25/6, p. 32.
15 *Toronto Star*, 15 March 1946, in Cockcroft's scrapbook, CKFTFAMILY.
16 Note of meeting on 12 March 1954, CAB 130/101, NA.
17 The 'hybrid' weapon was 'boosted' with lithium deuteride, of which about three hundred pounds was needed for each bomb.
18 Annex 1 to Defence Policy Committee, 'Russian capacity to produce and deliver thermo-nuclear weapons', CAB 134/808, NA.
19 Hermann et al. (1987: 107, 215–6, 431–8, 476, 486–7, 495, 498, 524).
20 Cockcroft, J. D., 'Nuclear physics since Rutherford', 6 November 1954, CKFT 4/21, pp. 7–9. Monk (2012: 537–47).
21 Chadwick to Bohr, 3 October 1952, NBI. The quotation is Chadwick's statement of Lindemann's view.
22 Cockcroft to his mother, 3 September 1954, CKFTFAMILY.
23 Cockcroft to his mother, 17 December 1954, CKFTFAMILY.
24 Jenkins (2001: 890).
25 Moran (1966: 508).
26 'Ismay Pledges Political Reign on Atom War', *Washington Post*, 16 December 1954, p. 2; 'NATO Has Cast the Die for an Atomic Defense', *New York Times*, 19 December 1954, p. E5.
27 WSC to Eisenhower, 12 January 1955, Boyle (ed.) (1990: 184–6).
28 Cockcroft to his mother, 1 January 1955, CKFTFAMILY. See also 'Churchill Goes to Atom Works', *Daily Mail*, 31 December 1954, p. 1.
29 Hartcup and Allibone (1984: 196).
30 'Churchill Goes to Atom Works', *Daily Mail*, 31 December 1954, p. 1.

Churchill's nuclear swansong

1 *Daily Telegraph*, 3 March 1955.
2 Jenkins (2001: 875–6); Shuckburgh (1986: 158).
3 Lindemann to WSC, 13 April 1954, PREM 11/785, NA.
4 'Return to Sanity in Warfare?', *Manchester Guardian*, 14 May 1954, p. 9.
5 Lanouette (1992: 317).
6 Lindemann to WSC, 20 May 1954, LIND J146/47.
7 FRUS 1952–4, Vol. 6, Memorandum of meeting between WSC and Eisenhower, 25 June 1954, pp. 1085–6. Larres (2002: 338–40).
8 Colville (1985: 653); Moran (1966: 573).
9 Hennessy (2006: 346–53).

10 Catterall (ed.) (2003: 326–8); CHBIO, Vol. 8, pp. 1020–1.
11 Catterall (ed.) (2003: 342).
12 Eisenhower to WSC, 22 July 1954, and the subsequent correspondence in Boyle (ed.) (1990: 162–8).
13 Larres (2002: 363).
14 Haslam (2011: 149–50).
15 Larres (2002: 218).
16 Moran (1966: 628).
17 Hennessy (2010: 163–5).
18 WSC comment, 12 December 1954, on briefing by Brook, 10 December 1954, DEFE 13/45, NA.
19 Butler (1971: 176).
20 Addison (1992: 420).
21 WSC to Commons, 23 February 1955, HNSRD.
22 Goodwin, P. C. (2005) 'Low Conspiracy? Government Interference in the BBC', *Westminster Papers in Communication and Culture*, Vol. 2 (1), pp. 96–118. See pp. 101–4.
23 Moran (1966: 633).
24 Diary entry for Clementine Churchill, 1 March 1955, CSCT 4/4.
25 This account of his speech is taken mainly from Moran (1966: 635–7) and from the reports published on 2 March 1955 in the *Manchester Guardian*, *The Times*, the *Los Angeles Times* and the *New York Times*, which published the speech in full.
26 WSC to Shaw, 18 August 1946, CHBIO, Vol. 8, p. 254.
27 Wheeler-Bennett (1969: 119–20); Colville (1985: 596).

Epilogue 1: Churchill's nuclear scientists

1 Bohr (1961: 1115).
2 Moss (1987: 227–31).
3 Russell to Einstein, 11 February 1955: http://www.spokesmanbooks.com/Spokesman/PDF/85russein.pdf; Clark (1975: 540–1).
4 Oppenheimer (1964a).
5 BMFRS of Oliphant (2001); *Biographical Memoirs of the Australian Academy of Science*, Vol. 14, No. 3, 2003, http://www.science.org.au/fellows/memoirs/oliphant.html.
6 Quoted in Cockburn and Ellyard (1981: 134).
7 See 'Patrick Blackett' by Michael Howard, in Baylis and Garnett (eds) (1991: 153–63).
8 BMFRS of Blackett, November 1975, pp. 74–6.
9 Edgerton (2006: 216–20).
10 Hill to Blackett, AVHL II 4/10.
11 'The Hydrogen Bomb', Third Programme talk, 14 March 1950, GPT

F169/13, see p. 3. For Churchill's version of the analogy, 4 April 1926: CHAR 1/188, f. 25.

12 Brown (1997: 340–53), see pp. 347 and 349.
13 Skemp (1978: 492n).
14 BMFRS of Gowing (2012), pp. 76–8.
15 Chadwick, J., 'Comments on Mrs Gowing's 2nd Volume', CHAD IV 14/11, p. 12.
16 Oppenheimer lecture 'Niels Bohr and His Times', OPPY B247 F3, p. 11.
17 Frisch to Genia Peierls, 30 April 1978, FRISCH F94/9.
18 Peierls (1985: 204–5).
19 Moss (1987: 230).
20 Lee (ed.) (2009: 756–7).
21 Groves (1962: 407–8).
22 Chadwick to A. C. Todd, 31 January 1951, CHAD IV 13/1.
23 Chadwick to Conant, undated, CHAD IV 13/1.
24 Howard (1974).
25 BMFRS of Hinton, December 1990, p. 226.
26 Priestley, J. B., 'Britain and Nuclear Bombs', *New Statesman*, 2 November 1957, pp. 554–6; Clark (1975: 557).
27 Cathcart (1994: 276).
28 Arnold (2012).
29 *Pugwash Newsletter*, Vol. 44, No. 2, October 2007 – this edition gives the full list of participants from 1957–2007 on pp. 39–155.
30 Rowe, A. P., review of *Science and Government* by C. P. Snow, marked 'Published by Time and Tide', ROWE.
31 Clark (1965: 416).

Epilogue 2: Churchill and his Prof

1 Lees-Milne (1994: 54).
2 Lindemann to WSC, 1 March 1955, LIND J146/18, 21, 24.
3 Moran (1966: 699).
4 The figure concerns his books and articles, not the speeches. 'Datelines', *Finest Hour*, number 129, Winter 2005–6, p. 9.
5 WSC to his wife, 30 January 1956, reproduced in Soames (ed.) (1998: 603).
6 Montague Browne (1995: 217, 230).
7 Montague Browne (1995: 220, 284).
8 Payn and Morley (1982: 323).
9 Montague Browne (1995: 220–1).
10 Montague Browne (1995: 222).
11 Birkenhead (1961: 329).
12 Fort (2004: 334–5).

13 Lindemann, letter to *The Times*, 21 May 1957.
14 Birkenhead (1961: 333–5).
15 Moran (1966: 729). See also WSC to Alan Lennox-Boyd, 3 July 1957, CHUR 2/214.
16 Birkenhead (1961: 335); Harrod (1959: 276).
17 Moran (1966: 703).
18 Oppenheimer (1964b).
19 Lees-Milne (1994: 54). The script for *Dr Strangelove*: http://www.visual-memory.co.uk/amk/doc/0055.html.
20 Lees-Milne (1994: 54).
21 WSC to Clemmie, *c.*11 October 1957, Soames (ed.) (1998: 621).
22 Plowden to WSC, 29 July 1957, CHUR 2/531.
23 Montague Browne (1995: 310).
24 Walsh (1998: 170–5).
25 Quoted in Cockcroft's review of Birkenhead (1961), *Sunday Times*, 5 November 1961.
26 Cockcroft, 'The Foundation of Churchill College', CKFT 12/78, p. 2.
27 CHSPCH, Vol. 8, pp. 8704–6. WSC visited the Churchill College site on 17 October 1959.
28 Correspondence and press coverage of the debate is in RVJO B.382 and AVHL I 2/5.
29 Snow (1962: 35–6).
30 Gilbert (2005: 446).
31 Sir Burke Trend to Montague Browne, 22 July 1963, CHUR 2/506.
32 Montague Browne to W. J. McIndoe, 8 August 1963, CHUR 2/506.
33 W. J. McTudor to Montague Browne, 20 September 1963, and Sir Burke Trend to Montague Browne, 22 July 1963, CHUR 2/506.
34 Montague Browne to W. J. McIndoe, 21 September 1963, CHUR 2/506.
35 Montague Browne to W. J. McIndoe, 21 September 1963, CHUR 2/506.
36 Montague Browne (1995: 326); Jenkins (2001: 911–12).
37 Coote, C. R. (1965) Obituary supplement in the *Daily Telegraph*, pp. ii–xv, see p. xi.
38 Muller (ed.) (2009: 259–66).
39 WSC to H. G. Wells, 9 October 1906, WELLS c.238.2.

Acknowledgements

1 WSC speaking at Grosvenor House, London, on receiving the *Sunday Times* Book Prize for the first two volumes of his memoir of the Second World War, CHUR 5/28A, f. 7.

Index

Works by Winston Churchill (WSC) appear directly under title; works by others under author's name.

Titles and ranks are generally the highest mentioned in the text.